MW01118025

The prehistory of flight

The prehistory of flight

Clive Hart

University of California Press

Berkeley   Los Angeles   London

University of California Press
Berkeley and Los Angeles, California
University of California Press, Ltd.
London, England

1 2 3 4 5 6 7 8 9

Library of Congress Cataloging in Publication Data

Hart, Clive.
The prehistory of flight.

Bibliography: p.
Includes index.
1. Aeronautics—History. 2. Flight—History.
I. Title.
TL516.H27 1985 629.13′009 84-8677
ISBN 0-520-05213-7
Printed in the United States of America

Karissimae Amoris Ymagini

Contents

Illustrations

Black and White Figures

Colour plates
(following page 210)

Preface

This book explores the development of ideas about flight and the air from preclassical times to the beginnings of successful balloon ascents at the end of the eighteenth century. I have focussed on four centres of interest: first, early concepts about the nature of the air and its place in the cosmos; second, the growth of ideas about the creatures of the air and about their mode of flight; third, changing attitudes to the desirability and possibility of manned flight; fourth, a group of seventeenth and eighteenth century experiments with flying machines. The first three are included in Part 1, 'Theory,' while the chapters devoted to the fourth make up the shorter Part 2, 'Practice.' A concluding chapter, 'Flying Ships,' covers the whole period of the book. My earlier work, *The Dream of Flight* (1972), dealt in the main with practical attempts at flying. This study, by contrast, is concerned with the conceptual background: discussions of the nature of the air are dominated by metaphysics, theology, and cosmology; descriptions of bird flight are sometimes empirical, more often rational and logical (though always wrong); the earliest visions of manned flight call forth zany and ingenious inventiveness.

Although they do not form a simple linear sequence, the ideas about the air explored here fall into three broad groups. One is the notion of an animated cosmos, beginning at least as early as Hesiod (probably eighth century B.C.) and repeatedly reinvigorated by Platonising writers until the eighteenth century. In such a concept of the world not only is the air, along with everything else, in some sense alive and active, but it stimulates the imagination because of its invisibility, its unpredictability, and its special position between heaven and earth. A second is the set of Aristotelian tenets about the physical structure of the world which, although often modified and sometimes vigorously questioned, remained generally dominant until the seventeenth century and was especially important in European natural philosophy from about 1100 A.D. While not necessarily incompatible with the notion of an animated world, Aristotelian physical concepts were more readily associated with the idea of a self-governing, self-sustaining, deanimated universe; the earlier world view was gradually supplanted by the more mechanical model. Deprived, by this evolutionary process, of its own personality, the air nevertheless long remained inhabited by truly aerial creatures—demons and fallen angels—which made it a hazardous region for man to explore. The remaining group of ideas emerged with the so-called New Science of the seventeenth century, during which knowledge of the nature of the air expanded rapidly. Not all of the new ideas are, however, directly relevant to the theory of flight. My very brief and selective account of the investigation of the air in the seventeenth and early eighteenth centuries entirely omits some topics—especially Cartesian cosmology—that are of central importance to the history of science but need no rehearsal here.

Coincident with the rapid expansion of post-Aristotelian science there arose a good deal of speculation about whether man might use his knowledge to learn how to fly. Views on this matter are best understood by comparison with contemporary commentaries on the mechanics and iconological significance of bird flight. Man's middle position in the universe led to ambivalent attitudes to flight: Surrounded by an intermediate element, he could use the air, at least in imagination, to help his ascent to Heaven, or he might fall through it, like Milton's Satan, to his damnation. Birds themselves were viewed in an ambiguous light. Superior to men in their capacity to fly, they could function as symbols of beauty, peace, and the power of the upward-tending intellect and will. Since many of them spent their time on or near the earth or water rather than soaring like the eagle, they could also represent the failure of spiritual aspiration and a capitulation to worldly temptation.

Discussions of bird flight before 1800 A.D. grew gradually less concerned with what could be directly observed and more concerned with ingenious mechanical theory. Despite the flourishing of empiricism, very few commentators attempted to gather firm data about the way birds move. The discussion offered by Mauduit de la Varenne in the eighteenth century is much less in touch with physical reality than that of Pierre Belon in the sixteenth. There were, of course, exceptions to the trend: while Johann Silberschlag and François Huber made little advance on the mechanical theories of Borelli, they showed a refreshing capacity to *see,* as Frederick II and Leonardo had done. In describing the many theories of bird flight my purpose is less to assess the degree of approximation to the truth than to register some of the concepts and concerns of each period.

Although in my chapter on the history of 'flying ships' I have given some attention to early speculations about flight based on the proposition that air is inherently light, I have not attempted to write a detailed account of aerostation before the Montgolfiers (1783). Francesco Lana, Lourenço de Gusmão, and Jacques Charles have often been discussed, and in any case my principal concern in the later chapters is with heavier-than-air flight. In describing the practical aeronautical work carried out in the seventeenth and eighteenth centuries I have been selective. The ornithopters of Besnier and the Comte de Bacqueville have long been famous, and it has not been my intention to write another general survey of aviation during that period. I have chosen a relatively small number of flying machines both to exemplify the typical work of heavier-than-air experimenters and to illuminate the later history of ideas discussed in the early chapters.

Although I make occasional reference to flight imagery in myth, folktale, and the arts, I have not attempted to explore these areas in depth. I hope to focus on them later in a study of the iconography of flight.

For quotations from most classical literature, except Aristotle, I have used Loeb. For Aristotle I have used the 'Oxford translation.' References to Aristotle in the footnotes are made by means of the conventional Bekker numbers. For most Patristic literature I have used Migne. Titles of Greek works and the names of the less familiar authors are usually given in the Latin form by which they were most commonly known to mediaeval and Renaissance readers. Except for Aristotle and the Loeb texts, I have included in the footnotes brief indications of the sources of quotations from published translations. Fuller details will be found in the bibliography. Other translations are my own.

For valuable assistance of various kinds I wish to thank John Bagley of the Science Museum; Peter Collier, a truly great flying instructor; Philip Gaskell; the late Charles H. Gibbs-Smith; Mrs Thomas Gilliland, at whose bright breakfast table I first read Burattini's manuscript; Professor Manfred Gordon; Helen Linthorne Hart, who is skilled at reading strange hands and who collaborated in an earlier version of Chapter 9; Dr J. A. Huisman; Philip Jarrett; Hubert C. Johnson; Dr and Mrs R. Kiess; Dr A. M. L. Knuth; Michael, Robin, Phil, and Jo, who flew with me; A. W. L. Nayler, Librarian of the Royal Aeronautical Society; Esen Piskobulu; Allan Ranius; Michael Rosen; Robin D. Smith; Kay Gilliland Stevenson, who taught me much about the feminine air, and who helped me not to stop; John W. R. Taylor; Sylvia, who helped me to see birds; Philip Titheradge, Custodian of Down House; Lynn White, Jr; Josette, Brian, and Simon Willis, who smoothed my way to a borrowable Migne; Jay Wolff, editor of *The Aeronautical Journal,* in which earlier versions of Chapters 6–10 and of Appendix I were first published.

Parts of this book were written with the aid of research grants from the British Academy and the Royal Society.

Clive Hart
Colchester, Essex
Feast of Saint Catherine
of Alexandria, 1984

Acknowledgments

I am grateful to the following libraries and institutions for permission to reproduce copyright material: Biblioteca Ambrosiana, Milan: Figs. 17, 18, 19, 20, 21, 22, 24; Biblioteca Mediceo-Laurenziana, Florence: plate XII; Biblioteca reale, Turin: Fig. 26; Bibliothèque municipale d'Amiens: plate III; Bibliothèque nationale, Paris: Figs. 32, 33, 53, plates IV, V, VII, X; British Library: Figs. 5, 10, 13, 14, 15, 16, 39, 50, 51, 54, 55, plates I, II, VI, VIII, XI; British Museum: Figs. 1, 2; the Syndics of Cambridge University Library: Figs. 7, 8, 9, 12, 40; Haags Gemeentemuseum, The Hague: Fig. 3; the Royal Aeronautical Society: Fig. 25; Österreichische Nationalbibliothek: Fig. 6; Staatsarchiv, Weimar: Figs. 42–49; Stifts- och Landesbibliotek, Linköping: Figs. 34, 35; Swedenborg Scientific Association: Fig. 36; Tekniska Museet, Stockholm: Fig. 37; Universitätsbibliothek, Heidelberg: plate IX.

As a long-winged hawk, when he is first whistled off the fist, mounts aloft, and for his pleasure fetcheth many a circuit in the air, still soaring higher and higher till he be come to his full pitch, and in the end, when the game is sprung, comes down amain, and stoops upon a sudden: so will I, having now come at last into these ample fields of air, wherein I may freely expatiate and exercise myself for my recreation, awhile rove, wander round about the world, mount aloft to those ethereal orbs and celestial spheres, and so descend to my former elements again.

The Anatomy of Melancholy, II.ii.3

Part One Theory

The nature of the air 1

The air is female. Or so the ancients thought. We have it on the best of authority: alluding to the idea that the region of the second element is the domain of Hera, Plato points out in passing that her name, ῞Ηρα, anagrams the Greek word for air, ἀήρ.[1] A familiar concept throughout Greek and Roman times, the idea of the air as feminine largely disappeared from mediaeval Christian cosmology and meteorology, but was readopted in attenuated form by Renaissance humanists. As Cicero says in his *De natura deorum,* it was most clearly formulated by the Stoics:

> The air, lying between the sea and the sky, is according to the Stoic theory deified under the name belonging to Juno, sister and wife of Jove, because it resembles and is closely connected with the aether; they made it female and assigned it to Juno because of its extreme softness.[2]

A similar comment is made by Macrobius in his commentary on Cicero's *Somnium Scipionis* (c. A.D. 400). After a passage about Jupiter, he writes: 'So also Juno is called his sister and wife, for she is air: she is called sister because the air is made of the same seeds as the sky, and she is called wife because air is subordinate to the sky.'[3]

That the air was made of the same seeds as the sky was an idea deeply rooted in the earliest Greek philosophy. By 'air' in this sense was meant the cloudy, damp, lower atmosphere; by 'sky,' or *aether* (αἰθήρ), the bright, dry upper regions beyond the clouds. In poetical cosmologies that appear to have been written in the sixth century B.C., both the lower air and the *aether* were given a common origin in a primaeval *aer.* In association with the Night that preceded the world, *aer* was the potential generator of the cosmos.[4] From the dark, moist, and hazy *aer,* generation proceeded both 'upwards' and 'downwards.' Rarified and energised, it produced the fiery *aether* and the lightning; condensed and thickened, it was transformed into water, earth, and stones.

Despite the fragmentary nature of the evidence, it is clear that the concept of evolution from the primaeval *aer* was the common property of many pre-Socratic writers and philosophers.[5] In the earliest poetic uses of the word, *aer* retains strong traces of its primordial origins. In Homer, where it is contrasted with the blazing brilliance of the *aether,* it is always an obscuring mist or haze; in Hesiod it is used as a part of the scenery in a quietly evocative rural passage:

> At dawn a fruitful mist is spread over the earth from starry heaven upon the fields of blessed men: it is drawn from the ever flowing rivers and is raised high above the earth by windstorm, and sometimes it turns to rain towards evening, and sometimes to wind when Thracian Boreas huddles the thick clouds. Finish your work and return home ahead of him, and do not let the dark cloud from heaven wrap round you and make your body clammy and soak your clothes.[6]

It was some time before these vestiges of an obscure, vaporous, generative matrix evolved into the subordinate though powerful female principle of middle and late classical antiquity. Only in about the sixth century B.C. did the range of meaning of *aer* expand from the idea of a dark mist to embrace the invisible atmosphere in general, from the surface of the earth to the sky.[7] As this extension of reference grew increasingly common, men began to think of themselves as living in the lower and relatively darker regions of a continuum manifesting itself in a variety of ways. More mysterious than solids or liquids, the all-pervasive, inescapable air aroused intense responses that, even at their most positive, were often tinged with fear. Before Aristotle reduced the air to one of the four primary 'bodies,' its generative power sometimes led to radical deification. Relating macrocosm to microcosm, Anaximenes (sixth century B.C.) considered air the mystical force informing and sustaining all things. In the only surviving fragment attributed directly to him, he expresses its powerful psychological and spiritual appeal: 'As our soul, being air, holds us together, so do breath and air surround the whole universe.'[8] Reporting these views at greater length, Hippolytus (second century A.D.) reveals in his *Philosophoumena* how the air may generate an attitude of mystical celebration:

> When it is of a very even consistency, it is imperceptible to vision, but it becomes evident as the result of cold or heat or moisture, or when it is moved. It is always in motion; for things would not change as they do unless it were in motion. It has a different appearance when it is made more dense or thinner; when it is expanded into a thinner state it becomes fire, and again winds are condensed air, and air becomes cloud by compression, and water when it is compressed farther, and earth and finally stones as it is more condensed. So that generation is controlled by the opposites, heat and cold. And the broad earth is supported on air; similarly the sun and the moon and all the rest of the stars, being fiery bodies, are supported on the air by their breadth.[9]

Of those who followed Anaximenes, Diogenes of Apollonia (fifth century B.C.) wrote most forcefully in the same vein: 'And it seems to me that that which has intelligence is what men call air, and that all men are steered by this and that it has power over all things.'[10] Diogenes goes on to develop his sense of air as universal life-giving breath or spirit by contrasting it with the principle of liquid or fluid. The more the air has developed from its original dark, warm matrix, the more it can contribute to the life of the mind. Clear, dry air is beneficial to thought, while damp air near the ground may assist physical growth but is harmful to the understanding. As animals breathe air closer to the soil than do men, their intelligence is less. Aware of a possible objection in relation to birds, Diogenes suggests an answer:

> This fluid the animals breathe in with the air from the ground, and they also eat wetter food. If it be asked why birds, which breathe purer air, have less intelligence than Man, the answer is that their bodies are so constituted that the breath, that is, air, does not pass right through them, but stays in the intestines. Hence they have rapid digestion but little thought.[11]

That water and soul or spirit were in some sense antithetical principles was a notion shared by a number of pre-Socratic philosophers, especially Heraclitus (sixth–fifth centuries B.C.);[12] as I discuss in Chapter 2, it becomes a commonplace of hexaemeral and Patristic writers.

Inherent in Diogenes' fragment is the Stoic polarity of a male principle associated with the brighter, fiery upper air and a female principle whose true home was the darker, moist air near the ground. When Empedocles (fifth century B.C.) established the system of the four elements or 'members' of the world, he put the deification of the air below the sky into the context of other spiritual forces:

> Hear first the four roots of all things, bright Zeus and life-bearing Hera and Aidoneus, and Nestis, who moistens the springs of men with her tears. Now by Zeus [his commentator goes on to explain] he means the seething and the aether, by life-bearing Hera the moist air, and by Aidoneus the earth; and by Nestis, spring of men, he means as it were moist seed and water.[13]

Although inferior in power to Zeus, the fiery *aether,* the lower air still represented something endowed with inherent integrity and purity. Attributing a similar deification to the Egyptians, Diodorus Siculus (first century B.C.–first century A.D.) writes of air as the highest part of the universe, that is, the region immediately below the *aether:*

> The air, they say, they called Athena, as the name is translated, and they considered her to be the daughter of Zeus and conceived of her as a virgin, because of the fact that the air is by its nature uncorrupted and occupies the highest part of the entire universe; for the latter reason also the myth arose that she was born from the head of Zeus.[14]

While belief in the divinity of the air persisted in attenuated form in Roman times, it became a less significant part of the world system and sometimes came under attack. Commentators on pre-Socratic ideas, no longer sensitive to the good cosmological reasons for the deification of the vaporous air in the sixth and fifth centuries B.C., tended to look on such propositions as exaggerated philosophical oddities. Considering himself to be bathed in a far from divine element, Cicero (106–43 B.C.) writes critically of Anaximenes in the *De natura deorum.* Into the mouth of his Epicurean speaker he puts a blunt summary of Anaximenes' position, which is as bluntly rejected:

> Anaximenes held that air is god, and that it has a beginning in time, and is immeasurable and infinite in extent, and is always in motion; just as if formless air could be a god, especially seeing that it is proper to god to possess not merely some shape but the most beautiful shape; or as if anything that has had a beginning must not necessarily be mortal.[15]

Softness and virginity, associated with the air by the Stoics and the Egyptians, are not, of course, the only traditional attributes of femininity. While the elemental air might be essentially incorruptible, a point about which there was to be a good deal of controversy, it was also liable to rapid and unpredictable change. Seneca (?4 B.C.–A.D. 65) describes the lower regions:

> The atmosphere is . . . especially variable, unstable, and changeable in its lowest region. The air near the earth is the most blustering and yet the most exposed to influences since it is both agitating and being agitated. Yet it is not all affected in the same way. It is restless and disturbed in different parts and in different places.[16]

The air's fickleness, variability of mood, and liability to sudden storms seemed especially female, while its mixed nature, ranging from brilliant clarity to dark and unpleasant fog, suggested the troubling duality of virgin and whore. Fallen man's necessary involvement with woman found a parallel in his inescapable dependence on the air. In place of a true deification, weather phenomena in general were put, during late antiquity, under the symbolic command of Juno, or her Greek equivalent Hera, who in Renaissance times became, by extension, the presiding goddesses of flight.

Corruptible or incorruptible?

Greek philosophers were repeatedly exercised by the nature of the enduring matter that must underlie the changing appearances of everyday reality. In proposing a fresh solution, Aristotle readopted, with major modifications, the now-familiar tetralogy of fire, air, water, and earth, which had previously figured in the work of Empedocles and Plato, in particular. Although the true substratum, 'primary matter,' is indeterminate and unobservable, these four 'elements' are the simplest corporeal manifestations of it, the simplest forms

1. Sexuality and the air: winged phallus, hung in the garden or used as a door-chime. Bronze. Pompeian. First century A.D. Horizontal length 131 mm. (Reproduced by permission of the British Museum)

2. More winged phalluses. The example second from the right is probably an eighteenth or nineteenth century copy. The others, of uncertain date and provenience, are probably Pompeian or Roman, first century A.D. The feathered phallus on the far right appears to represent a bird of prey. (Reproduced by permission of the British Museum)

of matter that can enter into our experience. Since they are irreducible to simpler bodies and are continually transforming themselves into one another, they are to be thought of as, in an important sense, a single 'stuff' in four different guises. In Aristotelian terms air is 'corrupted' and water 'generated' when air is condensed to become water, but the question whether air is 'corruptible,' in the sense in which one says that dead animal tissue corrupts, would for him have been meaningless. Although in most of the rest of this book I shall be concerned with the growth of post-Aristotelian scientific thought, Aristotle's was by no means the only voice heard before the Renaissance. For several centuries after the collapse of classical civilisation, most of his scientific works were lost to the West, and even after their full retrieval many competing ideas—Platonic, neo-Platonic, Pythagorean, sometimes Eastern—were developed in parallel. In such a context it is by no means surprising to find discussions about whether or not the air might be prone to fundamental decay.

Although the compiler of the pseudo-Aristotelian *Problemata* (fifth century A.D.) assumes that air is incorruptible, he formulates his question in a very non-Aristotelian way: 'Why is it that water and earth become corrupt, but air and fire do not?' The answer, he thinks, may have something to do with the heat of the upper two elements.[17] In a commentary on Hippocrates' *De aere, aquis et locis* (fifth century B.C.), Hieronymus Cardanus (1501–76) supported the idea by mentioning the common observation that air sealed into a container remains permanently fresh.[18] As late as 1773, Antoine Baumé asserted that air is 'indestructible.'[19] In meteorologies and encyclopaedias from classical times to the Renaissance the opposite view is, however, much more commonly found. The frequent association of air with fogs, miasmas, and stenches led to its being thought of not only as variable but also as subject to contamination. The writer of the pseudo-Aristotelian *Meteorologica* Book IV says that 'everything, except fire, is liable to putrefy; for earth, water, and air putrefy, being all of them matter relatively to fire.'[20] Whether this putrefaction

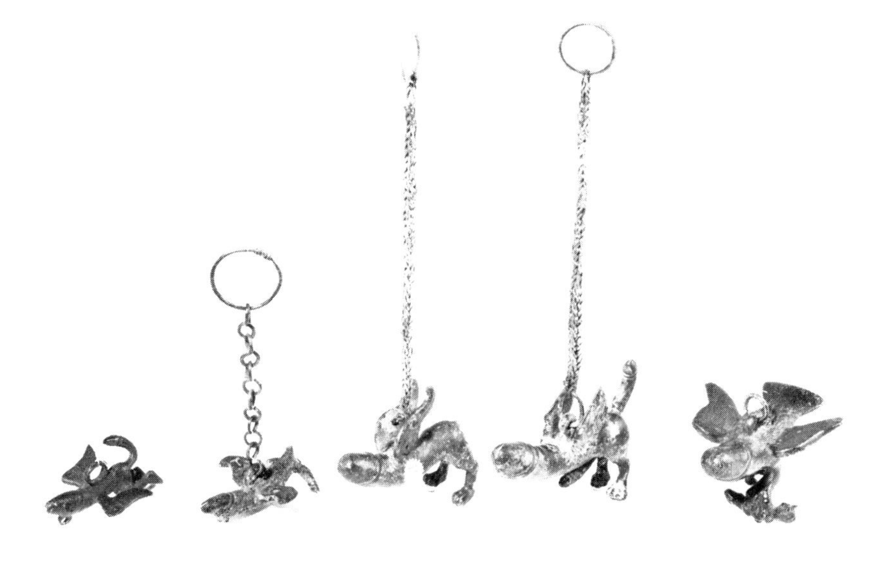

meant a change in the secondary, 'accidental,' qualities of air—for example, by increasing its heat to an unhealthy degree—or indicated a fundamental corruption of the element itself was often left unresolved.

In the many discussions of 'good' and 'bad' air a moral tone is often introduced, as for example in Seneca's *Naturales quaestiones:*

> The closer all air is to the earth, the thicker it is. Just as in water and in all liquids the dregs are at the bottom, so in air all the thickest particles settle downward. . . . The higher the air is, the farther it withdraws from the pollution of the earth, the less contaminated and purer it is.[21]

The upper air might nevertheless be too pure, too tenuous, to suit the needs of fallen human nature. Thomas of Cantimpré (thirteenth century) asserts that the best air is found in the middle regions, above the easily corruptible air near the soil and below the excessively subtle and penetrable regions surrounding the mountains.[22]

Moral considerations

Expanded from its obscure origins to occupy a broad band of the universe, the air held ambiguous moral connotations until well into the late Renaissance. As is indicated by Seneca, who was much respected by Renaissance writers, it played a highly important role in the structure of the world as the medium of connexion between lower and upper, mortal and divine, earth and the heavens:

> Thus, atmosphere is a part of the universe and, in fact, an essential part. This is what connects heaven and earth and separates the lowest from the highest in such a way that it none the less joins them. It separates because it intervenes midway; it joins because through it there is a communication between the two. It transmits to the upper region whatever it receives from the earth. On the other hand, it transfuses to earthly objects the influences of the stars.[23]

Similar ideas were developed outside the main stream of classical thought. Interpreting the story of Jacob's dream (Genesis 28:12), Philo Judaeus (early first century A.D.) stresses the role of the air as an access way to heaven. Offering an allegorical interpretation of Jacob's ladder, he says that when applied to the universe it 'is a figurative name for the air; whose foot is earth and its head heaven.'[24] As one ascends towards heaven, the purity of experience increases both in a spiritual and in a physical sense: 'It is a fine thought that the dreamer sees the air symbolized by a stairway as firmly set on the earth; for the exhalations given forth out of the earth are rarefied and so turned into air, so that earth is air's foot and root and heaven its head.'[25] Writing analogously about the range of purity in men, Philo makes a direct comparison between the air and man's soul. Just as the air connects the fallen

3. The goddess Fama (Rumour), who spreads ill reports, flying by night on foul wings midway between earth and sky (see *Aeneid* IV.173ff). From the reverse side of a medallion by Sperandio of Mantua, 1479. The motto, *fama super aethera notus* ('I am known by my fame in the heavens above') is from a speech by Aeneas to Venus, *Aeneid* I.379. (Collection Haags Gemeentemuseum, The Hague)

world to the heavens, so the soul connects our grosser, physical nature to our angelic potential: 'Such then is that which in the universe is figuratively called stairway. If we consider that which is so called in human beings we shall find it to be soul. Its foot is sense-perception, which is as it were the earthly element in it, and its head, the mind which is wholly unalloyed, the heavenly element, as it may be called.'[26]

In contrast with the vertical links imagined by Seneca and Philo, supernatural connexions in the lateral direction could be a source of psychological and spiritual unease. A celebrated passage of the *Aeneid,* immediately following the first sexual encounter of Dido and Aeneas, recounts in vivid language how news of the event was spread by the detestable flying goddess *Fama,* representing the most dangerous and unattractive aspects of the air's power:

> Then *Fama* rushed through the great cities of Libya—*Fama* who is swifter than any to do harm. She thrives on speed, is fattened by activity: at first she is small and frightened; but soon she rises into the air and moves with her head among the clouds. They say that she was borne by Earth, to spite the gods, as last sister to Coeus and Enceladus. Swift of foot and foul of wing she is a huge, appalling monster. Under each feather of her body there are, wonderful to relate, a watchful eye, a tongue, a loud mouth, and an attentive ear. By night she flies shrieking through the dark, midway between heaven and earth, never pausing for sleep. By day she perches on roof-ridges or high towers and terrifies whole cities, being as tenacious in lying as she is in truth-telling. (IV.173–88; my translation. See Fig. 3.)

A parallel passage in Ovid's *Metamorphoses* (XII.43–63) presents an equally disagreeable portrait, with less stress on aerial imagery but with a similarly painful exploration of the meeting of good and evil.

The idea of the air as a morally ambivalent region survived in mediaeval times partly through the influential number symbolism of the pagan encyclopaedist Martianus Capella (late fourth century–early fifth century A.D.). Governed by Juno, the second of the classical deities, the air, the second element, necessarily exhibited the characteristics of the number 2. Two was feminine, the numerological symbol for the powers of discord and separation. Sharing in both good and evil and serving to connect them, it was also, paradoxically, the symbol for justice, harmony, mediation, and generation.[27]

That the air should have been characterised as a divine generative power, as an external representation of the possibilities of corruption and infidelity, and as a source of mediation between impure human experience and the godhead manifests the general predisposition of ancient philosophers to look on the world as in some sense animate. Propositions of the kind range from the notion that matter itself is alive to vestigial analogies with no more than a vaguely anthropomorphic flavour. The later deanimation of the universe, more or less complete by the seventeenth century, was by no means a continuous development, but proceeded with many interruptions and reversals of direction. Even those mediaeval philosophers who tended to treat the processes of nature as mechanically self-regulating, the 'unnatural' acts of God being things of the remote past, often used language which showed the continuing appeal of the idea that the cosmos is animate.[28] It remains detectable as late as the sixteenth century in the emphatic tone of several passages that Paracelsus (1493–1541) devotes to celebrations of the air's power. Always treating the air as the most important of the elements, he associates it with the concepts of vitality, order, and organic growth:

> You should know that the air encloses within itself all other elements and all created things and ensures that they remain in an orderly arrangement as they were created. The air is that which has power over the three other elements. . . . the elements are set in the air as a house is set on foundations.[29]

Although by the seventeenth century the moral connotations of the air are less clearly defined, they continue to find indirect expression, as when Erasmus Francisci compares different qualities of air to aspects of human nature: good, evil, industrious, pious, and so on. He strengthens the connexion by going on to point out that with care we can purify bad air, while good air can be polluted by the effects of evil living.[30]

Active or passive?

Traces of animation are also found in frequent discussions about the possibility that the elements should be divided into two groups, active and passive. In *De generatione et corruptione* Aristotle refines and clarifies earlier ideas to evolve an elegant and symmetrical formulation. According to circumstances, the characteristics or powers of each of the four (hot, cold, moist, dry) could overcome, or be overcome by, those of another, so allowing transformation of the elements in any direction. Elsewhere, however, he echoes pre-Socratic ideas, considering hot and cold intrinsically active, moist and dry intrinsically passive, and air and fire in some sense superior to water and earth. Although many later writers attempted to sustain the clarity of Aristotle's harder-headed, purely physical account, the older anthropomorphic idea that air shared with fire an active principle dominating the heavy passivity of water and earth never lost its attraction. The difference of view is discussed by Plutarch (c. A.D. 36–120), who rightly attributes the more animist explanation to the Stoics: 'Air and fire because of their intensity are self-sustaining and to the former two, when blended with them, impart tension and permanence and substantiality.' Scoffing at this, he complains that such a distinction makes it impossible to consider water and earth as in any real sense primary elements.[31]

Despite Plutarch's reservations, and despite the primitive feminine associations of the air, the idea that it was an active force grew common, especially when it was identified with the nourishing spirit of God. Seneca celebrates its power:

> What song can be sung without the tension of air? Horns, and trumpets, and instruments which by water pressure form a sound greater than that which can be produced by the mouth—do they not accomplish their function by tension of the air? Let us examine things which produce great hidden force, such as very tiny seeds whose thinness finds a place in the joints of stones, and grow so powerful that they dislodge huge rocks and demolish monuments; while very thin, minute roots split crags and cliffs. This is nothing but the tension of air, without which nothing is strong and against which nothing is strong.
>
> That there is unity in air can be realized also from the cohesiveness of our bodies. What holds them together? Air. What else is it that puts our soul in motion? What is the air's motion? Tension. What tension can there be except from unity? What unity could this be unless it were the unity in the air? Moreover, what produces leguminous crops and slender standing grain-crops and forces up green trees and spreads out their branches or lifts them on high if it is not tension and unity in air?[32]

An interesting resolution of the difficulty is attributed by Seneca to the Egyptians, whose civilisation so fascinated the Romans:

> The Egyptians established four elements, then formed a pair from each one. They consider the atmosphere male where it is windy, female where it is cloudy and inactive. They call the sea masculine water, all other water feminine. They call fire masculine where a flame burns, and feminine where it glows but is harmless to touch. They call male the firmer earth, such as rocks and crags. They assign the term female to our soil that is tractable for cultivation.[33]

Although these ideas are somewhat alien to him, it may be possible, Seneca adds, to agree with some part of them.[34] The theory is generally coherent with the classical ideas that lay behind Seneca's world view.

The activity or passivity of the elements was a theme frequently developed by writers whose view of the nature of the physical world reflected recurrent themes of post-Platonic Greek cosmology. This was broadly true of Saint Augustine (354–430), who assumed the division into two pairs of elements, dismissed by Plutarch, to be almost self-evident. Speaking of the ethereal bodies of demons, he says that 'there prevails in them an element which is better suited to act than to be acted upon, placed above two others, viz. water and earth, and below one more, which is sidereal fire.'[35] In the twelfth century Thierry of Chartres (d. c. 1155) returns to the Platonic idea of air as a mediating element (see Chapter 2). Developing the idea of relative activity, he writes that while air is passive in relation to fire, it is, as it were, the active mediator and vehicle of the power of fire, transmitting it to the lower elements.[36] Similar concepts are revived in the fifteenth century by the brilliant young humanist Pico della Mirandola (1463–94), who writes of traditional attempts to account for weather phenomena. Although the air is Juno's realm, Jupiter may be considered as a part of the air when, in the form of rain, he descends through it to mate with the earth.[37]

The ambiguity between male and female, active and passive was implicit in a late classical 'Pythagorean' doctrine that extended the association of the air with the power of the gods by establishing a relationship between the elements and the seven planets. Treating the lowest, the moon, as another earth, one progressed upwards through the spheres of supramundane water (Mercury), air (Venus), and fire (the sun). Three planets remained, with the stars above them. The sequence of attributions was accordingly repeated, this time in reverse order, associating Mars with fire, Jupiter with the air, Saturn with water, leaving the stars as the new celestial earth. In his summary of this doctrine Pico says that among the upper planets 'the air is attributed to Jupiter whose nature is similar to that of Venus.'[38]

Powerful reinforcement of the mixed emotive connotations of the air was provided by an association of the elements with the long-lived mediaeval doctrine of 'humours.' Each of the four humours, the mixture of which determined a man's temperament, was linked to one of the elements. Nothing very surprising might be found in an association of black bile, the melancholy humour, with the heavy earth, nor in that of phlegm with water or of choler with fire. More complex emotional possibilities lay, however, in the relationship of blood and air. For mediaeval and Renaissance thinkers, physical links were not hard to find: Both blood and air were hot and wet; blood absorbed air from the lungs and foamed when pouring from arteries; both were essential to life. Robert Fludd (1574–1637) even pointed to the liver, the alleged source of blood, saying that it is of an airy constitution.[39] The immensely rich symbolic and iconological connotations of blood, which could well form the basis of a study in themselves, vary, like those of the air, from good to bad and from active to passive.

The implication of the air in systems of cosmic anthropomorphism has been remarkably persistent. In the early eighteenth century Petrus Clausenius, wanting to write crisply about physical facts, felt obliged to begin by clearing away all talk of male/female, Jupiter/Juno as superstition.[40] The tradition has nevertheless refused to die and writers with a strong interest in the occult have continued it into modern times. A powerful example is offered by Yeats who, appealing as so often to mediaeval concepts, allows the association of air and blood to underlie and strengthen his vivid image of Leda, 'so mastered by the brute blood of the air.'

Light or dark?

Related to the polarity of active and passive was a curious uncertainty as to whether the air was inherently dark or bright. Echoes of the obscurity of the primaeval *aer* are often heard, as in Seneca's proposition that 'by itself air is cold and dark.'[41] The point is more fully developed by Seneca's Alexandrian contemporary, Philo, who conceived of the air as inherently black: Together with the void, or 'abyss,' it 'completely filled the original immensity and desolation . . . of all that reaches from the zone of the moon to us.' Only with the creation of light did the air lose its blackness, after which it helped to play a mediating but also law-giving role in separating the constantly warring opposites of light and dark.[42] Remnants of its original blackness may be seen on the moon, which 'is not an unmixed mass of ether, as each of the other heavenly bodies is, but a blend of ethereal and aerial substance; and . . . the black which appears in it, which some call a face, is nothing else than the commingled air which is naturally black and extends all the way to heaven.'[43]

In his dialogue *De facie quae in orbe lunae apparet* Plutarch allows this Stoic idea to be treated with a good deal of ironic contempt:

> Whereat Lucius said: 'Nay, lest we give the impression of flatly insulting Pharnaces by thus passing over the Stoic opinion unnoticed, do now by all means address some remark to the gentleman who, supposing the moon to be a mixture of air and gentle fire, then says that what appears to be a figure is the result of the blackening of the air as when in a calm water there runs a ripple under the surface.' 'You are very nice, Lucius,' I said, 'to dress up the absurdity in respectable language. Not so our comrade; but he said what is true, that they blacken the Moon's eye defiling her with blemishes and bruises, at one and the same time addressing her as Artemis and Athena and making her a mass compounded of murky air and smouldering fire neither kindling nor shining of herself, an indiscriminate kind of body, forever charred and smoking like the thunderbolts that are darkling and by the poets called lurid.'[44]

In commentaries on the Bible, the equivalent of the obscure primaeval Greek *aer,* which is given little further attention after the early centuries A.D., is the darkness before creation in Genesis 1:2. Although Genesis speaks in detail of heaven, the earth, and the waters, exegetes were exercised by its total silence on the creation of air. A solution of the difficulty was proposed in the late second century by Theophilus of Antioch, who identified air with the spirit of

God moving on the face of the waters, a life-giving, nourishing medium that offered protection against the preexistent dark:

> The 'spirit borne over the water' was the one given by God to give life to the creation, like the soul in man, when he mingled tenuous elements together (for the spirit is tenuous and the water is tenuous), so that the spirit might nourish the water and the water with the spirit might nourish the creation by penetrating it from all sides. The unique spirit occupied the place of light and was situated between the water and the heaven so that, so to speak, the darkness might not communicate with the heaven, which was nearer to God, before God said: 'Let there be light.'[45]

An eloquent justification of the omission was offered in the fourth century by Saint Gregory of Nyssa (c. 335–94), who appealed to the special characteristics of air distinguishing it from the other elements: Neither inherently brilliant nor dark, it is capable of taking on any shape or quality dictated by its surroundings. It will accommodate itself to anything, move anywhere. Since man is in constant contact with it, it needs no special mention in Genesis.[46]

In the sixth century Philoponus developed further Theophilus's idea that the air is 'the place of light.' Suggesting that it is in a sense the fundamental substrate of light itself, he several times wrote of the air's capacity to be filled with shining brilliance. At other times it is dark only as a result of absence, through a lack of stimulation of its capacity to shine.[47] Although some mediaeval and Renaissance expositors assumed, as did Cardanus, for example, that 'by itself air is most obscure,'[48] the natural affinity of light and air became a commonplace. Partly due to interpretations of Genesis, this also owed something to Aristotle, who had stressed both the transparency of the air and its capacity to be brightly illuminated by reflected light. At such times, he said, it can glow with colours, usually reds and purples.[49] The natural whiteness of the air is again mentioned in the pseudo-Aristotelian *De coloribus,* where darkness, even the dark blue of the sky, is placed in opposition to its bright purity:

> Simple colours are the proper colours of the elements, i.e. of fire, air, water, and earth. Air and water when pure are by nature white, fire (and the sun) yellow, and earth is naturally white. . . . Air seen close at hand appears to have no colour, for it is so rare that it yields and gives passage to the denser rays of light, which thus shines through it; but when seen in a deep mass it looks practically dark blue. This again is the result of its rarity, for where light fails the air lets darkness through. When densified, air is, like water, the whitest of things.[50]

From the twelfth century until at least as late as the sixteenth, the idea of the air's inherent brilliance, transparency, and colourlessness was modified by the growth of a belief that when pure it shone with a greenish light. An anonymous twelfth century writer on the elements suggests that this is one of the ways in which common men might seek to deny that air cannot be perceived by the senses. The writer also appears to imply, in an interesting paradox, that its invisibility is commonly accepted provided that it is to some degree impure. Refuting the idea of its inherent greenness, he explains that the glow arises because light is reflected from the small suspended bodies that are too distant for the eye to perceive.[51] The popular belief in the greenness of the air is reported at about the same time by Thierry of Chartres, who uses similar language.[52]

The true colour of the air, the sky, and the heavens is several times discussed by Paracelsus. Although he stresses that the sky is essentially colourless, he says that it appears to be green: 'This firmament is transparent and looks green. . . .'[53] Elsewhere he refers to the common statement (which he is at pains to deny) that the sky is really green and that its blueness is merely an illusion caused by distance.[54] This popular belief may perhaps more readily be traced to the hue of some skies, and of summer shade, than to the traditional use of green, the colour of Venus and of femininity generally, to symbolize sexual passion and licentiousness.[55]

If the air were really endowed with some intrinsic colour, it would be less elusive, more obviously a part of the tangible, comfortable, physical world. While the blue of the everyday sky offered the happiest mean between the pre-Socratic dark *aer* and the transcendental place of light, the frequent speculations about green—so much more ambiguous a colour—reveal once again how uneasy men could feel about the mediating element.

The elemental character of air

Partly because of the vast extent of the atmosphere, partly because of its imprecise boundary with the sky, there could well be doubts as to whether it were composed of a single substance. In Aristotelian physics, the basic premises of which dominated late classical, mediaeval, and early Renaissance science, air was an intermediate element, sharing some of the characteristics of its neighbours, water and fire. As Aristotle himself points out, its physical nature is far from easy to determine:

> The first difficulty is raised by what is called the air. What are we to take its nature to be in the world surrounding the earth? . . . The question is really about that which lies between the earth and the nearest stars. Are we to consider it to be one kind of body or more than one?[56]

Arguing on the grounds of proportion and of the necessary balance between the four sublunary elements, Aristotle decides that the space above the earth must be filled with both air and fire, or something which is potentially fire: 'At the centre and round it we get earth and water, the heaviest and coldest elements, by themselves; round them and contiguous with them, air and what we commonly call fire.'[57] The pre-Aristotelian idea that the atmosphere was made up of several layers of air varying in density, the purest and thinnest at the top, the heaviest and thickest at the bottom, became closely allied to the fundamental Aristotelian concept of the transmutability of the elements. Not only were processes of transmutation going on all the time, water becoming air, air becoming fire or being changed back into water, but the problematical elemental fire, whose natural place was above the atmosphere, was often spoken of as if it were essentially a kind of superfine air. At the bottom of the atmosphere was a sort of mixed boundary layer, half water and half air, consisting of fog, mist, sea spray, and other evidence of humidity. While the doctrine of transmutability made all of the boundaries indistinct, the doubts surrounding the nature of elemental fire caused less attention to be given to the meeting of air and fire than to the readily observable merging of the spheres of air and water.

While earth was absolutely heavy and fire absolutely light, the one always tending towards the centre of the world, the other trying to move to the region contiguous with the sphere of the moon, water and air contained mixtures of heaviness and lightness in differing proportions and could thus be described in terms of what we would now call relative density. Aristotle's comments on the characteristics of these intermediate elements have an important bearing on the concept of ascension, and therefore of flight. Writing of transmutability, he twice compares the relationship of contiguous elements to that of form and matter. Each of the three lowest elements can be considered as matter that, when more fully formed, becomes the element next above it. This analogy he works out most completely in respect to water and air: 'For the one is like matter, the other form—water is the matter of air, air as it were the actuality of water, for water is potentially air. . . .'[58] Although value judgements play no very significant role in Aristotle's cosmology, form was for him a decidedly more powerful concept than matter. The analogy, suggesting a hierarchy of formal perfection, gives the higher elements some hint of a superior cosmological status that was to be greatly emphasised in mediaeval discussions of the nature and moral connotations of birds (see Chapter 2).

Aristotle attributes to the elements a natural tendency to move either up or down to their proper 'places,' unless they are forcibly prevented from doing so:

> Not only when a thing is water is it in a sense potentially light, but when it has become air it may be still potentially light: for it may be that through some hindrance it does not occupy an upper position, whereas, if what hinders it is removed, it realizes its activity and continues to rise higher.[59]

If the world were perfectly ordered, with the elements arranged in concentric rings, there would be no movement, no actualising of the potential to rise or fall, nor would anything be in a position to create a 'hindrance.' Although the tendency of the elements to move is indeed natural to them, all actual movement comes about because of the constant series of influences and changes brought about by the First Mover.[60] Transmitted downwards by chains of

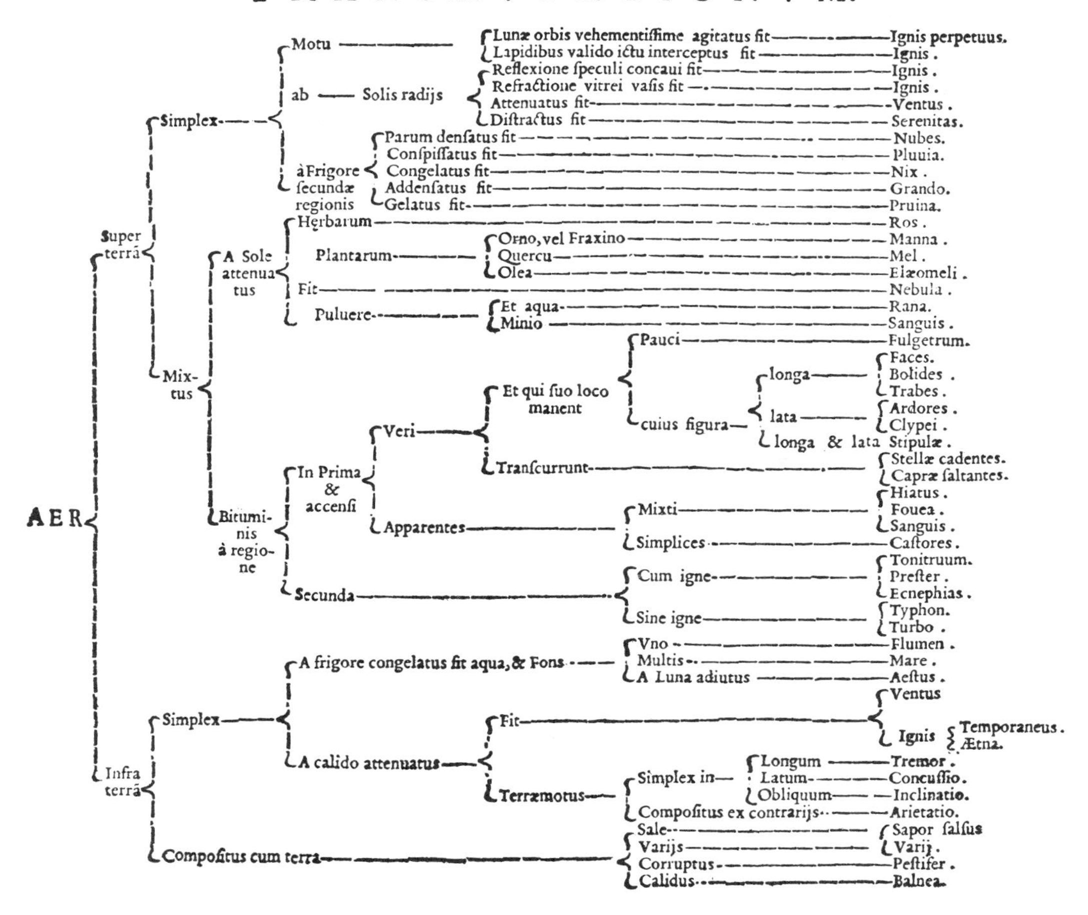

4. Table of the transmutations of the air. Giovanni Baptista della Porta, *De aeris transmutationibus libri iiii* (Romae 1614) before A[1].

contact, this movement continuously stirs the sublunary spheres, preventing the creation of homogeneous layers and causing the elements to realise their natural tendencies.

As the bodies of men and animals are made from the four elements, mixed in differing proportions, heaviness or lightness of spirit came to be associated almost as a matter of course with a preponderance of heavier or lighter elements, and with the balance of cold and hot, wet and dry. Among other contexts, this idea appears in comments prompted by the mediaeval fascination with dreams. Although a dream of flight might seem at first to suggest an admirable desire to rise above one's fallen condition, it was commonly interpreted as presumptuous, as an illegitimate desire to transcend the bounds of normal human activity.

5. The sublunary part of the Aristotelian cosmos, showing the three regions of the air. Gregorius Reisch, *Margarita philosophica nova* VII.I.41 (Ex Argentoraco veteri 1515). (Reproduced by permission of the British Library)

Regions of the air and the nature of wind

As a part of his argument to account for the origin and nature of wind, Aristotle proposed that the air was divided into three main regions, a theory that, with some elaboration and modification, became a commonplace of mediaeval and Renaissance meteorology. The lowest region, in which we live, was not only dense and moist but was also warmed by the sun's rays after their reflection from the earth's surface. Above this, at heights too great to be affected by the reflected warmth, was a cold moist region where the clouds and the usual weather phenomena originated. The second region extended as far as to the tops of the highest mountains, where 'the circumference . . . makes the earth a complete sphere.'[61] More than a philosophical speculation, this point was taken seriously by practical men, such as navigators, who lived outdoors. Christopher Columbus, for example, used the notional boundary between the first and second regions when estimating the height of new land. His son Ferdinand writes of 'seven little islands close to land which he said was the highest land he had seen up to that time; he believed that it rose above the zone of the air where storms are born . . . .'[62]

Above the region of storms the air was carried around by the rotation of the heavenly spheres, the movement preventing the formation of clouds in that region both because of the dispersal of vapours and because of the heat produced by friction. The clarity and warmth of the uppermost region were further increased by the proximity of the region of fire, which was to some degree mixed in with the air.[63] It is clear from many commentaries that the upper air was thought of as a transitional zone between the theatre of fallible earthly experience and the purity of the heavens. Despite the continuous circular movement of the upper region, it was commonly depicted as a region

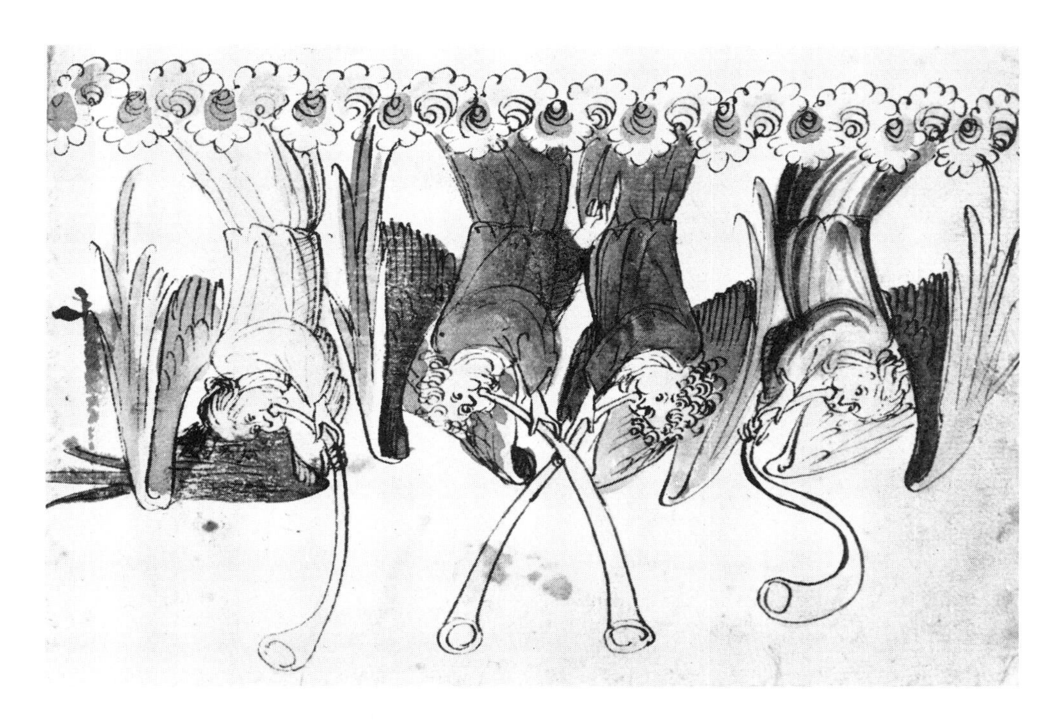

6. The four winds standing upside down on a line of clouds dividing the upper from the lower air. Vienna, Österreichische Nationalbibliothek, MS 3068, German, fifteenth century, f. 89[r]. (Reproduced by permission)

of great calm. While any attempt at human flight into such a place might be dangerous because of the supposed heat, it offered potential access to a state of something approaching bliss. Intimately relating the structure of the cosmos to central human concerns, this idea inevitably held great appeal for writers with a neo-Platonic tendency. In his account of the journey of Phronesis through the world, Alanus de Insulis, for example, sets the upper air in sharp contrast to the region of storms:

> Having traversed the expanse of Air to the place where the cloud-formations of heaven in their own night weave their mantle of darkness, where the hanging clouds collect water, where the hail hardens for a shower, where the winds contend, where the rage of the lightning swells, Phronesis passes on to Ether where the charm of ultimate peace prevails, where pleasing quiet and a breeze more pleasing still caress everything, where all is silence, where a clearer atmosphere smiles and banishes tears, where the veiled air sighs and shines throughout with a mystic light.[64]

The idea of a sharp division between the regions of the air is vividly realised in Figure 6 (fifteenth century). The sun and the moon shine in the upper atmosphere, which is divided from the lower by a line of clouds. Although the clouds are rendered with conventional formality, it is clear that they are imagined as forming a distinct line of demarcation amounting almost to a barrier. The spirits of the four cardinal winds stand upside down on the clouds, lending them still more an appearance of solidity.

Although inevitably wrong in some fundamental matters, Aristotle's description of the structure of the atmosphere is not so very far from the truth. His comments on density and temperature gradients were broadly true, and he was correct in supposing that the warmth of the lower region is mainly due to radiation from the earth rather than to direct heating by the sun's rays. For Aristotelian scientists and technologists, the material medium with which we were in direct contact and which might conceivably be used to enable us to rise to the heavens was thus, like mankind, of mixed nature. While its physical character was manifest, its mysterious invisibility led to its being frequently used to represent the soul, or spirit. Explanations of bird flight, and speculations about manned flight, invoked both the physical and the metaphysical aspects of these beliefs, flight away from the regions of impure human experience being equated with the search for the pure and serene upper air that so tempted Icarus. As the example of Daedalus and Icarus showed, caution was necessary if man were to dabble in experience outside his normal sphere of activity. The lower air, being almost water, offered a safer if less exciting and beautiful medium, any attempt to reach the upper air being not only physically but also spiritually hazardous.

Belief in the existence of discrete regions of the air continued on into the seventeenth century and even later. Writing to Marin Mersenne on 13 May 1634, Christophe de Villiers makes a passing allusion to manned flight, which was one of Mersenne's constant preoccupations. After saying that he thinks manned flight a real possibility, de Villiers continues:

> I leave aside the character of the upper air, not believing that for flight it is necessary to rise higher than the bottom region, in which this art, properly established and made familiar, could very well show its necessity.[65]

A modernised but still recognizably Aristotelian description of the three regions of the air is found as late as 1791 in Erasmus Darwin's *The Botanic Garden:*

> There seem to be three concentric strata of our incumbent atmosphere. . . . First, the lower region of air, or that which is dense enough to resist by the adhesion of its particles the descent of condensed vapour, or clouds. . . . The second region of the atmosphere I suppose to be that which has too little tenacity to support condensed vapour or clouds; but which yet contains invisible vapour, or water in aerial solution. . . . The second region or stratum of air terminates I suppose where the twilight ceases to be refracted . . . and where it seems probable that the common air ends, and is surrounded by an atmosphere of inflammable gas tenfold rarer than itself.[66]

The nature of wind, which we now understand to be merely the movement of the air, frequently puzzled early scientists. Although the idea that wind was air in motion was known to the Greeks, Aristotle rejected it in favour of a more complex theory that led to widespread controversy. An important part of Aristotelian meteorology had to do with the formation of 'vapours' and 'exhalations' arising from the effects of heat on parts of the earth's surface. The exhalations, drawn upwards by the sun and mixing with the spheres of air and fire, were responsible for thunder, lightning, and other 'hot' weather phenomena. When the exhalations moved about in the lowest region of the air, close to the earth's surface, they formed the winds. Although winds were occasionally thought to be found in the middle and upper regions, scientists normally adopted Aristotle's own view that they were characteristic of the lower region only. The cold middle region was turbulent and strife ridden but not because of the presence of wind in the ordinary sense. Here the cold air found itself besieged from both sides by the warmer air above and below, the resulting continuous battle being thought of as the immediate cause of the more violent and spectacular 'meteors' and storms.

Because of his assumptions about the nature of motion, Aristotle's explanation of the way exhalations caused wind was necessarily complex. Since natural movement was either straight up or straight down, he felt a need to explain how the horizontal, or transverse, movement of the earth's exhalations could come about. His account invokes the idea that air above the tops of the highest mountains is carried around in a circular direction by the heavens. After first rising straight up, the exhalation is deflected sideways by the circular motion, so causing the horizontal wind to appear to originate from above:

> The course of winds is oblique: for though the evaporation rises straight up from the earth, they blow round it because all the surrounding air follows the motion of the heavens. Hence the question might be asked whether winds originate from above or from below. The motion comes from above: before we feel the wind blowing the air betrays its presence if there are clouds or a mist, for their motion shows that the wind has begun to blow before it has actually reached us; and this implies that the source of winds is above. But since wind is defined as 'a quantity of dry evaporation from the earth moving round the earth,' it is clear that while the origin of the motion is from above, the matter and the generation of wind come from below. The oblique movement of the rising evaporation is caused from above: for the motion of the heavens determines the processes that are at a distance from the earth, and the motion from below is vertical and every cause is more active where it is nearest to the effect; but in its generation and origin wind plainly derives from the earth.[67]

In mediaeval and Renaissance times this explanation was supplemented by another based on a resolution of opposing movements and forces: The upward movement of the exhalation continued until it encountered the downward force of the sun's heat, which deflected it obliquely sideways. William Fulke described the process: 'The wynd is an *Exhalation* whote and drie, drawne up into the aire by the power of the sunne, & by reason of the wayght therof being driven down, is laterally or sidelongs caried about the earth.'[68] A more elaborate explanation appealed to the effects of the energy produced at the boundaries of the three main regions of the atmosphere. The rising warm exhalation continued upwards until it encountered the cold middle region, where the clash of hot and cold caused it to be repelled and to move violently sideways.[69]

The difficulty experienced by mediaeval and Renaissance scientists in trying to establish a rational distinction between air and wind is well revealed in the extensive notes of Leonardo da Vinci (1452–1519). While treating the winds as having special characteristics and a generally Aristotelian genesis, Leonardo takes issue with many of the details of the standard account:

> You say that the vapour which generates the wind is carried upwards by heat and pressed down again by cold; which having been said, it follows in course that this vapour, finding itself between two contrary motions, escapes to the sides; and this lateral movement is the wind, which has a tortuous movement because it cannot descend to the earth because the heat pushes it up, and it cannot move very high up because the cold presses it down; hence this necessity gives it a latitudinal and tortuous movement. Now many drawbacks will follow from this theory of yours, of which the first is that the wind will never descend to the plain, and secondly that the cold in being driven down by such a vapour would be acting contrary to its inert nature.[70]

Retaining many of the Aristotelian categories, Leonardo nevertheless fails to free himself entirely from the air/wind distinction:

> Every movable thing continues its movement in the shortest way and either shuns the obstacle or is bent by the obstacles; therefore the wind curves in penetrating the thick air, and bends upwards towards the light air.[71]

Even when, a few years later, Leonardo virtually contradicts the basis of that comment, the distinction is not wholly discarded:

> The air moves like a river and carries the clouds with it; just as running water carries all the things that float upon it. This is proved because if the wind were to penetrate through the air and drive the clouds these clouds would be condensed between the air and the moving force and would take a lateral impress from the two opposing extremities, just as wax does when pressed between the fingers.[72]

Although Renaissance poets do not often comment expressly on the distinction between air and wind, a conceptual difference may sometimes be felt, especially when they use personification. Imagery is in any case sharpened if an implicit difference is understood, the current of air being conceived of as a discrete entity. Wyatt's simile 'That fleith as fast as clowd afore the wynde'[73] contains an image that for readers in the sixteenth century was more clearly focussed than it is for us.

Apart from the familiar dangers of strong winds, exhalations were thought to create other hazards. Not only were they the cause of violent storms, but they were also, in Aristotelian meteorology, responsible for earthquakes. Struggling to break free of the earth's surface, they sometimes pushed the ground violently aside. Although this idea may at first seem far fetched, it will not appear so ridiculous if considered in relation to the violent explosion of gases from volcanoes, the eruptions of which are often accompanied by earth tremors. In view of its association with many violent natural phenomena, the air was obviously to be treated with respect.

Two further misapprehensions about the nature and effects of wind caused difficulty to mediaeval and Renaissance Aristotelians when they attempted to develop insights into the principle of flight. The first was the failure of most observers of soaring birds to understand that the winds move not only horizontally but also vertically, and that birds can support themselves on columns or waves of rising air. In the absence of upcurrents, the continuous flight of gliding birds would indeed seem to be a mystery requiring special explanation. The second has to do with the relationship between the horizontal force of the wind and the downward-tending force of weight. The idea that, after the effects of inertia have been overcome, an object moved by two such forces at right angles to each other may be thought of as experiencing a single force compounded of the original two was slow to develop. The assumption underlying many commentaries on bird flight before the late nineteenth century is that the line of descent followed by a bird when gliding downwards in a steady horizontal wind is somehow held firm against the effect of the wind, almost as if the bird were sliding down a fixed wire. Even when flying downwind, therefore, the bird always feels air pressure from the windward quarter. Although some commentators such as Leonardo understand that the 'wire,' or flight path, is moved with the wind, they almost invariably think of the flight path as moving more slowly than the wind, so that some relative air pressure continues to be felt.[74]

For mediaeval and Renaissance thinkers the effects produced by the invisible air provoked continual surprise and speculation. Even when not disturbed by wind, it was more than a passive, transparent medium and could influence our assessment of things perceived through it, such as the size and nature of celestial objects. Since they were too distant to be measured directly, the stars and planets, and even the clouds, had to be interpreted on the basis of visual information that, having to traverse the air, was sometimes thought liable to change en route. As sight was traditionally the most fallible of the senses, it was easy to suppose that the fickle female air could lead one into error. In his rimed disquisition on Savoy, Jacques Peletier takes up this point. Knowing from experience that the cloud that seems shapely from a distance is found to be formless when one climbs up the mountain into it, he writes:

> The clouds themselves are convex only in appearance and not in truth. For the intervening air, made heavy or subtle by the power of the sun, and the space across which one perceives, makes the sense which sees, or thinks it sees, easiest of all the senses to deceive.[75]

Changing ideas

Fundamentally new ideas were generated in the seventeenth century when the great upsurge of interest in experimental science led to an especially intense concern with the nature and properties of air. Making the point aptly and succinctly, Marie Boas Hall says: 'Pneumatics was to the seventeenth century what the theory of light and colours was to the eighteenth century: the supreme example of science which could be popularized, because it could be made to fire the imagination, and because it illustrated the philosophic basis of the new experimental, mechanical natural philosophy.'[76] Important experiments were conducted in many centres of learning, the most significant and influential work in England being that of Robert Boyle (1627–91), who invented a simple and efficient air pump. Fully reported in his many publications and letters, his discoveries became rapidly known to an international audience.

The study of the air had a direct bearing on many branches of natural science. Relevant not only to chemistry and biology, it was intimately involved in late Renaissance discussions of cosmology, gravity, and light, and was frequently invoked in arguments about the propositions of Cartesian mechanics. The degree to which it stirred the imagination of seventeenth and eighteenth century scientists may be gauged from the plea voiced by Stephen Hales in the final paragraphs of a chapter on air in his *Vegetable Staticks:*

> May we not with good reason adopt this now fixt, now volatile *Proteus* among the chymical principles, and that a very active one, as well as acid sulphur; notwithstanding it has hitherto been overlooked and rejected by Chymists, as no way intitled to that denomination?
>
> If those who unhappily spent their time and substance in search after an imaginary production, that was to reduce all things to gold, had, instead of that fruitless pursuit, bestowed their labour in searching after this much neglected volatile *Hermes,* who has so often escaped thro' their burst receivers, in the disguise of a subtile spirit, a meer flatulent explosive matter; they would then instead of reaping vanity, have found their researches rewarded with very considerable and useful discoveries.[77]

I shall make no attempt even to summarise the range of topics relevant to this new interest in the air but shall confine my attention to a few matters most immediately connected with the idea of flight.

Although by the seventeenth century the air was no longer universally considered to be an 'element' in the full Aristotelian sense, it was commonly thought of as essentially a single substance in which, as in water, other substances might be dissolved or mixed. A simple, 'true' air absorbed the various vapours and exhalations that were generated from time to time out of the earth. The character of these propositions is revealed in the following representative quotations from Boyle's *The General History of the Air,* an incomplete compendium published in 1692, the year after his death:

> By the *Air* I commonly understand that thin, fluid, diaphanous, compressible and dilatable Body in which we breath, and wherein we move, which envelops the Earth on all sides to a great height above the highest Mountains; but yet is so different from *Æther* [or *Vacuum*] in the intermundane or interplanetary Spaces, that it refracts the Rays of the Moon and other remoter Luminaries . . . .
>
> According to my Thoughts, the Air may be taken *either* for that which is *Temporary,* (if I may so call it;) *or* in a *Transient* State; *or* that which is *Lasting,* and in a *Permanent* State. . . . if you sufficiently heat an Eolipile furnished with Water, and stay a pretty while to afford time for the expulsion of the *Aerial* Particles by the *Aqueous* Vapours, you may afterwards observe, that these last named will be driven out in multitudes, and with a noise, and will emulate a Wind or Stream of Air, by blowing Coals, held at a convenient distance, like a pair of Bellows, and by producing a sharp and whistling Sound against the edg of a Knife, held in a convenient Posture almost upon the Orifice of the Pipe, whence they issue out. But this vapid Stream, though in these, and some other things it imitates true Air, whilst the vehement Agitation lasts, which the Vapours it consists of, received from the Fire; yet in a very short time, especially if the Weather, or the Vessels it enters into, be cold, loses the *temporary* Form it seemed to have of Air, and returns to Water, as it was at first. . . .
>
> It seems then not improbable to me, that our *Atmospherical* Air may consist of *three* differing Kinds of Corpuscles. The *first* is made of that numberless Multitude and great Variety of Particles, which, under the form of Vapours or dry Exhalations, ascend from the Earth, Water, Minerals, Vegetables and Animals, *&c.* and in a word, of whatever Substances are elevated by the Celestial or Subterraneal Heats, and made to diffuse themselves into the Atmosphere. The *second* sort of Particles that make the Air, may be yet more subtile than the former, and consist of such exceeding minute Parts, as make up the *Magnetical* Steams of our Terrestrial Globe, and the innumerable Particles, that the Sun and other Stars, that seem to shine of themselves, do either emit out of their own Bodies, or by their Pressure thrust against our Eyes, and thereby produce what we call *Light* . . .
>
> I shall add a *third* sort of Atmospherical Particles, compared with which, I have not yet found any, whereto the Name of Air does so deservedly belong. And this sort of Particles are those, which are not only for a while, by manifest outward Agents, made Elastical, but are *permanently* so, and on that account may be stil'd *Perennial* Air.[78]

Not until well into the eighteenth century did scientists begin clearly to formulate the idea that 'true' or 'perennial' air is essentially a mixture of different 'gases.' Even at the end of the eighteenth century the view that the air is a single substance continued to be advanced. In Thomas Henry's translation of Lavoisier's *Opuscules physiques et chimiques,* published in 1776, readers were still offered Antoine Baumé's assertion that 'Air . . . is one and the same: there is only one species of air. This element may, and in fact does enter into an infinite number of combinations; but when it is disengaged from bodies with which it was combined, it recovers all its properties, and when it is properly purefied it in no wise differs from that which we respire.'[79]

Beliefs about the air and about the mechanics of flight were influenced by opinions on the existence or nonexistence of vacuum, which for centuries had been a matter of controversy.[80] Aristotle's strongly argued opinion that vacuum was impossible was followed by many physicists, and, although the question was continually raised in mediaeval and Renaissance treatises on the natural world, the standard position was to deny that a vacuum could exist. Strong competition against this view was nevertheless offered by atomist theories of matter, adopted by, among others, Hero of Alexandria (first century A.D.). Although there is no self-evident contradiction between the compressibility of the air and Aristotle's belief that, together with all other matter, it is a continuous substance, Hero found it easier to account for compressibility by thinking of atoms with empty space around them being squeezed closer together:

> When any force is applied to it, the air is compressed, and, contrary to its nature, falls into the vacant spaces from the pressure exerted on its particles: but when the force is withdrawn, the air returns again to its former position from the elasticity of its particles, as is the case with horn shavings and sponge, which, when compressed and set free again, return to the same position and exhibit the same bulk.[81]

Atomic and continuum theories of matter coexisted until well into the eighteenth century, the choice of approach having some influence on ideas about how movement through the air is possible. While the pushing aside of atoms surrounded by vacuum is relatively easy to understand, it is not at first clear how a continuum may be penetrated. The arguments, forcefully put by Lucretius in the first century B.C., are implicit in many later discussions about the mechanics of flight.[82] The need for local disruption of a continuum accounts for a number of comments about the birds' use of sharp beaks and leading edges in order to 'cut' the air and so create a passage through it. In chapter I of *The Compleat Angler* (1653), Auceps, celebrating the flight of the falcon, says that she 'makes her nimble Pinions cut the fluid air, and so makes her high way over the steepest mountains and deepest rivers, and in her glorious carere looks with contempt upon those high Steeples and magnificent Palaces which we adore and wonder at.'

When Galileo's pupil Evangelista Torricelli (1608–47) created an apparently empty space on top of a column of mercury, many 'plenists' continued to deny the existence of true vacuum by recourse to the idea that a superfine *aether* was still to be found there. Variously used by classical writers to refer sometimes to the upper air, sometimes to the celestial 'fifth element' above the region of fire, sometimes to fire itself, the word *aether* has a complex history.[83]

Given vigorous new life in the seventeenth century, it was used as the name of a hypothetical substance permeating the universe and so maintaining an unbroken continuum of matter. Robert Boyle and others attempted to provide experimental demonstration of its existence, but since they were unsuccessful many scientists grew increasingly sceptical of the hypothesis. It nevertheless continued to play an important role in scientific thought, particularly for those who, like Isaac Newton, found the idea of 'action-at-a-distance' difficult to accept. If the world were filled with *aether,* that would provide a medium through which forces such as gravity might act. Without it, and in the absence of the 'field theories' of the nineteenth and twentieth centuries, many scientists found the mode of operation of such forces incomprehensible. The 'new philosophy' having called so much in doubt, the element of fire was taken less seriously than had been the case in the Middle Ages, and by the seventeenth century little credence was given either to its role in the composition of physical objects or to the existence of a fourth elementary sphere above the air. The revived *aether* theories nevertheless provided a partial substitute to fill the gap in nature. The domestication of the fifth element, bringing the cosmic brightness down to play an all-pervasive role in the sublunary world, allowed something like fire to survive in hiding, so to speak, within the material universe.

While much mediaeval science had concerned itself with forces of affinity and attraction, objects tending to move towards their natural places, seventeenth century science was more prone to think in terms of pressures, of 'push' rather than 'pull.' Not only had the pressure of the air become inescapably evident, but hypotheses were formulated about '*aether* pressure' and '*aether* winds.' From this time on, discussions of bird flight tended to be dominated by analyses of pressure systems.

Of particular interest to those who theorised about the mechanisms of flight was the growing evidence that air has weight. While in the many mediaeval controversies about relative and absolute heaviness and lightness the proposition that the air has absolute weight was sometimes advanced, the Aristotelian view that air tends naturally upwards, being relatively heavy only in comparison with fire, was generally accepted. In the seventeenth century, however, the experiments of Torricelli proved conclusively that air has weight and that its accumulated layers in the atmosphere create a sufficient pressure to sustain a column of mercury about 30 inches high.

Experiments concerned with the weight of the air led to further investigations of its compressibility. That the air was compressible had long been understood. Simple experiments designed to demonstrate this were described by Hero of Alexandria, who implied that the results would merely confirm his reader's expectations. Again, however, it was not until the seventeenth century that the phenomenon of compressibility was investigated by the use of fully quantitative techniques. With the development of Boyle's air pump it became possible for the first time to measure the forces required to compress a given quantity of air, and therefore to appreciate what large amounts of energy could be stored in this way. Measurements by Robert Hooke and others established that at a constant temperature there is a simple inverse relationship between the volume of a body of air and the pressure to which it is subjected (a relationship now known as 'Boyle's Law'). Extrapolating from this, and imagining a column of air compressing itself under its own weight, Hooke reasoned that the atmosphere must be increasingly tenuous with increasing height, and that it probably extends indefinitely.[84] The Aristotelian idea of an upper region of subtle, purer air was thus transformed into the concept of an atmosphere that grew ever thinner as one rose but that in all other respects remained essentially similar.

Closely related to experiments with the compressibility of the air were considerations of its 'elasticity.' Independently developing the concept that Hero had included in his analogy with horn shavings and sponges, Boyle, Borelli, and others attempted to account for the air's resistance to compression by describing the particles of which it is composed as if they were little coils or springy cylinders:

> Of the Structure of the Elastical Particles of the Air, divers Conceptions may be framed, according to the several Contrivances Men may devise to answer the *Phenomena:* For one may think them to be like the *Springs* of *Watches,* coil'd up, and still endeavouring to fly abroad. One may also fancy a Portion of Air to be like a Lock or Parcel of curled *Hairs* of *Wooll;* which being compressed by an external Force, or their own Weight, may have a continual endeavour to stretch themselves out, and thrust away the neighbouring Particles, and whatever other Bodies would hinder them to recover their former State, or attain their full Liberty. One may also fancy them like extreamly *slender Wires,* such as those of Gold and Silver, that Tradesmen unwind from some Cylindrical Bodies of differing Sizes, on which they were rolled; which Pieces of Spiral or curled Wire may be, *as* of differing Substances and Consistences, *so* of very differing Lengths and Thicknesses, and have their Curls greater or lesser, nearer each other, or more distant, and be otherwise diversified; and yet all have Springiness in them, and (notwithstanding) be, by reason of their Shape, readily expansible on the score of their native Structure, as also by Heat, Girations, and other Motions; and compressible by an external Force into a very little room. I remember too, that I have among other Comparisons of this kind, represented the springy Particles of the Air, like the very *thin Shavings of Wood,* (that Carpenters and Joiners are wont to take off with their Plainers;) for, besides that these may be made of differing Woods, as Oak, Ash, Firr, *&c.* and thereby be diversified as to their Substance, they are usually of very various Breadths, and Lengths, and Thicknesses.[85]

Although not everyone was happy with such models, the notion of the air as composed of particles in which elastic energy was stored became commonplace. The pressure of the atmosphere and the powerful forces which it could exert when subjected to artificial compression or rarefaction gave renewed interest to the theory and to the embryonic practice of flight. From the middle of the seventeenth century onwards it was understood that the air offered much greater sources of energy than had usually been suspected; while man's muscles might be weak, the air itself might be manipulated and harnessed so as to make flight possible. Birds were increasingly thought to fly by using the elastic power of the air; man, it was expected with confidence, would one day learn to do the same.

The creatures of the air 2

Birds

Of all animals, the birds have most persistently been associated with divinity. Their beauty, their melodious song, and above all their capacity to move upwards, at will, towards the heavens, have given them a central place in the iconography of spiritual life. In Western thought the tradition goes back at least as far as Plato, who writes of bird wings and bird flight in a lyrical passage of the *Phaedrus:*

> The natural function of the wing is to soar upwards and carry that which is heavy up to the place where dwells the race of the gods. More than any other thing that pertains to the body it partakes of the nature of the divine. But the divine is beauty, wisdom, goodness, and all such qualities; by these then the wings of the soul are nourished and grow . . .[1]

The tradition remains alive through early Christian times and on into the religious symbolism of the Renaissance. Writing in his highly individual and characteristic manner about Genesis 1:20, Origen (A.D. 185 or 186–253) constrasts the birds with the 'moving creature that hath life,' or reptiles (*reptile animae viventis*):

> If we look on a woman to lust after her, we have become a venomous reptile; but if we have a sense of restraint, even when an Egyptian casts her eyes upon us we shall become a bird, leaving between her hands our Egyptian clothing and flying from her infamous snares. If we are tempted to steal, we are in danger of acting as a detestable reptile; but if, even though we have only two mites, we think of casting them in unto the offerings of God, we shall act as a bird, freed from thoughts of earthly things and guiding ourselves with our wings towards the firmament. If we allow ourselves to think that we need not go so far as to endure the sufferings of a martyr, that is to be as a venomous reptile; but if we are moved by the idea that we must fight to the death for truth, that is the thought of a bird which will free us from the earth and raise us to heaven. The same reasoning may be applied to any other sins and virtues. And thus it is necessary to separate the creeping beings from the birds, since the waters in us received from the sight of God the order to produce them that they might be distinguished.[2]

Two passages about birds by Saint John Chrysostom (c. 345–407) have a decidedly Platonic character:

> For this reason wings are given to birds; that they may avoid snares. For this reason men have the power of thinking; that they may avoid sin.[3]

> The eagle has his light pinion; but I have reason and art, by which I am enabled to bring down and master all the winged animals. But if thou wouldest see my pinion too, I have one much lighter than he; one which can soar, not merely ten or twenty stadia, or even as high as heaven, but above heaven itself, and above the heaven of heavens; even to 'where Christ sitteth at the right hand of God!'[4]

As the second of these passages indicates, the limited intelligence of birds, together with their mixed nature and middle position between heaven and earth, led to their being viewed with a good deal of ambivalence. In *De divisione naturae,* Eriugena (c. 810–c. 877), who stressed the unity, purity, and remoteness of God, described them as far from perfect creatures. It is because no physical creature—not even the birds—could participate in the wholly subtle and spiritual quality of the heavens that they are commanded in Genesis 1:20 to fly 'below' the firmament (*sub firmamento caeli*).[5] Associated since the earliest times with the soul, symbolic birds could of course fly higher. While bird souls are usually depicted on their way towards salvation in the heavenly regions, evil bird souls are occasionally encountered. Both good and bad kinds were seen by Saint Anthony of Egypt (c. A.D. 250–356) in a remarkable vision reported by Palladius (A.D. 419 or 420). A great giant stood with his feet in a lake the size of a sea and with his hands stretched up to heaven, while around him flew the bird souls of the dead. Those birds that flew higher than his head and hands were attended by angels, to symbolise their salvation, while those that flew low enough to be struck by the giant's hands were hurled down to the lake and to damnation. The low-flying birds represented souls attracted by the desires of the flesh.[6]

When, at the end of the *Timaeus,* Plato himself writes of birds, he does so in a dismissive manner, associating them with a superficial approach to participation in the divine. Along with women and the lower animals, birds are spoken of as if they were degenerate men. While superior to the land and water animals, they represent essentially unsatisfactory aspects of experience:

> Birds were made by transformation: growing feathers instead of hair, they came from harmless but light-witted men, who studied the heavens but imagined in their simplicity that the surest evidence in these matters comes through the eye.[7]

Having told 'the story of the universe as far as the generation of man,' and having given an account of the spiritual existence that corresponds with the element of fire, Plato concludes the *Timaeus* with a brief description of the creatures that inhabit the other three sublunary elements: 'second, winged things whose path is in the air; third, all that dwell in the water; and fourth, all that goes on foot on the dry land.'[8]

While Plato does not question the assignment of the birds to the air, 'solider Aristotle' finds it difficult to define what naturally belongs to each of the elemental regions. Part of Aristotle's argument proving the divinity of the cosmos depends on the presence of living creatures in all of the elements, including the fifth and extraterrestrial element, the *aether.* Although the stars and planets were clearly conceived of as living creatures vastly superior but also analogous to the more familiar inhabitants of the earth, it was not obvious in what sense each of the four lower elements could be said to be fully inhabited. Aristotle does not attempt to make a simple division of the lower animals such as Plato had included in the *Timaeus,* but says: 'Plants may be assigned to land, the aquatic animals to water, the land animals to air.'[9] Thus the inhabitants of the air include not only the birds but all animals that breathe.

While some winged creatures were more closely adapted to the air than were others, there being a wide range from flightless birds to 'certain flying serpents in Ethiopia' that were said to lack feet,[10] most, if not all, flying creatures of which we have direct knowledge had much in common with the heavy earth. Insect flight, which amused and puzzled Aristotle, seemed far from truly aerial:

> The flight of insects is slow and frail because the character of their feathery wings is not proportionate to the bulk of their body; this is heavy, their wings small and frail, and so the flight they use is like a cargo boat attempting to make its voyage with oars.[11]

Aristotle is careful to draw structural parallels between birds and land animals, not only pointing out that birds have feet, but also alleging that they 'cannot in fact fly if their legs are removed.'[12] Elsewhere he takes these qualifications about the 'airiness' of flying creatures much further, saying in *Historia animalium* that 'no creature is able only to move by flying, as the fish is able only to swim.'[13] In the spurious Book IV of the *Meteorologica* the point is made more directly:

> Of the elements earth is especially representative of the dry, water of the moist, and therefore all determinate bodies in our world involve earth and water. Every body shows the quality of that element which predominates in it. It is because earth and water are the material elements of all bodies that animals live in them alone and not in air or fire.[14]

Aristotle nevertheless conceived of all creatures as having some degree of natural affinity not only with the element from which they were mainly constituted, but also with the regions in which they moved. This assumption is expressed in *De respiratione,* in which he speaks of the constitution of different classes of animals:

> For some have a greater proportion of earth in their composition, like plants, and others, e.g. aquatic animals, contain a larger amount of water; while winged and terrestrial animals have an excess of air and fire respectively. It is always in the region proper to the element preponderating in the scheme of their constitution that things exist.[15]

The idea of a close relationship between living creatures and their medium of movement is at the basis of a beautiful passage in the pseudo-Aristotelian *De mundo,* in which the author, more inclined to mystical lyricism than was Aristotle himself, discusses the natural order of the universe:

> For one thing, moved by another, itself in due order moves something else, each acting according to its own constitution, and not all following the same course but different and various and sometimes even contrary courses . . . It is just as though . . . . one should hold in the folds of a garment a water-animal, a land-animal, and a bird, and let them go; clearly the animal that swims will leap into its own element and swim away, the land-animal will creep away to its own haunts and pastures, the bird of the air will raise itself aloft from the earth and fly away, though one original cause gave each its aptitude for movement. So is it with the universe.[16]

Fire animals

The most immediate difficulty associated with the assigning of classes of animals to the elements had to do with the region of fire above the atmosphere. Were living creatures to be found there, and if so, were there any animals that could survive in ordinary fire of the kind with which men are familiar? The question puzzled Aristotle, and continued to puzzle cosmologists until well into Renaissance times. The curious and vague remarks in *De respiratione* about air and fire, and the avoidance of specific comment on the nature of the habitat of flying creatures reflect the difficulty of attaching any notion of reality to such creatures. Aristotle appears to have felt unsure whether life and fire were compatible. On one hand he was attracted by the harmony and symmetry of a theoretical system that would allow all four of the sublunary spheres to be inhabited: in *De generatione animalium* he plainly wants to believe in the existence of fire animals, saying that 'there certainly ought to be some animal corresponding to the element of fire, for this is counted in as the fourth of the elementary bodies.'[17] On the other hand Aristotle's empirical common sense led him to doubt the physical reality of the proposition: elsewhere in *De generatione animalium* he says bluntly that 'fire generates no animal.'[18] Book IV of *Meteorologica* is still more explicit, asserting that animals exist only on land and in the water, and not in air or in fire.[19]

Although Aristotle had little to say about creatures that might be found above the region of the air, he continued to ponder the possibility that some animals might be able to survive in terrestrial fire. When he came to write *Historia animalium,* in which he included notes on various commonly held beliefs about the animal world, he was willing to give indulgent credit to the idea that there existed certain insects and 'salamanders' whose natural home was fire:

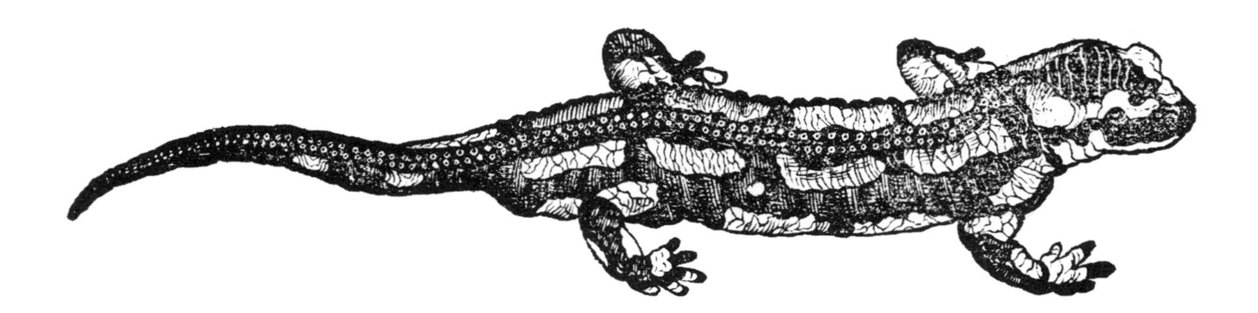

7. Salamander, said to have been able to live in fire. Conrad Gesner, *Historiae animalium* II (Tiguri 1555) part 1, p. 74. (Reproduced by permission of the Syndics of Cambridge University Library)

> In Cyprus, in places where copper-ore is smelted, with heaps of the ore piled on day after day, an animal is engendered in the fire, somewhat larger than a bluebottle fly, furnished with wings, which can hop or crawl through the fire. And . . . these latter animals perish when you keep [them] away from the fire . . . Now the salamander is a clear case in point, to show us that animals do actually exist that fire cannot destroy; for this creature, so the story goes, not only walks through the fire but puts it out in doing so.[20]

The inconsistency inherent in the notion of a creature that both lives in fire and puts fire out, popularised by the *Physiologus* and the bestiaries, was explained away by some of the later commentators with the suggestion that the salamander is endowed with an exceptional degree of protective natural coldness. While it can therefore survive in a large fire, a small one is likely to be extinguished by the cold. The process of rationalising is plainly seen in a fifteenth century bestiary that quotes the sixth century encyclopaedist, Isidor: 'It lives in the midst of the flames without pain and without being consumed, and not only is it not burned up, but it puts the fire out.' The writer of the bestiary goes on to say, however, that the fire is put out 'if it is small and cannot prevail against the natural qualities of the salamander.'[21] As he also judiciously points out, 'various opinions have been expressed about salamanders.' Some, like Apuleius (2nd century A.D.), were prepared to give full credit to the passage from *Historia animalium,* while others, like Aelian, who wrote a generation or so later,[22] denied that such creatures could be generated from flames however close their association with fire. Throughout the Middle Ages the compatibility of salamanders with fire received a great deal of attention, although the process of their generation was usually left unresolved.

The salamander's imperviousness to fire, coupled with its traditional reputation as sexless, led to its adoption, along with the turtledove, as an emblem of chastity. The writer of the Latin version of the *Physiologus* (4th–5th century A.D.) explores the salamander's symbolic potential further. Relating it to the story in Daniel about the boys who escaped unhurt from the fiery furnace, he finds it expressive of the ideas of ascension and indomitable spirit.[23]

Aristotle's difficulties with fire animals sprang from his fundamental desire to conceive of the world as a unified and continuous whole. Their nature was a puzzle not so much because of the problem of how they could survive in an environment of elemental fire, which was not necessarily believed to cause combustion when 'in its own place,' but because of the apparent incompatibility of an ethereal habitat and a physical creature that must have some earthy attributes. Trying to avoid at least a part of this objection, he offers an ingenious suggestion in *De generatione animalium.* After assigning creatures to the other three elements, he continues:

> The fourth class must not be sought in these regions . . . Such a kind of animal must be sought in the moon, for this appears to participate in the element removed in the third degree from earth [i.e., in the region of fire].[24]

Although salamanders and fireflies might exist on earth, the true home of the fire animals was thus seen to lie at a midpoint between the earth and the heavens.

Truly aerial creatures

While the problem of the fire animals was serious, the true character of the creatures of the air soon created still greater difficulties. Despite Aristotle's readiness to assign the breathing land animals to the air, the absence of any creatures that spent the whole of their time in the air, on the analogy of the fish in the water, destroyed the symmetry of the system. The existence of purely airy creatures could nevertheless be admitted if one were to posit a middle order of beings whose nature enabled them to mediate between men and the gods. Such beings, the daemons, had been proposed in a famous and seminal passage of Plato's *Symposium.*[25] A transformation of the idea of an animate atmosphere, itself a mediating element, this proposition was developed in detail by Apuleius, whose *De deo Socratis* became the primary source for many later discussions of the matter. Like Aristotle, Apuleius thinks it inappropriate to call the birds truly aerial:

> For you have every reason to pronounce his opinion false who assigns the birds to the air; for not one of them raises itself above the summit of Mount Olympus, which, though it is said to be the highest of all mountains, yet if you measure its height in a straight line, the distance to its summit is not equal, according to the opinions of geometricians, to ten stadia; whereas the immense mass of air extends as far as the nearest portion of the cycle of the moon, beyond which æther takes its rise in an upward direction. What, then, are we to say of such a vast body of air, which ranges in extent from the nearest part of the revolutions of the moon as far as the highest summit of Mount Olympus? Will that, pray, be destitute of its own appropriate animated beings, and will this part of nature be without life, and impotent? Moreover, if you attentively consider the matter, birds themselves may, with greater propriety, be said to be terrestrial than aërial

> animals; for their whole living is always on the earth; there they procure food, and there they rest; and they only make a passage through that part of the air in flying which lies nearest to the earth.[26]

Suggesting how the gap may be filled, Apuleius paraphrases earlier philosophy[27] that argues for the existence not only of salamanders and other fire animals but also of truly airy creatures. Spirits and angels of the kind that he describes do not in any way depend on the earth for their survival:

> They are then by no means animals of an earthly nature, for such have a downward tendency, through their gravity. But neither are they of a fiery nature, lest they should be carried aloft by their heat. A certain middle nature, therefore, must be conceived by us, in conformity to the middle position of their locality, that so the nature of the inhabitants may be conformable to the nature of the region.[28]

Apuleius follows this with a long passage in which he describes the properties essential to such middle-region creatures if they are to remain easily suspended in the air. After postulating an equilibrium of heaviness and lightness, he draws an analogy with the clouds:

> That I may not appear to you to be devising things that are incredible, after the manner of the poets, I will just give you an example of this equipoised middle nature. We see the clouds unite in a way not much different from this tenuity of body; but if these were equally light as those bodies which are entirely devoid of weight, they would never cap the heights of a lofty mountain with, as it were, certain wreathed chains, depressed beneath its ridges, as we frequently perceive they do. On the other hand, if they were naturally so dense and so ponderous that no union with a more active levity could elevate them, they would certainly strike against the earth, of their own tendency, just like a mass of lead and stone. As it is, however, being pendulous and moveable, they are guided in this direction and in that by the winds amid the sea of air, in the same manner as ships, shifting sometimes in proximity and remoteness; for, if they are teeming with the moisture of water, they are depressed downward, as though for the purpose of bringing forth. And on this account it is that clouds that are more moist descend lower, in dusky masses, and with a slower motion, while those that are serene ascend higher, and are impelled like fleeces of wool, in white masses, and with a more rapid flight.[29]

Although *De deo Socratis* became the most important source, Apuleius was not alone in advancing such arguments. A briefer but similar case for the existence of aerial creatures had been made a century earlier by Philo:

> The universe must needs be filled through and through with life, and each of its primary elementary divisions contains the forms of life which are akin

> and suited to it. The earth has the creatures of the land, the sea and the rivers those that live in water, fire the fire-born, which are said to be found especially in Macedonia, and heaven has the stars. . . . And so the other element, the air, must needs be filled with living beings, though indeed they are invisible to us, since even the air itself is not visible to our senses . . . Nay, . . . if all the other elements produced no animal life, it were still the proper function of the air to do what none other did and bring forth living beings, since to it the seeds of vitality have been committed through the special bounty of the Creator.[30]

In another argument for the existence of such creatures, Philo stresses the nourishing role of air, which he calls 'the best of earth's elements':

> It is more to be expected that air should be the nurse of living creatures than that land and water should, seeing that it is air that has given vitality to the creatures of land and water, for the Great Artificer made air the principle of coherence in motionless bodies, the principle of growth in bodies which move but receive no sense-impressions, while in bodies that are susceptible of impulse and sense-impression He made it the principle of life.[31]

That the air is inhabited by beings of a physical though not earthy constitution became an established doctrine adopted in various forms by most Christian thinkers, both Catholic and Protestant, until well after the Reformation. In a provocative and characteristically idiosyncratic treatise on the inhabitants of the four elements—nymphs (water), sylphs (air), pygmies (earth), salamanders (fire)—Paracelsus offers encouragement to those who hope one day to fly. Pointing out that our environment is most like that of the creatures of the air, he draws a series of parallels between men and sylphs: like us, sylphs live in air and die without it; like us, they may be burned by fire; like us, they may be drowned in water or trapped in earth.[32]

Belief in the existence of the creatures of the air, especially favoured by writers of a neo-Platonist tendency, was reinforced by the mystical concept of the scale of nature or chain of being. As no links could properly be thought missing from the chain joining God to the humblest oyster, a whole genus of middle-region beings, more spiritual than man but more physical than God, had to be posited. As late as the seventeenth century, Sir Thomas Browne (1605–82) appealed to the doctrine to support his belief in witches:

> It is a riddle to me . . . how so many learned heads should so farre forget their Metaphysicks, and destroy the Ladder and scale of creatures, as to question the existence of Spirits: for my [owne] part, I have ever beleeved, and doe now know, that there are Witches; they that doubt of these, doe not onely deny them, but Spirits.[33]

Platonist writers arranged the aerial beings on the ladder of nature not only according to the degree to which they were ethereal or material, but also according to a scale of moral worth, the most virtuous spirits being those that lived nearest to God, the most evil those that lived close to man's fallen world. A forthright account of the character of these various spirits is given by the twelfth century Platonist Bernardus Silvestris:

> In the sublunar atmosphere the higher portion differs from the lower more by its climate than by its location. The highest levels are more refined, and somewhat warmer, being affected by the contingency of the fiery condition of the ether, so far as mean things can be influenced by great, or sluggish life by that which is more highly animated. The class of spirits who dwell in the atmosphere, but in serenity, maintain calm of mind, as they live in calm. Second in rank to these is the genius which is joined to man from the first stages of his conception, and shows him, by forebodings of mind, dreams, or portentous displays of external signs, the dangers to be avoided. The divinity of these beings is not wholly simple or pure, for it is enclosed in a body, albeit an ethereal one. For the creator drew forth the distilled essence of ethereal calm and ethereal fluidity, and adapted divine souls to a material which was, so to speak, unmixed. Since their bodies are virtually incorporated, and subtler than those of lower creatures, though coarser than those of higher powers, the feeble perception of man is unable to apprehend them. Below the midpoint of the teeming air wander evil spirits, and agents of the lord of cruelty. They cannot avoid the taint of earthly foulness, for they hover close to the surface of earth. These beings, having been only slightly cleansed of the ancient evil of matter, have been contained within extremely narrow bonds by the great foresight of God. And since they persist in wickedness and the desire to do harm, they are often empowered by divine decree to inflict torment on those stained with crime. Often, too, they decide for themselves, and inflict injury of their own accord. Often they insinuate themselves invisibly into minds at rest, or concerned with their own thoughts, through the power of suggestion. Often, assuming bodily existence, they assume the forms of the dead.[34]

Just how these essentially pre-Christian creatures might best be accommodated in the rest of Christian doctrine presented a problem. While it became usual, from as early as Saint Augustine, to assimilate them in some way to the angels and devils, the Bible's complete silence on the matter of the angels' origin left much room for doubt and argument. Were they coeternal with God? Were they created before the world? Were they created after it? The spiritual nature of the angels, and of their counterparts the demons, was largely dependent on the answers to such questions. Although Origen believed that all the angels were fallen creatures, their distance from God and the light reflecting the seriousness of their transgression,[35] it was more usual to suppose that, having been created at some time, the angels subsequently formed two groups: those that remained with God in a state of grace and those that, having fallen, lived somewhere much nearer to the world.[36]

The particular abode of the evil demons, or fallen angels, was the lower, thicker, more sullied part of the atmosphere. The idea may have begun with Saint Augustine, who once again raised, by implication, the issue of the ambiguity of the air. Although it may not seem possible to us, he says, that a creature could be locked away in the air, it can only serve the demons, 'having regard to their nature,' as a prison until the Day of Judgement.[37] An original and powerful variation appears in an early work by Eriugena, *De divina praedestinatione.*[38] Commenting on the nature of the eternal fires prepared for the devil and his angels (Matt. 25:41), he says that after the Resurrection we shall all in fact experience it. The bodies of the blessed, however, will have been transformed into an ethereal substance unharmed by fire, while the damned will be burned because they will have received bodies of air like that with which Satan was clothed after his fall. In this passage Eriugena explicitly relates the scale of moral worth to the hierarchy of potencies in the physical world, asserting that the higher elements can affect the lower, but not vice versa.

The more conventional image of the air as a prison was repeated and developed many times by later writers on hexaemeral themes. Taking up the point about relativity of perception, Alexander Neckam (1157–1217) of St Alban's, who in his early twenties became a distinguished professor in Paris, explains that the lower air in which the demons are imprisoned is 'an infernal region' in relation to the upper layers.[39] Anders Sunesøn (c. 1167–1228), Archbishop of Lund and another frequenter of the Parisian scholastics of the twelfth century, echoes Saint Augustine but adds torture to imprisonment: 'The air is a prison in which, on account of their crimes, pain will never cease to be meted out until the end of time.' Meanwhile they maliciously enjoy exercising their power to distress humankind.[40] The point was most forcefully made by Alanus de Insulis (twelfth century) who, in *Anticlaudianus,* depicts the evil of the demons as a perversion of the soul. In the demons' unhappy state, normal apprehensions are inverted so that their habitat, the air, which for others may be synonymous with liberation and joy, is perceived as a place of misery:

> Phronesis examines in more detail the hidden approaches of Air, its secret places and coverts and pursuing her investigation with keen perception, sees with clearer vision the wandering denizens of Air, for whom Air is a prison, the abyss a punishment, joy is sorrow, life is death, triumph is failure. Their minds, impaired by the degrading poison of envy, cause this venom to overflow on the human race so that it may suffer from a like illness and the same disease. They are the ones who ever arm themselves against us, lay low the unarmed, overcome the armed, rarely do they themselves surrender but once defeated they cannot renew the battle.[41]

Augustine, Anders, and others write of the 'cloudy air,' associating the evil spirits with a home more suited to their nature than is the clearer upper atmosphere. Alanus, by contrast, offers no qualification to the clarity of the air, so ironically and sadly felt to be a prison. While his metaphysical vision of the incongruity between dark, distorted perceptions and bright, transcendent reality has more poetic power, it was much less common. Not only Shakespeare's witches but innumerable other evil or dangerous creatures 'hover through the fog and filthy air.' From the earliest times the demons of the air had been held at least partly responsible for fouling their own nests, using their perverse powers to create the storms.[42] In his famous handbook of exemplary church activities, Thomas of Cantimpré gives ecclesiastical authority to this old pagan idea. God has granted the demons certain magic powers unknown to man. These they use to impel and stir the air, causing thunder, storms, and lightning.[43] The idea survived long into the Renaissance. Including in his *Hierarchie of the Blessed Angells* an acerbic description of the malicious demons, Thomas Heywood (d. c. 1650) directly attributes to them the creation of the corruptible middle region of the atmosphere:

> *Spirits of th' Aire are bold, proud, and ambitious,*
> *Envious tow'rd Mankinde, Spleenfull, and malicious:*
> *And these (by Gods permission) not alone*
> *Haue the cleare subtill aire to worke vpon,*
> *By causing thunders and tempestuous showr's,*
> *With harmefull windes: 'tis also in their pow'rs*
> *T'affright the earth with strange prodigious things,*
> *And what's our hurt, to them great pleasure brings.*[44]

Controversy was generated not only by the question of where the demons had their natural home, but also by whether their constitution were physical or incorporeal. While on one hand some Christian writers found Apuleius's attribution of a tenuous physical body difficult to accept, others were not sure that wholly immaterial beings could properly be said to exist below the heavens. Although conceding that many philosophers had adopted Apuleius's position, Saint Augustine remained doubtful.[45] Writing in *Civitas Dei* about the punishment of evil demons, he says that they may perhaps

> have a kind of body made of that dense and humid air which we feel strikes us when the wind is blowing . . . But if any one maintains that the devils have no bodies, this is not a matter either to be laboriously investigated, or to be debated with keenness. For why may we not assert that even immaterial spirits may, in some extraordinary way, yet really be pained by the punishment of material fire, if the spirits of men, which also are certainly immaterial, are both now contained in material members of the body, and in the world to come shall be indissolubly united to their own bodies?[46]

Despite Augustine's impatience with the matter, the debate continued. Three main positions could be adopted: first, that the spirits were wholly incorporeal and could therefore make no physical contact with mankind; second, that they were always endowed with some form of fine body, as Apuleius described; third, that they were essentially incorporeal but could at will assume or fashion a body for themselves.[47] That the demons had a corporeal existence of some kind was the most usual position, readily understood by ordinary people and accepted by many churchmen. A commonly held theory was that of Bishop Fulgentius (A.D. 467–533), who proposed that while the body of the unfallen angels was made of *aether,* which he equates with fire, that of the demons was formed from the lower element, air.[48] This convenient contrast is prompted and supported by Ps. 103 (104):4: 'Who maketh his angels spirits; his ministers a flaming fire.' The more sophisticated theory proposed that, although wholly spiritual, the middle creatures could adopt a body in order to communicate with men or to give the illusion of physical presence, and it was especially favoured by many Renaissance angelologists and demonologists. In his influential book on the angels, Henry Lawrence argued that the creation of a body from air was entirely possible:

> If you object that the ayre is improper to take figure or coulour, because it is so thin and transparent? The answer is, that although the ayre remaining in its rarity doth not reteyne figure or coulour, yet when it is condenced and thickened, it will doe both as appeares in the clouds.[49]

The extreme position, which made the angels and demons not only incorporeal but also essentially unable to communicate directly and physically with men, was less convincing to ordinary people and appeared to go counter to many well-attested accounts of visions and other supernatural experiences. It remained usual to believe that bodies could be made of air.

Because of the awe-inspiring power of all aerial creatures, it was not easy to persuade nervous people, as Calcidius tried to do, that there was no need to be frightened by the name 'demon,' which is, he says, 'indifferently fixed upon the good and the wicked, because neither does the name "angel" cause us to fear, although some angels are the servants of God . . . while others, as is well known, are the accomplices of the hostile power.'[50] As was doubtless natural, poets and philosophers gave far more attention to the malevolent spirits than to their more virtuous brethren whose normal habitat was the upper regions. Pierre Ronsard (1524–85) offers a comparatively rare exception. Writing at a time when the creatures of the air were viewed with superstitious terror, he speaks of the demons in the original Platonic sense as messengers of God. Not only do they inspire no fear, but they can be wholly tamed by the power of female beauty:

> *Legers Daimons, qui tenez de la terre,*
> *Et du haut ciel justement le milieu:*
> *Postes divins, divins postes de Dieu,*
> *Qui ses segrets nous apportez grand erre.*
> *Dites Courriers (ainsi ne vous enserre*
> *Quelque sorcier dans un cerne de feu)*
> *Razant noz champs, dites, a' vous point veu*
> *Cette beauté qui tant me fait la guerre?*

Si de fortune elle vous voit ça bas,
Libre par l'air vous ne refuirez pas,
Tant doucement sa douce force abuse,
Ou comme moy esclave vous fera,
Ou bien en pierre ell'vous transformera
D'un seul regard, ainsi qu'une Meduse.[51]

More common than Ronsard's easy acceptance is an ambiguous and self-protective playfulness. When the demons are transformed into semimalicious but attractive beings such as Puck and Ariel, whose origins lie both in Platonism and in pagan folklore,[52] their role in human affairs can be contemplated without undue disquiet.

Endowed with their delicate bodies, the aerial creatures were truly in their element. They were at home among the winds and clouds, their mixture of levity and gravity making flight not only a natural but also an effortless form of motion. Unhampered by man's earthy constitution, they could both intervene in mortal life and, when their moral character permitted, ascend at will towards the heavens. Not surprisingly, they could fly at great speed:

> Because of the greater mobility of their airy bodies they not only outpace men and wild animals, but are also incomparably faster than the flight of birds.[53]

Continuing his account of their skills, Saint Augustine had pointed out that as a consequence of their speed of movement, the spirits of the air are able to convey information so rapidly that they appear to be able to predict events.

Recalling their pre-Christian origins, the creatures of the air sometimes appeared partially disguised as messengers of another kind, the neoclassical gods of Renaissance art. The transformation was easily made, the gods themselves having been admitted into Christian iconography as a band of ancillary superhuman powers akin to the angels though more secular in character. Since in this form the creatures of the air are relieved of much of their heavy moral burden, the flight imagery associated with them tends to lay stress on mechanical matters to do with the size and shape of wings and the strength and direction of the winds. The absence of a physical barrier to the flight of such beings is implicit in an amusing passage concerning the god Mercury in Aurelian Townshend's masque *Albions Triumph,* staged by Inigo Jones in 16$\frac{31}{32}$. His wings having been singed by a lady's eyes, Mercury finds to his surprise that he is unable to fly:

MERCURY *What mak's me so vnnimbly ryse,*
That did descend so fleete?
There is no vp-hill in the skyes;
Clouds stay not feathered feete.[54]

The means by which truly aerial creatures, both good and evil, might communicate with man generated a long-lived controversy. Whatever their true constitution, how much physical contact was possible? Could a demon seize a man and carry him through the air? Commentaries on the Bible commonly included the remark by Saint Augustine (adopted in the *Glossa ordinaria*) to the effect that since the bodies of demons were made of the tenuous upper elements, they were more suited to act than to be acted upon. In mediaeval times it thus became usual to believe that aerial creatures could establish fully physical contact with man and that there was little that man, in his turn, could do in retaliation. Continuing his description of the evil spirits, Alanus de Insulis writes of their capacity to deceive us by their assumption of bodily form:

> These confounders of truth with falsehood, equipped with a kind of aerial body, counterfeit a likeness to us and with many phantoms delude dull man. They feign light in darkness, agreement in contention, they conceal treachery under the mantle of peace, they represent the bitter as sweet, they give noxious draughts under the pretext of doing good.[55]

After the Renaissance, belief in the essential or potential physicality of the creatures of the air slowly waned, and flying demons were increasingly looked upon as wholly spiritual beings rather than as material creatures of the 'middle region.' It grew increasingly difficult to credit the power of spirits to possess or assume a bodily form,[56] and by the end of the eighteenth century Apuleius's cloudlike, airy beings were virtually extinct. It is by no means accidental that their extinction should have coincided with the first true conquests of the air. Once man had penetrated the hitherto unknown regions, its original inhabitants withered and died.

Truly aerial birds

A quite different and simpler resolution of the apparent imbalance between the degrees of habitation of the elements was offered by a long and non-Aristotelian tradition about footless birds. In contrast to Aristotle's implication that birds are really land animals, some writers on natural history believed that there existed a species of bird that entirely lacked feet. These *apodes* were described in two distinct forms: some believed they were small grain eaters, while others identified them as long-tailed birds of paradise (*manucaudiata*). They were held to fly continuously and to fall to earth only after their death, when the earthy parts of their constitution prevailed over their upward rising nature. These birds were thus truly aerial because, as Buffon put it, they could remain 'suspended in the air as long as they breathed, just as fish are sustained in the water.'[57]

The identification of the *apodes* as a species of small grain eater may have originated because of the way such birds take occasional 'dust baths,' with their bodies and wings outstretched on the ground, showing no readily apparent sign of their small feet. The origin of the alternative identification with the birds of paradise is found in their curled tail feathers, which suggested that they were prehensile (hence *manucaudiata,* 'hand-tailed') and could serve to hold them on a perch in an emergency. The most likely source of the whole tradition, however, is probably the capacity of some swifts to fly for days, weeks, or even months without alighting, so giving rise to the idea that they spend their entire existence in the air. A fairly elaborate life cycle was worked out for these birds, including the notion that the females laid their eggs in a hollow on the male's back, where the eggs remained until they hatched and the young were fully fledged. In India, according to Buffon, the bodies of footless birds were occasionally found by the inhabitants, their beaks spiked into the ground like tent pegs.

Although occasionally disputed by Renaissance naturalists,[58] belief in the existence of footless birds was still current as late as the 1770s. Not only does Buffon report the widespread acceptance of the idea, but in his introductory discourse on the nature of birds he himself appears to believe the story. In a short passage about modes of flight he describes the range of birds from the flightless struthious species through those that both swim and fly but cannot walk ('like the penguins, sea parrots, etc.'), to those that, like the birds of paradise, 'neither walk nor swim, and can move only by flying.'[59] Later in his book, when writing about the bird of paradise, he nevertheless changes his mind, treating the story of perpetual flight as a myth.[60] After Buffon the idea fades rapidly, being mentioned in only a few of the subsequent books of ornithology.

With the exceptions of angelology and the minor but significant question of footless birds, most comments on flying creatures from the twelfth century to the late Renaissance show the influence of Aristotle's work on natural history, while many premediaeval writers, such as Pliny, Aelian, and Isidor, echo and embellish Aristotelian remarks and adopt his categories. Not only Aristotle's interests but also his prejudices and omissions are often reflected. While he devotes some space to the varying flight patterns of birds and insects, taking a stronger interest in the latter, he nowhere shows any concern for the phenomenon of flight itself, which apparently does not strike him as in any way mysterious. Nor does he appear to ponder the problems that would be posed by human attempts at flight. Although the pseudo-Aristotelian *Problemata,* probably based on a lost book of problems by Aristotle himself, makes in passing the interesting comment that of all animals birds 'most resemble man in form,'[61] the *De incessu animalium* asserts that wings are wholly incompatible with the erect posture of the human body:

> It is also evident from these considerations that a bird cannot possibly be erect in the sense in which a man is. For as it holds its body now the wings are naturally useful to it; but if it were erect they would be as useless as the wings of Cupids we see in pictures. It must have been clear as soon as we spoke that the form of no human nor any similar being permits of

8. Footless bird. Conrad Gesner, *Historiae animalium II* (Tiguri 1555) part 2, p. 612. (Reproduced by permission of the Syndics of Cambridge University Library)

> wings; not only because it would, though Sanguineous, be moved at more than four points, but also because to have wings would be useless to it when moving naturally. And Nature makes nothing contrary to her own nature.[62]

This discouraging attitude to the aspirations of potential aviators undoubtedly helps to account for the scanty attention given to manned flight by mediaeval philosophers. When the subject is implicitly raised, the Aristotelian background is usually felt, as when Pierre Belon strikingly echoes the comments in the *Problemata* and the *De incessu animalium* by his detailed comparison of human and avian skeletons (Figure 10).

Birds, air, and water

> *Who . . . could ever have thought, that out of water, a nature could be produced, which should by no means endure water? But, the WORD of God* speaks, *and, in a moment, out of water are created* birds.
>
> —*Luther,* The Creation

Throughout the Christian era attitudes to bird flight were strongly influenced by interpretations of Genesis 1:20–22, which reinforced classical Greek ideas about the close relationship of water and air. Correctly understood, the brief description of God's creation of the birds and fishes is not remarkable: 'Let the waters teem with countless living creatures, and let birds fly above the earth across the vault of heaven' (*NEB*). The Old Latin versions, the Vulgate, and most subsequent translations, misconstruing the Hebrew text, offered readings containing the sense of the King James Bible: 'Let the waters bring forth abundantly the moving creature that hath life, and fowl that may fly above the earth in the open firmament of heaven.' The suggestion that birds were created from water thus indirectly echoed the Aristotelian idea that they were not to be thought of as truly aerial creatures. The point was further reinforced by the words of Genesis 1:22: 'And God blessed them, saying, Be fruitful and multiply, and fill the waters in the seas, and let fowl multiply in the earth.'

That the Scriptures should name water as the primary substance of birds was an oddity that Patristic commentators and later writers on the hexaemeron sometimes explained away by invoking analogies with swimming. The idea may have arisen partly owing to the indirect influence of Plato, who mentions the matter briefly in *The Sophist.* During a discussion of methods of hunting, the Elean Stranger asks if it is not the case that 'of swimming creatures we see that one tribe is winged and the other is in the water?'[63] The analogy appears very early in the history of hexaemera, being invoked by Saint Basil (A.D. ?330–379),[64] and fully developed by Saint Ambrose (A.D. ?340–397):

> Birds seem to be primarily related to the fish species, since each has a certain element in common, that of being able to swim. The second element

which fishes and birds also share lies in the fact that the art of flying is an aspect of that of swimming. As a fish cuts through the water in the act of swimming, so a bird 'cuts the air' in his swift flight. Both species are provided in a similar way with tails and 'with the oarage of wings.' So the fish directs himself forward and advances to distant points by the aid of his wings [fins]. He uses his tail as a rudder in order to guide himself or change his route by a sudden movement from one area to another. Birds also exercise their wings in the air as if they were floating on water, using them in the way one would use one's arms. By use of their tails they are able to direct themselves upward or downward at will.

Hence, while all of these species follow the same pattern, they are but complying with the divine precept that places the origin of both in water. For God said: 'Let the waters bring forth reptiles, living creatures according to their kind, and winged creatures flying above the earth along the firmament of heaven, each according to its kind.' Not without reason, therefore, do both species have the innate faculty of swimming, since both have their origin in water.[65]

Others, including Saint Augustine in particular, appealed to the three regions of the air. In an early, unfinished draft of a book on Genesis he points out that birds are not to be found in the upper atmosphere but fly only in the lower, thicker regions that may well be considered merely very fine water. By this association he hopes to demonstrate the birds' essential wateriness:

Clouds are formed from water, as anyone will feel who has occasion to walk in the hills among the mists or on the plains through fog. And it is in air of this kind that birds are said to fly. But in the rarer and purer form, which is properly called air by everybody, they are not to be found, since owing to its tenuousness it will not support their weight. In this, moreover, it is said that no clouds are formed, and that no turbulences exist. There, indeed, no wind blows, as is the case on the top of Mount Olympus, which is said to rise beyond the humid air.[66]

Returning to the subject in his later, fuller treatise *De genesi ad litteram,* Saint Augustine considers the silence of the Scriptures about the origin of the air. This is to be explained, he says, by its division into upper and lower parts. Although a little inclined to doubt the truth of the generally accepted Aristotelian model of the atmosphere, he says that if the upper region really exists, it is included in the Scriptural word 'heaven' (*caelum*). Similarly, the word 'earth' (*terra*) should be understood to include everything from 'fire [lightning], and hail; snow, and vapour; storming wind' (Ps. 148:8) down to the dry earth in its usual meaning. Receiving exhalations from the sea and the land, the lower air is so thickened as to be a form of water:

Thus whether we speak of water which flows in fluid waves, or of water in tenuous form suspended as vapour—the first the domain of the moving

> creature that hath life, the second that of the fowl—in either case we are concerned with the same humid element.[67]

He thus counters the objection that the birds should be said to have been made from air rather than from water. The birds are well suited to this thickened air, its humidity sustaining them and allowing them to use their wings to fly just as fish swim by the use of their fins.[68] This way of rationalising Genesis 1:20 grew very popular with later commentators. In the ninth century it was repeated, with variations, by Eriugena, who, more original than most theologians of his day, added further evidence of a more radical kind. There are, he says, some creatures that mutate twice a year: for six months they inhabit the air, and for another six they live in the water. In Migne's text the example given by Eriugena is the coot (*fulica*), but a variant reading that seems to provide a more likely origin for the idea is *luligo,* or *loligo,* the cuttlefish, whose 'sail' suggests that it belongs half to the air, half to the water.[69] In the twelfth century William of Conches proposed an elaborate theory of the origin of the world as we know it, distinguishing between the birds and the fish on the basis of the amount of warmth the water received from its contact with heated air:

> After their creation, the heavenly bodies, since they are of a fiery nature, began to move, and from their motion to heat the underlying air. And from the air the water was heated. Out of the heated water various kinds of animals were created, among which those containing more of the upper elements are the birds. Whence because of the action of the lighter elements birds are sometimes to be found in the air while sometimes, because of the action of the heavier, they descend to earth. The other animals, which have more water in them, are the fish, which can live only in this element and in no other.[70]

In the *Summa theologiae* (thirteenth century) the explanation based on two forms or states of water is again taken up by Saint Thomas, who adds the further comment that air is omitted from special mention in Genesis because, unlike the other elements, it is not immediately apprehendable (*insensibilis*).[71]

The emphasis on the relationship between water and the lower air, and thus on the close kinship and biological similarity of birds and fish, persisted throughout the Middle Ages and well on into the Renaissance. The explanation of Genesis 1:20, based on Basil, Ambrose, and Augustine, gained widespread currency as a result of its inclusion in the *Glossa ordinaria,* so frequently recopied in mediaeval manuscripts. Subtle variations of the argument were sometimes offered, as when Ludovicus Buccaferrea suggested that one should consider the structure of flying animals according to the familiar twofold distinction of matter and form. Recalling Aristotle's discussion of water and air as the matter and form of the same entity, Buccaferrea writes of the double nature of birds. In relation to their substance, he says, they can be deemed to belong to water (or to the earth in or on which God said they should multiply). In relation to their form, on the other hand, one may view them differently:

> Not all of them belong to the water or to the earth, but some to the air, because from their spirit they derive their will to motion, since they have their motion thanks to the imagination and the will. And therefore some of them are attracted to the air: and thus it is that by their form they live in the air while by their substance they live on earth or in the water.[72]

Form and substance are manifestations, respectively, of potentiality and actuality. The tension between the two had already grown familiar in scholastic commentaries on the corporeality of angels and demons. Resolutions of the problem of angels' bodies were sometimes achieved by positing a twofold nature, a combination of the actual and possible, of matter and form. In differing degrees both birds and angels were endowed with the form of pure flight, but as neither was a species of pure being, only God being pure form, the potentiality was in both cases compromised by imperfect physical embodiment.[73]

Popular encyclopaedists of the Middle Ages frequently appeal to the idea of thickened air in relation to flight. One of the best-known compendia of general information was *The Book of Sydrach,* a collection of questions and answers compiled in French or Langue d'Oc in about 1250. Translated into many languages, variant versions of the text were widely read from the thirteenth century until well into the Renaissance. The answer to the question 'How do birds fly?' is usually a crude shorthand account of the main points in Aristotle and Saint Augustine, taking the form 'Birds fly in the air by means of the thickness of the air. When the air is thick and moist it thereby sustains the birds when they fly.'[74] A version more interesting than most is contained in an English verse translation published in about 1510:

The .C.xviij. questyon

The fowles that in the ayre flye
How are they borne on loft a hye
The thycknesse of the ayre ahye
Is the chefe cause truly
That the fowles a lofte are borne
Els ware theyr flyght forlorne
The ayre is thycke and moyst also
And wyl dysclose and close agayne to
And whan the fowle is a lofte
His wynges styreth he ofte
With the warpyng of his wynge
He doth the ayre a sondre mynge
Which waxeth thycke and he is lyght

And that hym holdyth in his flyght
And that the ayre is thycke aboue
A man may wel proue
Who so in his hande wyl take
A yarde and it smertly shake
It shal bowe in the shakyng
And that doth no nother thyng
But thycknes of the ayre withstandys
The yard in shakyng of thy handys
And so it withstandyth the fowles al
In theyr fleyng that they ne fall.[75]

Although the thickened air near the surface of the water was the natural home of birds, it could also be perceived as a potential obstruction. In 'Upon Appleton House' Andrew Marvell writes of the 'modest *Halcyon*', or kingfisher. Flying at twilight, 'betwixt the Day and Night' (line 670), the kingfisher participates in the worlds of light and dark, above and below, good and evil. Embodying the mediating powers of the air and of flight, it is associated with the establishment of a system of equilibrium in the natural order:

And such an horror calm and dumb,
Admiring Nature *does benum.*[76]

An intermediate creature, the predatory kingfisher itself becomes a prey to the elements as it hunts:

The viscous Air, wheres'ere She fly,
Follows and sucks her Azure dy;
The gellying Stream compacts below,
If it might fix her shadow so.[77]

A theory with a decidedly mystical content, transcending such appeals to common experience, coupled the problem of the wateriness of birds with another puzzle contained in Genesis 1:6–7, which speaks of a strange celestial region of water:

> And God said, Let there be a firmament in the midst of the waters, and let it divide the waters from the waters.
>
> And God made the firmament, and divided the waters which were under the firmament from the waters which were above the firmament: and it was so.

Hexaemeral writers had often pointed out that whatever the true nature of the waters above the firmament, they must presumably be endowed with some special form of lightness.[78] Speculating about this in his *Physices christianae pars altera,* the Calvinist theologian Lambert Daneau (1530–95) tentatively proposed a theory of the creation of birds, only to dismiss it as mistaken. Birds clearly have an affinity with the mysterious upper waters. Both are subtle, agile, tenuous, warm, and light. It is therefore tempting, Daneau suggests, to suppose that while fish were made from the ordinary water of the sea, God used the finer celestial water from which to fashion the birds. However attractive this idea may seem, Daneau is confident that it is wrong. Birds must be made from the lower waters since otherwise they would not need to seek the upper regions in which they were, so to speak, born but would be able to live there permanently. Instead, they have continually to strive upwards. Writers who assert that birds stay in the air all the time are mistaken. Birds are unable to hang suspended in the air like clouds but must descend to the water or the earth as soon as they stop flapping their wings. Making the conventional point that birds and fish are strikingly similar, Daneau says that God made them both from the same substance in order to demonstrate his power. From the one substance he formed creatures that in some respects, especially the degree of heat that they contain, are at opposite ends of the scale of possibilities.[79]

The matter is further complicated by Genesis 2:19, where the creation of the birds in the second, Yahweh, version suggests, this time, that they were made from the earth: 'And out of the ground [Vulgate: *de humo*] the Lord God formed every beast of the field, and every fowl of the air.' While implicitly treated by the commentators as subsidiary to Genesis 1:20 and not, after careful consideration, incompatible with it, this statement did nothing to promote an association of birds and the air.

The difficulty encountered by exegetes attempting to explain the Biblical account of the birds' creation led to a further strengthening of doubts about their moral connotations. In addition to the problematical matter of their physical constitution, there was the question of priority. Since terrestrial animals are higher on the scale of being, Saint Thomas considers the objection that they should have been created before the birds. While agreeing that birds are indeed inferior to quadrupeds, he argues that that which is better should appear later:

> *Obj. 5.* Further, land animals are more perfect than birds and fishes, which appears from the fact that they have more distinct limbs, and generation of a higher order. For they bring forth living beings, whereas birds and fishes bring forth eggs. But the more perfect has precedence in the order of nature. Therefore fishes and birds ought not to have been produced on the fifth day, before land animals . . .
> *Reply Obj. 5.* The order in which the production of these animals is given

has reference to the order of those bodies which they are set to adorn, rather than to the superiority of the animals themselves. Moreover, in generation also, the more perfect is reached through the less perfect.[80]

The mixed response to birds and bird flight is more strongly felt in another passage from the *Summa theologiae* concerned with emblems and symbols. Along with other birds including the ostrich, whose flightlessness represents a form of earthbound life, Saint Thomas mentions the gull (*larus*):

> The gull, which both flies in the air and swims in the water, signifies those who are partial both to Circumcision and to Baptism: or else it denotes those who would fly by contemplation, yet dwell in the waters of sensual delights.[81]

The inherent antithesis here is consistent not only with the implications of Genesis 1:20, but also with the hint of value judgements in Aristotle's discussion of water and air in the *Physics* (see Chapter 1). The ambiguity spills over into Saint Thomas's comments on the idea that man might try to fly. While spiritual flight on the wings of contemplation was a way of approaching God, it is clear that Saint Thomas had no sympathy with the thought that man might one day fly in physical fact. He states several times that human flight is impossible, and indeed in a number of passages in which he speaks about simple logic he uses the statement as the paradigm of self-evident truth: 'Nothing, indeed, prevents a conditional being true, the antecedent of which is impossible, such as in this conditional: If man flies, he has wings.'[82] As Saint Thomas indicates in his passage about the gull, the necessity for birds to spend much of their time either on the ground or on water, and their frequent choice to remain there rather than to fly, made the flight of birds a symbol of the need to escape from the sufferings of post-Edenic life rather than of essential freedom from fallen nature. The theme of flight as the avoidance of corruption rather than as the enjoyment of unassailable purity is taken up in a later tradition that birds avoid flying in unclean air.[83] Although man might fly above the physical miasmas, his spiritual uncleanliness was more difficult to escape. The theme is echoed in many prophetic comments about the horrors that might be expected from successful aeronautical experiments, the old cautionary tale about Icarus often giving way to predictions of theft, rapine, and murder. Meanwhile the associations of birds with the lower elements rather than with the air continued to find expression in early scientific treatments, including that of Buffon, who presented a somewhat tortuous argument about the affinity of birds and water:

> It appears that the element of water belongs more to the birds than to the quadrupeds; for, with the exception of a small number of species, all terrestrial animals avoid water and swim only when they are forced to do

so by fear or the need for food. Among birds, on the other hand, there is a large group of species that are happy only on the water, and seem to go on land only for special reasons, such as laying their eggs away from the reach of the water, etc. And what demonstrates that the element of water belongs more to birds than to terrestrial animals is that there are only three or four quadrupeds that have membranes between their toes, whereas one can count more than three hundred birds provided with these membranes which give them the capacity to swim. . . . Birds, whose feet are like oars, the form of whose body is oblong, rounded like that of a ship, and so light that it sinks only far enough to sustain itself, are, for all these reasons, almost as well suited to swimming as to flying; and this faculty of swimming even develops first, for one sees ducklings moving about on the water long before they take flight in the air.[84]

Flying fish

Probably because of their relatively unfamiliar, exotic character, flying fish, which provide a link between birds and the water from which they were said to have been created, were discussed only rarely by the naturalists. Aristotle mentioned them primarily in relation to the sound they make when flying:

The scallop, when it goes along supporting itself on the water, which is technically called 'flying', makes a whizzing sound; and so does the sea-swallow or flying-fish: for this fish flies in the air, clean out of the water, being furnished with fins broad and long. Just then as in the flight of birds the sound made by their wings is obviously not voice, so is it in the case of all these other creatures.[85]

A later tradition concerning the flight endurance of flying fish suggests a more interesting relationship between their watery nature and the Biblical origins of birds. In an attempt to account for the brevity of the glides of which most flying fish are capable, it became traditional to state that they could remain airborne only for so long as their wings remained wet. In his *De subtilitate* Cardanus claimed to speak about the matter on the basis of personal experience:

I have seen them: they are small and winged, with wings which are attached near their gills and in the case of these fish are longer than a span. When wet they sustain the fish in the air, but let it fall when they dry out.[86]

Responding negatively to this, as to the whole of *De subtilitate,* Cardanus's vigorous antagonist Julius Caesar Scaliger scorned the idea, offering more routine reasons for the brief duration of the flight:

When the flying gurnard and the sea-swallow have returned to the water after the completion of a short flight, no one can know whether their wings

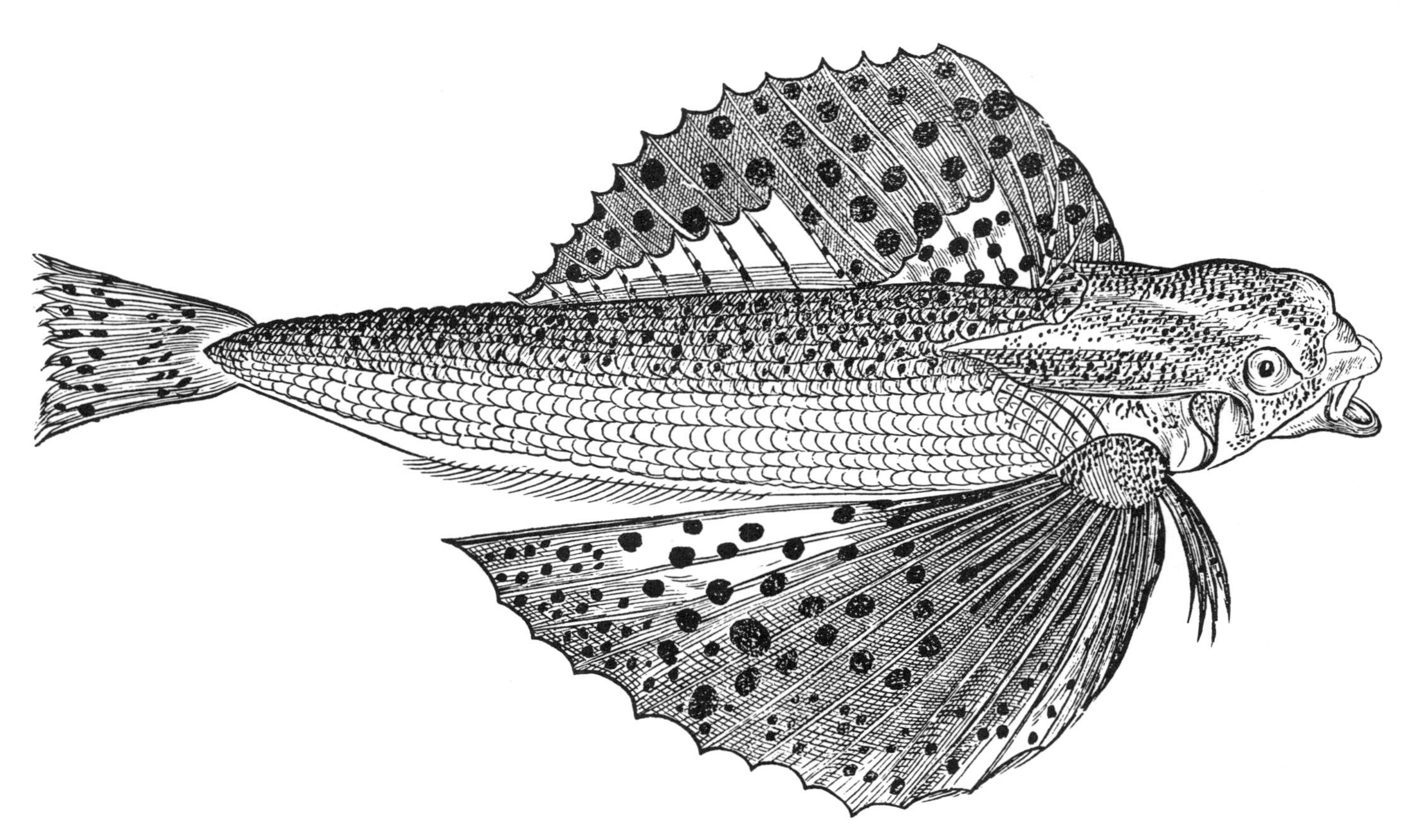

9. Flying fish. Conrad Gesner, *Historiae animalium* III (Tiguri 1558) 1279. (Reproduced by permission of the Syndics of Cambridge University Library)

> have dried out. I thought you would have known the reason for this vulgar error. When they have just emerged from the water we see drops shaken off by the motion of the wings; towards the end of the flight these grow fewer, and finally there are none. It is nevertheless not because of the absence of humidity that the flight ceases, but because the fish is tired, or because it has had sufficient enjoyment, or because it thinks that it is far enough from the enemy that is pursuing it.[87]

Scaliger's objection nevertheless failed to eradicate the idea. In 1599 it is mentioned in passing by John Nashe ('that fish neuer flies but when his wings are wet').[88] A generation later, John Swan, relying perhaps on Cardanus, makes the same point in his *Speculum mundi,* in which he calls the flying fish *hoga,* a name elsewhere given to a kind of sea calf:[89]

> This fish hath wings . . . Neither can it flie high or farre, or longer than her moistened wings keep wet; nor yet swimme fast, having exchanged finnes for wings.[90]

This 'vulgar error' probably gained some of its tenacity from the relationship between water and flight established by the authority of the Bible and supported by Aristotelian cosmology.

However beautiful, the 'sea-swallow' was a hybrid, a monstrosity, whose poor performance both in the air and in the water was a natural result of its mixed and impure nature. The point recurs in the bestiary tradition, though the moral emphasis could be varied. Thomas of Cantimpré's *De natura rerum* speaks of the flying fish in terms similar to those used by Saint Thomas in his passage about the gull:

> These fish most aptly signify those who, having because of the cares of office been concerned for a while with secular matters, nevertheless recollect themselves at the appropriate time and place and, recalling the sweetness of heaven, are devoutly lifted to more spiritual thoughts.[91]

The twelfth century bestiary translated by T. H. White makes the same point the other way round. Although the flight in this case is attributed to one of the traditional sea monsters, the *serra* (sawfish), the passage surely refers, as White says, to flying fish:

> There is a beast in the sea which we call a SERRA and it has enormous fins. When this monster sees a ship sailing on the sea, it erects its wings and tries to outfly the ship, up to about two hundred yards. Then it cannot keep up the effort; so it folds up its fins and draws them in, after which, bored by being out of the water, it dives back into the ocean.
>
> This peculiar animal is exactly like human beings today. Naturally the ship symbolizes the Righteous, who sail through the squalls and tempests of this world without danger of shipwreck in their faith. The Serra, on the contrary, is the monster which could not keep up with the righteous ship. It symbolizes the people who start off trying to devote themselves to good works, but afterwards, not keeping it up, they get vanquished by various kinds of nasty habits and, undependable as the to-fro waves of the sea, they drive down to Hell.[92]

Bats

Considered monstrosities or sports of nature, and viewed with even more distaste than were flying fish, bats were unjustly scorned and neglected by most classical, mediaeval, and Renaissance writers. Although Aristotle mentions them several times, he is less interested in their magnificent flying skills than in their 'mutilated' or 'misshapen' structure, which follows from their being half quadruped, half bird.[93] Treated as appendages to nature's catalogue of flying creatures, bats were commonly included by later encyclopaedists at the end of the list of birds. This position proved convenient in any case because of the nomenclatural accident that placed them at the end of the alphabet (Latin *vespertilio*), where, after alphabetical ordering had begun to supplant the older hexaemeral arrangements of material, they kept company with the hoopoe (*upupa*), traditionally the dirtiest of birds. Apart from their apparently hybrid nature, the characteristic of bats most often noticed by the encyclopaedists was the leathern quality of their wings,[94] which in the Middle

Ages and the Renaissance became the standard iconographic flying equipment of devils, in contrast to the feathers of the angels.[95]

Before their adoption in the late eighteenth and nineteenth centuries as Gothic symbols of the dark,[96] few poets chose to give bats an important place in their range of imagery. Ovid is a rare exception. Accounting for their origin as a transformation of the daughters of Minyas, he describes a terrifying scene:

> Suddenly unseen timbrels sounded harshly in their ears, and flutes, with curving horns, and tinkling cymbals; the air was full of the sweet scent of saffron and of myrrh; and, past all belief, their weft turned green, the hanging cloth changed into vines of ivy; part became grape-vines, and what were but now threads became clinging tendrils; vine-leaves sprang out along the warp, and bright-hued clusters matched the purple tapestry. And now the day was ended, and the time was come when you could not say 'twas dark or light; it was the borderland of night, yet with a gleam of day. Suddenly the whole house seemed to tremble, the oil-fed lamps to flare up, and all the rooms to be ablaze with ruddy fires, while ghostly beasts howled round. Meanwhile the sisters are seeking hiding-places through the smoke-filled rooms, in various corners trying to avoid the flames and glare of light. And while they seek to hide, a skinny covering overspreads their slender limbs, and thin wings enclose their arms. And in what fashion they have lost their former shape they know not for the darkness. No feathered pinions uplift them, yet they sustain themselves on transparent wings. They try to speak, but utter only the tiniest sound as befits their shrivelled forms, and give voice to their grief in thin squeaks. Houses, not forests, are their favourite haunts; and, hating the light of day, they flit by night and from late eventide derive their name.[97]

Compared to the birds, which have an important symbolic function throughout their frequent appearances in the Old and New Testaments, bats receive in the Bible only three mentions, all of them unflattering. In Leviticus 11:19 and Deuteronomy 14:18 they are listed among the unclean animals that are unfit to be eaten. In Isaiah 2:20 it is predicted that they and the moles will take into their custody the unworthy idols of gold and silver cast aside at the latter day. Doubtless as a consequence of their insignificance to the writers of Holy Scripture, most Patristic commentators accorded bats little respect, usually dismissing them as vile and ignoble animals. The loving Saint Ambrose is exceptional in finding some attractive characteristic in their mode of life:

> The bat is an ignoble creature, whose name is taken from the word for evening. They are equipped with wings, but at the same time they are quadrupeds. They are provided with teeth, in this respect differing generally from other birds. As a quadruped, too, the female brings forth her young alive and not in the oval stage. Bats fly in the air like birds but prefer

> to be shrouded in the dusk of evening. In flight they do not use the support of wings but rely on their webbed feet which serve as wings, both as a balance and as a means of propulsion. These common creatures have this faculty, too, of adhering one to another, assuming any position like a pendant bunch of grapes, so that, if the lowest in place gives way they all fall apart. Here we see the virtue of love in action—a virtue difficult to find among men here below.[98]

More typical are the interpretations of the *Glossa ordinaria,* which explain that bats are morally worthless because although they can fly about the earth they also walk on the ground, using their wings as feet. This is symbolic of a turning aside from the contemplative life and is no desirable example for mankind.[99]

The unhappy connotations of bats appear to have deterred mediaeval and Renaissance expositors of flight. Those who discuss the theory are almost wholly concerned with birds, while among the aspiring aeronauts only Leonardo thought bats worth imitating. Although artificial wings based on those of bats would always have been easier to build than simulated bird wings, the prejudice against bat flight as an awkward and unnatural form of locomotion was a more potent influence on the minds of lesser men.

How do birds fly? 3

The mechanics of flight

A thoughtful man must find the flight of birds as remarkable a thing as any other in nature.

—*Belon,* La nature des oyseaux

Delighted by a bird's mysterious power to remain suspended in air, Pierre Belon devoted an early chapter of his fine book *L'histoire de la nature des oyseaux* (1555) to an attempt to explain the processes at work.[1] Except for Leonardo, whose manuscripts remained almost unknown until comparatively recent times, few others before Giovanni Borelli (1608–79) proved thoughtful enough to seek clarification of the mystery. The dampening example of Aristotle, whose interest in the matter was slight, was certainly responsible in part, while those who did choose to write about bird flight were often concerned with such specific aspects of ornithology as falconry and augury.

Along with other commentaries on the nature of the physical world, the sporadic attempts to explain bird flight reveal the common assumptions and the state of scientific knowledge of their time. Classical writers tended to focus on the apparently natural capacity of some things to rise away from the centre of the world. The development of the study of mechanics in the Middle Ages and the Renaissance led to explanations in terms of levers, angles, weights, wedges, and action and reaction, culminating in the seventeenth century image of the bird as a complex machine. Experiments in the seventeenth and eighteenth centuries that demonstrated the elasticity of the air engendered a further class of explanations based on the notion that a bird was propelled upwards and forwards by the expansion of parcels of air that had previously been compressed by the downstroke of the wing. Developments in biological science, reflected in the long and detailed anatomical treatises of the eighteenth century, led to a very full account by Paul-Joseph Barthez,[2] who in 1798 proposed that the flight of birds could be explained in terms of the internal action and reaction of the muscles.

In addition to physical analyses, many accounts of flight invoke irrational causes or postulate the existence of unexplained forces. Such elements, often coexisting with propositions of a greater or lesser degree of scientific rigour, are indicative of the peculiar potency of bird and flight symbolism. Among the most frankly mystical is the commentary of Iamblichus (c. A.D. 300), who exerted a decisive influence on the development of neo-Platonist Christian thought. Speaking of observations made for the purpose of divination, he treats bird flight as entirely controlled by the will of the gods. The passage is in keeping with Iamblichus's view of the world as a tightly integrated hierarchy:

> With respect to birds . . . the impulse of their proper soul moves them, and also the daemon who presides over animals; and, together with these, the revolution of the air, and the power of the heavens which descends into the air, accord with the will of the Gods, and consentaneously lead the birds to what the Gods ordained from the first. Of this the greatest indication is, that birds frequently precipitate themselves to the earth, and destroy themselves, which is not natural for any thing to do; but this is something supernatural, so that it is some other thing which produces these effects through birds.[3]

While most compilers of natural histories wrote in a more matter-of-fact vein, the universe was known to be governed by mysterious laws. Although it is presented in a scientific context, there is more than a trace of the mystical in Galen's idea (second century A.D.) of buoyant energy arising directly from the rapid and sometimes imperceptible palpitation of a bird's wings. Nonrational elements continue to be found in theories of bird flight as late as the sixteenth century, when Belon is led to attribute the otherwise inexplicable phenomenon to some strange quality of the bird's form. Even in the eighteenth century the Chevalier de Vivens, postulating a whole new class of 'centrifugal' forces, is unable to account for lift in terms of the known mechanics of the universe.

Most attempts to explain how birds fly nevertheless appeal to fairly simple contemporary ideas about the nature of the physical world. The principal mechanisms invoked by the commentators discussed in this chapter fall into seven interrelated but distinct categories:

1. The effect of the inherent lightness of the upper Empedoclean and Aristotelian elements
2. 'Resistance': the application of simple action and reaction, the wing pushing against the air as against a rigid object
3. Action and reaction between the air and the wings on the analogy of the inclined plane
4. The capacity of a bird to sense and harness physical forces inherent in the air through which it passes
5. Movement analogous to that of a swimmer, the air being treated as a tenuous liquid through which the bird rows itself by the use of its wings

6. The application of buoyancy produced by the springiness of the air, assisted by the springiness of the flight feathers
7. An interplay of muscles, some pushing back while others draw the bird forward

Although an historical progress is discernible, the various kinds of explanation are found combined in the work of individual commentators, ancient prejudices often surviving in the company of sophisticated scientific thought. Except for Leonardo, to whom I devote a separate chapter, writers are discussed in approximately chronological order.

Aristotle

Keen to find common ground among all moving animals, Aristotle avoids singling out the distinctive qualities of flight. Wings, he points out, are analogous to the forelegs of quadrupeds, and the forelegs, he thinks, initiate movement.[4] The bending and straightening of the wings enable the bird to progress by a series of jumps, with the forward movement through the resisting medium made easier by the 'oblique' attachment of the wings, by which Aristotle appears to have in mind some degree of 'sweepback.'[5]

More significant than his casual comments on flight mechanisms are Aristotle's influential remarks about the nature and sources of animal motion in general. Some especially seminal passages, included in the crucially important and much-discussed eighth book of the *Physics,* contribute implicitly to many mediaeval and Renaissance descriptions of birds. The argument is presented in two steps. First Aristotle accounts for the apparent difference between inanimate objects, which need to be moved by an external agency, and animals that are self-movers. What distinguishes the animals is that they contain two separate parts, one that causes the movement and another that is moved: 'It would seem that in animals, just as in ships and things not naturally organized, that which causes motion is separate from that which suffers motion, and that it is only in this sense that the animal as a whole causes its own motion.'[6] And, a little later: 'In the whole of the thing we may distinguish that which imparts motion without itself being moved and that which is moved: for only in this way is it possible for a thing to be self-moved.'[7]

This interesting identification of a separate power plant within the animal nevertheless needs qualification. In the second part of the argument, Aristotle points out that an animal is a self-mover in only a secondary sense. As the mover within it, unmoved with respect to the animal itself, moves along with the animal during locomotion, it is far from being a true 'unmoved mover,' and in fact depends on activity outside itself, such as the state of the atmosphere, the cycles of respiration, digestion, and sleep.[8] Ultimately, the movement of animals, like all other movement in the universe, is derived, through chains of cause and effect, from the true First Mover. Birds, often perceived as the most independent of living creatures because of their freedom of flight, are thus firmly established as organic parts of an integrated system of action and reaction. Just how the physical connexions were established was the subject of much subsequent discussion.

Lucretius, Plutarch, and Pliny

While it may seem obvious to us that birds are sustained by the air, pre-Renaissance writers were often puzzled as to what role, if any, it played in flight. An early affirmation of the importance of the air is contained, by implication, in a passage of Lucretius (c. 99–55 B.C.) concerned with 'Avernian regions.' Deriving 'Avernus' from ἄορνος (birdless), Lucretius believes it to be a place that causes the death of birds by poisoning the air or by so disrupting it that flight becomes impossible. An emphatic supporter of the theory that the world is made up of atoms surrounded by void, Lucretius proposes that large areas of vacuum may be created by the exhalations over Avernus:

> It happens also at times that this power and exhalation of Avernus strikes apart all the air that lies between birds and the earth, so that an almost empty pocket is left here: and when they have come flying over against this place, the beat of their wings suddenly goes halting and ineffective, and all the effort of the pinions on either side is wasted. In this case when they cannot find rest or support on their wings, nature assuredly forces them to fall down to the earth by their own weight, and through this almost empty space they as they now lie disperse abroad their soul through all the pores of the body.[9]

A similar suggestion is offered a century later in two passages of Plutarch (c. A.D. 46–c. 120) relating how birds were made to fall to the earth by loud noises:

> And that which is often said of the volume and power of the human voice was then apparent to the eye. For [while an audience was loudly acclaiming] ravens which chanced to be flying overhead fell down into the stadium. The cause of this was the rupture of the air; for when the voice is borne aloft loud and strong, the air is rent asunder by it and will not support flying creatures, but lets them fall, as if they were over a vacuum, unless, indeed, they are transfixed by a sort of blow, as of a weapon, and fall down dead. It is possible, too, that in such cases there is a whirling motion of the air, which becomes like a waterspout at sea with a refluent flow of the surges caused by their very volume.[10]

Elsewhere Plutarch appears to be less sure that such a rupturing can occur and decides, instead, on the idea of a violent 'shock wave':

> The people were incensed and gave forth such a shout that a raven flying over the forum was stunned by it and fell down into the throng. From this it appears that such falling of birds is not due to a rupture and division of the air wherein a great vacuum is produced, but that they are struck by the blow of the voice, which raises a surge and billow in the air when it is borne aloft loud and strong.[11]

While most other commentators shared Plutarch's awareness that the air could exert forces on a bird, the relationship between the bird's weight, its wingbeats, and the forces of the air was formulated in a wide variety of ways.

A perceptive foreshadowing of later analyses of flight, especially those of Leonardo (see Chapter 4), is found in a brief passage of Pliny (A.D. 23–79). Although he is mainly concerned with the different styles of movement of the various species, he includes an interesting comment about the flight mechanism of those birds that follow a series of shallow curves, folding their wings between bursts of rapid flapping:

> There are some that fly with their wings for the greater part folded, and after giving one stroke, or others also a repeated stroke, are borne by the air: by as it were squeezing it tight between their wings, they shoot upward or horizontally or downward.[12]

Affirming that birds are 'borne by the air' (*aere feruntur*), Pliny offers an analysis that is not only entirely physical but also takes the active participation of air for granted. Some of his successors, such as Galen, are less clear on this point.

Galen

A mixture of the physical and the mystical is found in Galen's supposition that energy may be created, stored, and controlled by a kind of universal biological vibration or palpitation. Writing in particular about bird flight, he proposes a theory that, although quite different from Aristotle's description of organic self-movers, nevertheless depends on an analogous system of self-contained action and reaction. Although a soaring or hovering bird may appear to hold its wings still, it is in fact, according to Galen, maintaining a rapid vibration resulting from tension in the muscles, the energy of which counteracts the tendency to fall. Galen does not think of this movement as a series of ordinary wingbeats on a small and rapid scale, but writes of the energy as if it were applied in direct opposition to the force of the bird's weight that is tending to draw it down to the centre of the earth:

> Those things which are seen to move are not the only moving objects. Many things which are said to be unmoving, which remain in one and the same place, are nonetheless in motion. Birds are moving not only when they are carried up or down but also when they stay in one place in the air. If the body of a dead bird were put into that place, it would readily fall to the ground, impelled by its weight. From this it is plain that a body which remains in the same place aloft does so because of the application of some motion, as a consequence, that is, of the amount of motion which it would lack if it were brought down by its weight alone. For what you think of as immobility is a motion composed of two motions which can draw the body in contrary directions. If you take one of these to be raising the body upwards, you will readily see that the other draws the body down.[13]

Galen's idea was adopted by a number of writers and developed in various forms. Ulysses Aldrovandus makes it somewhat more explicit:

> Tonic motion is a certain firm and stable motion by which the bird is moved when, while making the maximum of effort, it appears to remain in one place rather than to fall from it. According to the doctors, it is brought about by a simultaneous contraction of the muscles which keep the limb [= the wings] as it were immobile. It is called tonic, that is to say firm, because it is the most steadfast form of flight and motion, to sustain which a great deal of strength and power are needed, especially when by this means birds rise high into the air with outspread wings, though remaining motionless, even resisting adverse winds and the weight of their own body so that they are neither forced downward by the latter nor driven hither and yon by the force of the former.[14]

The tension in the bird's muscles, allowing the imperceptible tonic motion to be produced, came to be attributed to 'animal spirits,' an effect of the life force itself. Since artificial wings would not be subject to such vital influences, attempts to imitate the soaring flight of birds were bound to be vain. On 13 May 1634 Christophe de Villiers wrote to Marin Mersenne:

> No amount of industry can create the means of flight such as that of kites when they fly around in a circle and more or less parallel to the horizon by a tonic flight and movement of their wings that are filled with animal spirits continually flowing in and being carried through them in order to produce so long an extension of them. A man's wings being extended like those of the kite remain motionless and would need other wings to support them. For the tonic movement of the kite's wings cannot be simulated in artificial wings that, being only imitations, cannot receive animal spirits creating motive force. And I would be more easily persuaded that one could raise or lift oneself from the horizontal by a continual movement and beating of wings, although the difficulty, as you say, is not slight.[15]

As late as the eighteenth century, bird flight was still occasionally explained by an appeal to such ideas. Even the great Abbé Nollet supposed imperceptible vibration to be the only possible explanation of soaring flight:

> There are some birds that can sustain themselves for a time at the same height without appearing to move their wings (which is known as *gliding*); it must nevertheless be supposed that their vibrations are so rapid and so short that they are imperceptible from a certain distance.[16]

The concept of semimystical 'tonic motion' was thus transformed into a purely physical phenomenon entirely explicable in terms of classical mechanical theory.

Frederick II

The most famous of all books on falconry, the *De arte venandi cum avibus* of Frederick II of Hohenstaufen (1194–1250) includes a sequence of passages describing, often in great detail, the difference between the flight patterns of the various species. Despite his fascination, amounting almost to an obsession, with details of the natural history of birds, he gives no indication that he finds the phenomenon of flight in any way remarkable. Although showing no serious interest in the physical principles, Frederick offers in passing a few remarks that reveal the general nature of his assumptions. At the start of his discussion of the wing feathers, of which he writes a detailed catalogue and description, he makes a comment that, though brief, is interesting because of its implied separation of the systems of lift and of propulsion. (Only after Sir George Cayley had fully understood the need for such a separation was it possible to make significant progress in artificial flight.) Frederick's comment includes one of the earliest anticipations of the idea:

> Air and wind support the bird while aloft, but progress through space is accomplished by flying in practically the same manner as beasts walk along the ground and fishes swim in the water.[17]

When describing the arrangement of the flight feathers, Frederick again shows that he has some understanding of the different functions of the inner and outer parts of the wings:

> The chief purpose of the secondary remiges is to support the bird in the air when the wing is expanded and in action. They also assist forward motion; and when the wing is folded they cover the 'knife-blade' feathers and, in fact, all ten outer quills (the primaries). These latter ten quills, when extended, help to sustain the weight of the bird in flight; but their chief function is to give a forward impulse. By the semicircular motion of all the primaries (parallel with each other) the weight of the body is raised. . . . the greatest lifting power and driving force is exerted by the feathers farthest from the bird's body. (p. 89a–b)

Returning to the common analogy between movement in the air and in water, Frederick later compares flying to rowing:

> Those that exhibit infrequent wing beats . . . have large wings and long flight feathers with which they are able to describe wide circles in the air and to make rapid progress after the manner of galleys furnished with long oars. (p. 93b)

It is nevertheless clear to Frederick that the wings in some way act to produce a lifting force that counteracts the bird's natural tendency to fall to the centre of the earth. Although some birds are obviously lighter than others, he does not, as do many other early writers on the subject, appeal to the idea of a natural quality of lightness. Working in normal Aristotelian terms, he thinks of them as earthy and heavy, and therefore as tending downwards. By implication their flight is a form of unnatural or 'forced' motion:

> Certain birds have longer wings and primary flight feathers than their body measurements seem to demand; others, again, have short wings with long quills. These are all able to float in the air and fly quickly forward without oft-repeated wing strokes. When, as stated, such birds have strong and well-fashioned quills with no vacant interspaces, they need for long and effective flight comparatively few wing beats. But if, with a similar arrangement of wings and pinions, the latter are soft and slender or with broken spaces, the bird cannot maintain a forward motion or remain long in the air without wing motion.
>
> Heavy birds with short, relatively soft wings must naturally employ rapid wing beats in flying, otherwise they would fall to the ground, as all weights are attracted to the center of the earth. (pp. 91b–92a)

Two paragraphs later, Frederick's son Manfred adds a further gloss, in which he specifically refers to Aristotle:

> Birds with long wings and perfect flight feathers sail along by backward strokes of their wings as if to set the air in motion. The longer the wings and the more nearly perfect the pinions, the more support is given by the spreading out of the wings and these backstrokes of the latter on the air, thus reducing the necessary number of beats. It has, in this connection, a resemblance to the relative movement of a broad and of a pointed piece of lead; the downward fall of the former being slower because of the greater resistance of the air, as noted by the philosopher. (p. 92a–b)

Continuing with his analysis of differing modes of flight, Frederick comments on their capacities to cope with head and tail winds. Failing, as did virtually everyone else until recent times, to understand that a steady horizontal wind has no effect on a bird's aerodynamic conditions, he says that certain kinds of wing are better suited than others to slow or fast wind speeds:

> Birds with long wings and good flight feathers fly better with the wind than those with short wings and pinions, even when they both fly equally fast without the aid of the wind. However, short-winged birds fly better against the wind than those with long wings, even if both species progress equally fast in the absence of a stiff breeze.
>
> Also, birds with rapid wing strokes fly better against the wind than those with a slow wing movement; and there are small birds of every species that make better time in defiance of an adverse wind than do larger birds. (p. 94b)

If there is any truth in Frederick's remarks, it applies to the general question of the airspeeds attainable by birds of different species, the matter of adverse or favourable winds being irrelevant. While his comments are based on an unfounded causal relationship between wind and airspeed, he includes passages indicating that he is well aware of the effects of wind on groundspeed:

> By the aid of favoring winds, such birds as quail, whose usually short flights are made by rapid wing strokes, undertake long journeys (as when migrating or returning to their nesting places). On such occasions, however, they take their time and husband their wing power as they fly from island to island. (p. 94b)

Although it is not clear whether Frederick intends it, there is an implication here that with a favourable wind a bird is able to sustain itself in the air with less muscular effort. If so, that implication is of course entirely untrue. Elsewhere Frederick refers to the matter of wind shear without falling into error:

> If birds wish to take advantage of a favoring breeze they often rise high in the sky; but if the wind is adverse they generally remain lower down, because at a great altitude wind has greater force. (p. 95a–b)

Although they are little more than ancillary to his main concern with falconry, Frederick's passing comments about the methods and principles of bird flight show him to be an exceptionally acute observer gifted with great insight. Despite the error about wind and airspeed, he is surpassed by no one except Leonardo until well after the end of the Renaissance.

Pierre Belon

Argument by analogy was the most common means of explaining how the air supports a bird: the descending wings were thought to encounter resistance under them, so allowing a series of aerial leaps to be made. Although unable to give a truly convincing account of the physical nature of flight in terms of the knowledge available to him, Pierre Belon (1517–64) attempts an explanation in which most of the obvious difficulties are confronted. Basing his ideas on Aristotelian physics, he begins by pointing out that it is not easy to see what a bird pushes against in order to urge itself forward. Rightly invoking the necessity of a system of action and reaction, he says:

> While flight is achieved through movement and all movement is made by means of a contrary which acts against it, nevertheless one finds nothing else that is contrary to the force of the flying bird than the air. And so what reaction does one find in the air opposed to the force of the bird?[18]

There follows a somewhat confused passage in which Belon discusses the Aristotelian ideas of 'forced' and 'natural' motion, giving them a special application. As Aristotle had done,[19] Belon pays particular attention to volun-

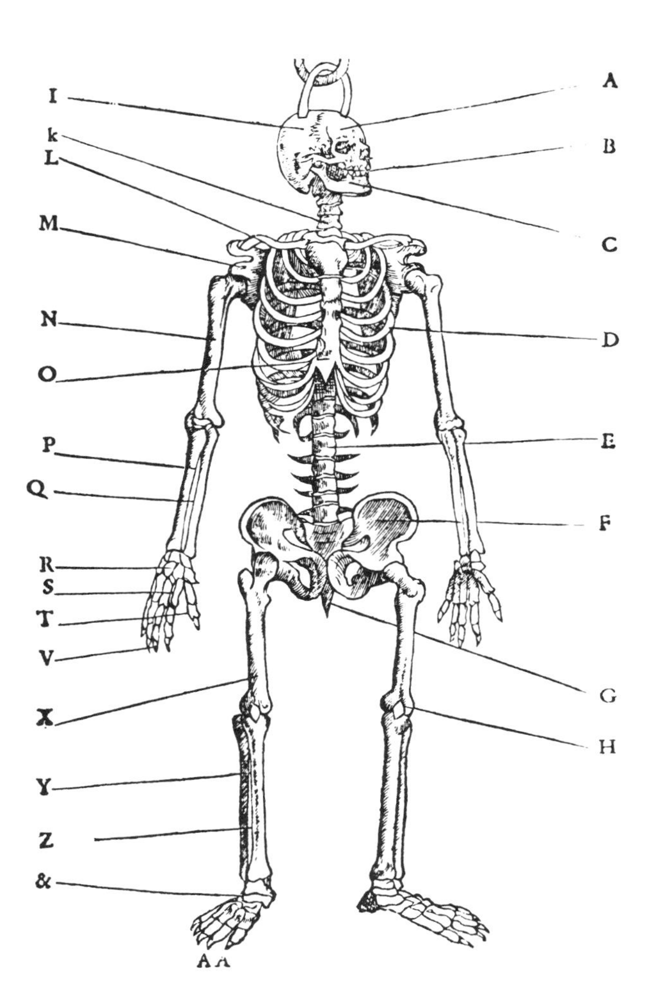

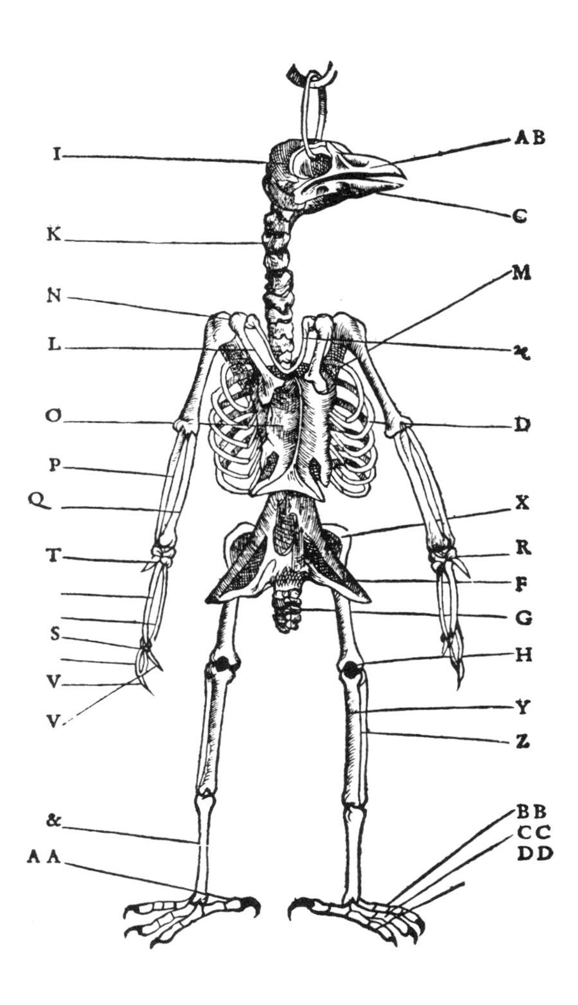

10. Comparison of the skeletons of (a) a man and (b) a bird. Pierre Belon, *L'histoire de la nature des oyseaux* (Paris 1555) 40, 41. (Reproduced by permission of the British Library)

tary natural motion, as in animals, also pointing out that in addition to their limbs animals have a further aid to motion, namely, their affections and passions. Although the bird moves voluntarily, and therefore naturally, Belon is still confronted with the difficulty created by his attempt to base an explanation on action and reaction. Mentioning Aristotle's argument that a land animal cannot move unless one limb is immobile,[20] he avoids suggesting an analogy for the reciprocal motion of walking and tries, instead, to specify what it is that the bird pushes against:

> It is necessary to think of their flight as of something light carried into the air and to attribute the movement to the reaction of the air against the lightness of the feathers that divide it as if by force. For the feathers that seize a great quantity of air because of the form of the wings, act in their place like our feet down here, walking on the earth.[21]

Aware that he has still given no clear indication of anything against which to push, he adds 'that which reacts against them is in the air, not on the earth,' and speaks of forces within the air itself. Energy in the form of violent turbulence, he says, can be observed acting independently of any source outside the air. He appeals particularly to experience at sea, when, as he himself had noticed, it is sometimes possible to witness squalls of great violence surrounded on all sides by calm air and water. Belon suggests that resistances against which birds may push their wings are smaller and less violent parcels of the same kind of energised air.

At this point a reader might suppose that Belon has explained the mechanism of flight to his own satisfaction. His sense of wonderment at the possibility of flying is nevertheless still alive, and he goes on to seek a fuller and more cogent account. Appealing, as his starting point, to another familiar idea, Belon attributes the power of flight to the bird's form. Although there is a strong implication of the metaphysical importance of form as discussed earlier in connexion with the dual nature of birds and angels, Belon's appeal to the concept is also partially dependent on an aspect of Aristotelian physics. Many postclassical discussions of lightness and heaviness include comments on the different effects produced when a given quantity of metal or other heavy matter is placed, in concentrated or flattened form, on the surface of a bowl of water. While a lump of iron will sink, the same quantity, beaten out and made slightly concave, will float. Still thinner sheets of metal will float, supported by the surface tension, even when quite flat; if carefully placed on the surface, thin needles will float. As the principles of displacement, buoyancy, and surface tension were by no means well understood, this capacity of thin metal to float was often attributed, as by Belon, to some lightening effect brought about by the process of beating it out.[22]

Developing further the general principle that the form of the object appears to affect its capacity to sustain itself, Belon suggests that not only the ability to rise but also the ability to move swiftly through a fluid medium may be attributable to subtleties of form. Referring once again to his shipboard experience, he writes of the astonishing speed of dolphins that he saw swimming before the prow. As anyone who has watched dolphins leading a ship will know, their effortless speed[23] is indeed a matter for amazement, and it is in no way surprising that Belon should have been able to find no obvious mechanical explanation. As he rightly says, their fins, or 'wings,' as he calls them, are far too small to provide the necessary thrust. Seeing no familiar source of power, he finds the cause of their speed in some characteristic of their shape that he does not, however, attempt to specify more closely:

> Now, making a comparison between the rapid movement of a bird cutting through the air, and the fish in the water, I should like to attribute the cause to their form. For form contributes greatly to making a movement either slow or fast; for just as lead, stone, and any metal can float on water if it is given a hollow shape, so birds according to their differing natures fly either more heavily or more lightly.[24]

Hardly an explanation, nor even a fully formulated hypothesis, Belon's invocation of 'form' nevertheless expresses his sense of the mystery of flight, transcending such mechanical concepts as rowing, swimming, or pushing back against the solid earth.

Fabricius

Half a century after Belon, the Italian anatomist Hieronymus Fabricius (1537–1619) wrote a monograph, *De alarum actione, hoc est volatu* ('Of the action of wings, that is, of flight'), which he appended to his book *De motu locali animalium secundum totum* (1618). Although he fails to avoid many contemporary misconceptions and makes startling, unfounded statements, his thoughts on the subject show a good deal of real insight.

Fabricius first considers the question of inherent lightness, which he thinks essential to flight. A heavy body, he says, can be sustained in the air only if aerial matter is mixed with its earthiness. In the belief that lightness is concentrated in feathers, he mentions experiments showing that birds cannot fly if the wing feathers are left intact while all the others are removed.[25] Although it is of course hardly surprising that a bird in such a deplorable state should be unable to fly, Fabricius treats the point as a demonstration that the feathers provide the lightening effect without which flight would be impossible. The matter is taken up again some decades later by Pierre Gassendi (1592–1655), who suggests a possible explanation. Perhaps, he says, the fine down feathers supply a lifting force by permitting warm, volatile spirits to flow through their tubelike shafts, so adding to the effectiveness of the wings.[26]

Pursuing further the concept of lightness, Fabricius considers the function of the curvature and extension of the wings, which he takes to be the gathering of air. In common with many others, he points out that a sail or cloth sheet falls more slowly if spread sideways and more rapidly if rolled into a tight bundle. This, he suggests, shows that the lateral extension of an object is itself a means of creating supportive forces. The wings need to be extended because in that way they gather beneath them the light air that sustains the bird. Observing that birds frequently vary the extension of their wings, he continues with the logical suggestion that in this way, as Leonardo had believed, they can make themselves lighter or heavier so as to rise or descend. Offering a further gloss on the supportive function of the gathered air, Fabricius suggests a comparison with the way in which a man who is learning to swim may help himself to remain afloat by the use of bladders of air attached to his body. Logic nevertheless fails to provide entirely satisfactory answers since it is clear that birds often (if not always) descend to land with their wings fully open. Attempting to solve this difficulty by taking a further logical step, Fabricius appeals, quite reasonably, to the need to distinguish between 'falling' and 'descending,' between involuntary and voluntary motion.[27]

Aristotle had left somewhat unresolved the relationship between, on one hand, the fundamental categories of 'natural' and 'violent' motion, and, on the other, the movement of an object, such as a top or mill wheel, spinning on a fixed axis. Among the many attempts to resolve the problem was the development by Cardanus, in 1570, of the idea of 'voluntary motion,' conceived as a kind of self-contained force, related directly neither to natural nor to violent motion. Going beyond Aristotle's comments in the *Physics* (see earlier), Cardanus speaks of the motion of living animals as similarly self-contained and self-regulating.[28] Considering the problem of compound motion, he nevertheless goes on to say that since when an animal moves it counteracts gravity, its motion is in fact a mixture of voluntary motion and of the thwarted natural motion towards the centre of the earth.

Without being able to formulate his analysis in a precise way, Fabricius, basing his ideas on Cardanus, is clearly aware of the difference between passive and active movement, between a glide under the influence of gravity and powered flight against it, between freedom and control. In general he conceives of bird flight as a combination of the control of lightness and the application of muscular effort. At this point Fabricius shows himself capable of a conceptual distinction beyond the grasp of most of his predecessors. He is able, in effect, to distinguish between lift and thrust, between the energy needed to maintain the bird in the air, preventing it from falling, and that needed to push it along in a horizontal line. The lift that keeps the bird up he attributes, once again, to the expansion of the wings, carefully adjusted at all times to maintain the bird at its chosen height. The forward thrust he sees generated by the backwards motion of the wings, pushing the air behind the bird. This he compares to the rowing action of oarsmen or of swimmers, or, better still, to that of a wader's webbed feet. A duck has, at it were, winged feet, and if one watches carefully it is possible to see the water retreating behind the bird.[29] In addition, the regular rowing action of the bird's wings causes the air to be alternately compressed and rarefied, the expansion of the compressed air leading to the creation of a further forward thrust to help the bird along.

Returning several times to the distinction between lift, for remaining airborne, and thrust, for change of place, Fabricius finds evidence for an additional contributory flight force. Giving special attention to matters of anatomy, he believes that since a bird's thorax is highly expandable, in contrast to the comparative rigidity of a man's, it makes possible a further lightening of the body by the inhalation of large volumes of air.[30]

At the end of his monograph Fabricius discusses an important matter to which few others except Leonardo devoted attention until recent times: the stability of a bird in flight. He first describes a simple toy that the local children called a *salta martino:* a word meaning a kind of grasshopper or, with particular relevance to this context, a doll so weighted at the base that it rights itself after having been pushed. Fabricius's *salta martino* consisted of a small body (*corpusculum*) made of pith or elder fruit, to the end of which was attached a thin piece of lead. When thrown into the air at any angle or in any direction, this toy would descend with the lead first. The same phenomenon could be observed in birds that, if thrown into the air upside down, always turned over immediately. This interesting process is caused, he believes, by the distribution of the bird's weight, the powerful wing muscles being placed on the chest rather than on the back so as to cause the centre of gravity to lie well below the line of the wings. Although Fabricius's analysis of stability leaves a lot unexplained, he has a clear sense of the value of pendulous stability in flying creatures. Considering, finally, the internal arrangement of birds in relation to the needs of motion through the air in various directions, he asserts that nature has been careful to fix all of the organs firmly in place, avoiding anything that might cause inconvenience by moving around inside. For similar reasons birds lack a bladder, since otherwise their urine might splash about or be ejected in flight, both unbalancing and soiling them.

Borelli

Until modern times, by far the most influential description of bird flight was included in the first volume of Giovanni Alfonso Borelli's posthumous work *De motu animalium.*[31] From the time of its publication and for more than a hundred years afterwards, Borelli's book was quoted, paraphrased, plagiarised,[32] and cited as the recognised authority. Its popularity was largely dependent on the modernity of its approach, the explanation of motor systems applying up-to-date mechanical knowledge and analysis, with careful appeal to anatomical information. In the forty-page section on bird flight, nevertheless, little is original or physically sound beyond Borelli's clear-sighted awareness that birds do not flap their wings backwards on the downstroke.

After describing the basic structure of a bird's wing, in the course of which he reiterates Belon's point about the basic similarity of wings and the forelegs of quadrupeds (see Figure 10), Borelli analyses the system of forces that accounts for the phenomenon of flight. While he does not share Fabricius's belief in the power of inherent lightness, he adopts as his primary mechanism one of the contributory forces that Fabricius includes towards the end of his monograph, namely, the elasticity of the air:

> The particles of air are condensed by the rapid beating of the wings and, springing back by virtue of their elastic power, resist the compression, just as if they were a hard surface. From which it arises that the whole avian machine springs up, making a new jump through the air. And thus flight may be seen as nothing more than a composite motion made up of frequently repeated jumps through the air.[33]

Attempting to quantify this process more fully than had any of his predecessors, he introduces a comparatively sophisticated analysis based on relative velocities. As a result of the beating process, the compressed air moves away from the bird as it reexpands. If the air moves away at the same speed as that of the wingbeat, the bird will neither ascend nor descend, but be sustained at a given height (I.302). If, on the other hand, the wingbeats are faster than the rate at which the air escapes, the bird will rise at a speed equal to the difference between the two velocities (I.303).

Glossing the 'jumping' action further, Borelli makes analogies with walking on solid ground but points out that jumping through the air has certain

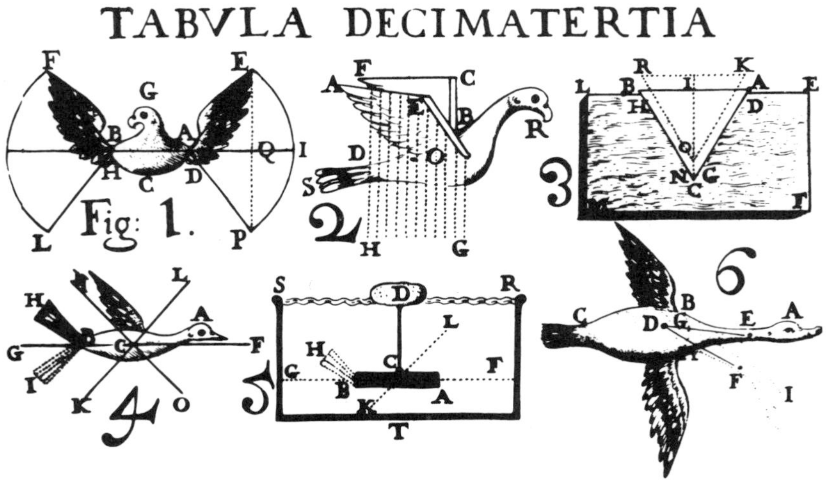

11. The mechanism of bird flight. Giovanni Alfonso Borelli, *De motu animalium* I (Romae 1680) plate 13.

mechanical advantages. On one hand, the absence of any encounter with hard surfaces leaves body tissues less fatigued and less damaged. On the other, the air provides a more continuous lifting force: when a man jumps on the earth, the impetus is quickly dissipated after the foot has made contact with the ground, but since flight takes place in a fluid medium, the impetus remains undissipated after the downstroke and can assist the subsequent impulse of the wings (I.307).

At this point Borelli introduces his most original and influential, though totally erroneous, contribution to the theory of flight. Pointing out that, as any observer may note, birds do not flap their wings backwards, he appeals to the flexibility of the wing behind the rigid line of the leading edge. As the leading edge is moved vertically downwards, the pressure of the air underneath causes the trailing edges to be bent upwards and inwards toward the centre line of the body. The trailing edges of the wingtips are thus brought together so that the surfaces of the wings form a wedge whose apex is at the rear and whose base is towards the bird's head. Appealing to the simple theory of wedges and inclined planes, Borelli points out that the compressed air, pressing against the sides of the wedge so formed, will cause the wedge to move in the direction of the base, that is, will cause the bird to move forward in the direction of the head (I.308–09). In Borelli's estimation, the true explanation of forward flight therefore has nothing to do with the alleged rowing action so often adduced, on which he pours, in passing, a good deal of scorn.

Offering an immediate physical expression of geometric principles, the wedge and the inclined plane appealed greatly to the mechanically minded seventeenth century. Borelli was not alone in adapting these ideas for application to aerodynamic matters. In one of his notebooks about physical problems,

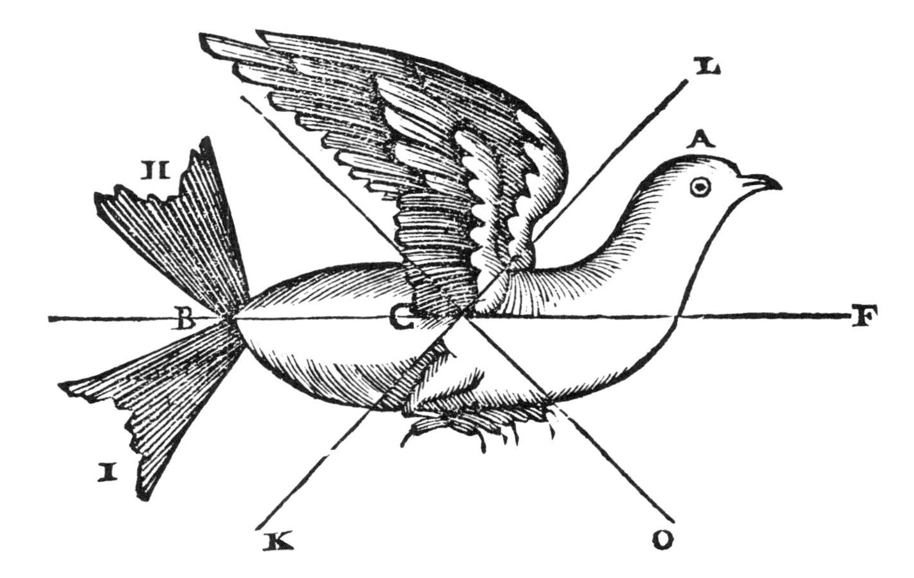

12. Redrawing of Borelli's plate 13, Fig. 4. Pierre Régis, *Cours entier de philosophie* new ed. II (Amsterdam 1691) 618. (Reproduced by permission of the Syndics of Cambridge University Library)

John Conyers, a London apothecary, used the theory of the wedge or cone to explain various kinds of motion in the air, beginning with the rising of a kite:

> Considerable is the reason of the Rise & motion of a paper Kite which boyes raise by adding more & more string untill it attaine a great height, which how much [*sic*] of the reason of the thickness of the ~~supperior~~ inferior ayre to bare it & the holdeing against the winde it should seeme to have a Power supporting Conickall as allso in the raisinge the same[34]

Although he does not attempt an explanation of flight itself, Conyers uses the cone to account for the sloping descent of gliding birds and for the sawtooth appearance of some cumulus cloud formations:

> that a pidgeon or other fowle cannot desend perpendicularly in flight is a truth, but flye sloapeing downewards somtymes beyond the marke it seemes the motion sloapeing is easier as allso the rise in flyeing cannot be other then sloapeing
>
> The Rise of clouds for raine is partly so the motion of wind caries the cloud on untill it comes to the end of the winds force then it tends downewards for the motion & Rise of clouds is like indentures cutt in & out as hills grow & the fashion of Bubbles[35]

Applying these popular ideas of the day, Borelli attempts, as Fabricius had done, to distinguish between lift and thrust. While the wedge action will cause

13. Kite, demonstrating the oblique action of the wind. British Library, MS Sloane 919, f. 17r. Notebook of John Conyers, apothecary of London. Entry for January or February 16$\frac{73}{74}$. (Reproduced by permission of the British Library)

the bird to move forward, the continuing downward beat of the wing is essential to maintain it in the air. This movement must, he says, be straight down, since only then will the bird generate a force acting vertically upwards to counter the weight. For equilibrium to be maintained, the centre of gravity of the bird must lie on a line joining the leading edges of the wings. Thus not only observation but also reason can tell us that a bird does not flap its wings backwards. If it were to do so, the centre of gravity would be too far forward and the bird, now unbalanced, would fall.[36] Since it is clear that some adjustments of balance are often necessary, both in the air and on the ground, Borelli points out that birds can in fact vary the relative position of their centre of gravity by extending or retracting their necks and legs. Although he does not give the matter as much weight as did his predecessor Fabricius, he is careful to mention the value to the bird of having the centre of gravity placed a good deal lower than the line of the wings (I.292, 297).

While Borelli's appeal to wedges merely confused the theory of flight, his authority had a decidedly salutory effect in helping to persuade people that birds do not 'row' themselves along through the air. Although that mistaken view continued to be given widespread currency, its dominance began to wane after 1680.

Towards the end of his discussion, Borelli attempts to provide some account of the nature of soaring flight. After denying the truth of earlier theories, such as the weakness of downward-tending forces in the upper air, Borelli correctly, if perfunctorily, explains soaring flight by recourse to the movement of the air in which the bird is flying. As we may see from the movements of clouds, he says, winds blow even at the great altitudes reached by soaring eagles. Although he does not attempt to analyse how it happens, he says that the bird, whose outstretched wings make it sink only slowly, is sustained by the force of the wind as thin iron plates are sustained by water. Having given itself sufficient lift and forward motion by initial flapping, it can by this means remain aloft, or at least sink very slowly (I.321). Borelli thus discards any appeal to mystical forces, to gross differences in conditions at altitude, and to such devices as 'tonic flight.' While going only a part of the way towards an explanation of soaring flight, his account is a good deal more modern than that of his immediate predecessors.

Ray and Parent

Among the best known of British naturalists, John Ray (1628–1705) wrote a much-quoted book, *Synopsis methodica avium & piscium,* which was posthumously published in 1713. More immediately interesting from the present point of view, however, are the comments on bird flight that he included in a book published more than twenty years earlier, *The Wisdom of God Manifested in the Works of the Creation* (1691). Noting that wings are convex above and concave below, he offers a half-truth based on the idea of 'resistance':

> The underside of them is also made concave, and the upper convex, that they may be easily lifted up, and more strongly beat the Air, which by this means doth more resist the descent of their body downward.[37]

Showing an intuitive awareness of the importance of drag, he goes on to speak of the streamlining of a bird's body, adopting Aristotle's idea that the beak 'cuts' the air:

> Then the Trunk of their body doth somewhat resemble the Hull of a Ship; the Head the Prow, which is for the most part small, that it may the more easily cut the Air, and make way for their bodies . . .[38]

In common with many ornithologists, Ray was more concerned with biology and taxonomy than with the mechanics of flight. Content to describe rather than to analyse, he leaves the matter there.

The mechanical emphasis of Borelli's pioneering attempt at a scientific account is increasingly felt in the work of eighteenth century writers, some of whom also offer quantitative measurements and estimates. Among these was Antoine Parent, who included a discussion of flight in his *Essais et recherches de mathematique et de physique,* published in 1713. Attempting a detailed analysis, he adopts an almost ludicrously complex geometrical and arithmetical presentation, supported by only a handful of basic propositions none of which goes very far towards advancing the general theory. Making the assumption that flight depends on the resistance offered by the air to the beating action of the wings, Parent takes issue with Borelli, pointing out the fallacy of the argument based on analogies with the wedge and with forces governing the

14. Two pages of calculations about bird flight. Antoine Parent, *Essais et recherches de mathematique et de physique* III (Paris 1713). (Reproduced by permission of the British Library)

392 *Recherches ou Eſſais de Phyſique*

ce qui donne pour cette hauteur $\pm \frac{\pi}{p\int y dx} \pm hcs \times \frac{C-D}{arC}$.

Et faiſant encore l'analogie de l'art. 2. Comme la racine de la hauteur d'où les corps tombent dans la premiere ſeconde de tems $\propto \sqrt{150}$, Eſt à la racine de la hauteur cy-deſſus $\pm \sqrt{\pm \frac{\pi}{p\int y dx} + \frac{hcs}{ar} \times \frac{C-D}{C}}$; Ainſi la viteſſe uniforme acquiſe par 150 pouces, ſçavoir 300 pouces par ſeconde a un quatriéme terme $\propto 10\sqrt{\pm \frac{6\pi}{p\int y dx} \pm \frac{6hcs}{ar} \times \frac{C-D}{C}}$, qui eſt la viteſſe v en pieds par ſeconde avec laquelle l'oiſeau montera ou deſcendra verticalement, & cela dépendamment de la valeur d'h & de p.

A l'égard de cette valeur d'h pareille à celle du ſecond article, elle eſt aiſée à trouver à cauſe de l'équation du ſecond article $hp\int y dx \propto$ l'effort direct de l'aile, ou au poids P qu'elle eſt capable de ſoûtenir en tirant perpendiculairement ſelon QLM, d'où l'on déduit $h \propto \frac{P}{p\int y dx} \propto$

& de Mathematique. 393

$\frac{1000P}{307\int y dx}$. Subſtituant donc cette valeur d'h dans la derniere valeur cy-deſſus, on aura pour la viteſſe v de l'oiſeau dans le ſens vertical $10\sqrt{\pm \pi \mp \frac{Pcs \times \overline{C-D}}{arC} \times \frac{6}{p\int y dx}}$, ou plutôt cette autre . . . . $100\sqrt{\pm \pi \mp \frac{Pcs \times \overline{C-D}}{arC} \times \frac{15}{307\int y dx}}$, qui ſe fera de bas en haut ſi π eſt plus petit que le membre qui le ſuit ſous le ſigne radical, & de haut en bas s'il eſt plus grand. Donc réciproquement ſi cette viteſſe v eſt donnée, on trouvera la force inconnuë P, ou le poids que chaque aile doit pouvoir ſoûtenir ſelon QL dans ſon centre L, ſçavoir $\overline{307v^2\int y dx \mp 600000\pi} \times \frac{arC}{600000cs \times \overline{C-D}} \propto P$, avec laquelle valeur de P on trouvera le nombre de coups que l'aile doit battre par ſeconde, en ſe ſervant de l'égalité cy-deſſus $h \propto \frac{P}{p\int y dx}$, & $10\sqrt{6h} \propto 10\sqrt{\frac{6P}{p\int y dx}}$, qui eſt la viteſſe que le

novement of solid objects in general. Being fully aware of the need to find a different approach to explain movement in a fluid medium, Parent points out in a number of ways that Borelli took insufficient notice of the relationship between centre of gravity and 'centre of resistance.' Offering little that was new to the subject, he nevertheless showed a clear understanding of the need to establish an equilibrium between weight and lift before straight and level flight could be achieved. Although the bird's wing encounters resistance as it rises, Parent explains, as Leonardo had done, that owing to its curvature and flexibility the resistance is smaller on the upstroke than on the downstroke. To arrive at the net resistance, one must therefore subtract the pressure on the underside, which he thinks is made greater by the concavity of the wing.[39]

When writing of the equilibrium of forces in the horizontal plane, Parent has rather less to offer. In common with virtually everyone at that time, he fails to distinguish between groundspeed and airspeed, writing about the increase in air resistance that results from a strong head wind. It is for that reason, he says, that birds of passage prefer to fly 'above the region of the clouds, where the course of the winds is more uniform and much gentler, not to say almost imperceptible, as is discovered by men who walk over high mountains' (p. 382).

Concerned more with lift than with drag, Parent concludes his analysis with a brief paragraph about the possibility of artificial flight. Taking issue again with Borelli, who had thought such a thing impossible, he says that the

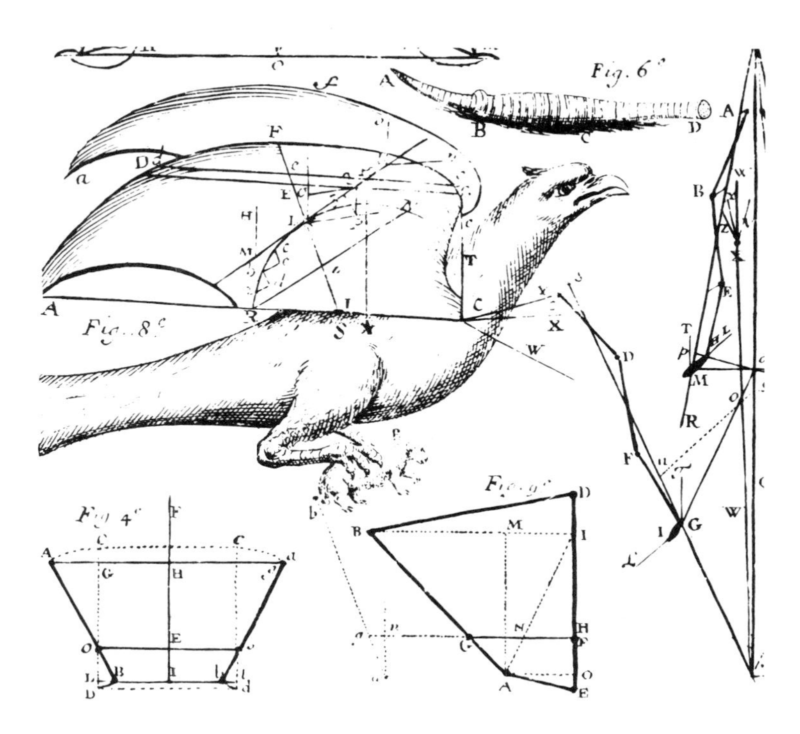

15. The mechanism of bird flight. Parent, *Essais* III (Paris 1713) plate 26. (Reproduced by permission of the British Library)

matter will be decided by the future: 'Experience will show whether it is not possible to make a spiral spring that may apply to the centre of resistance of each wing the force necessary to sustain the whole machine' (p. 400).

Le Chevalier de Vivens

A member of Montesquieu's circle, the engaging Chevalier de Vivens (1697–1780) proposed in 1742 an apparently outrageous but in fact quite rational new idea to account for flight. As Montesquieu himself showed a lively interest in the subject, it is likely that, in this as in other things, Vivens collaborated with Montesquieu and his son in the carrying out of observations and experiments. In his *Pensées* Montesquieu included some orderly notes on bird flight and the possibility of human flight,[40] but there is no reason to suppose, as Jules Duhem has done, that Montesquieu collaborated in the writing of Vivens's monograph, which uses concepts alien to Montesquieu's cast of thought.[41]

Vivens develops an imaginative theory that might be termed a transformation of the idea of 'tonic flight.' Unsatisfied with earlier attempts to explain the familiar gravitational anomalies of floating metal foil and sinking grains of cork, and unable to credit the alleged minute flapping movements of soaring birds, some of which he had been able to observe at close quarters, Vivens suggests the existence of a new physical force. When a body or particle of matter is divided or extended, it acquires something that Vivens calls centrifugal force at the same rate as it loses centripetal or compressive force.

Although he does not say so explicitly, Vivens's centripetal force is Newtonian gravity, while his centrifugal force is antigravitational. The centrifugal force has, among other things, the effect of reducing the specific gravity of particles of matter as they are either subdivided or beaten out into thin sheets.

Applying this idea to bird flight, Vivens supposes that birds make use of matter that shows an abundance of centrifugal force:

> I say that birds have in them and around them, and in a highly adhesive fashion, elastic substances whose centrifugal force, coupled with the resistance of the air, is equal to the weight of the birds, becomes greater when the bird wants to rise, and less when it wants to descend.[42]

Explaining where this matter is stored and how its volume can be varied so as to vary the centrifugal, lifting force, Vivens has recourse to the anatomical point that fascinated ornithologists throughout the seventeenth and eighteenth centuries: the variable air-filled cavities in a bird's body. He supposes that by expanding and compressing its 'elastic substances' the bird is able to vary its specific gravity on the analogy of the use that fish make of their ballast sacs. He even supports his argument by an appeal to the impossibility of a bird's flying if the feathers of its wings are seriously damaged:

> That does not diminish the force of its muscles. In damaging its wings it seems to me that all one is doing is diminishing their volume and introducing air into the pipes of the feathers through the ends that have been cut.[43]

As for the nature of the elastic substance, Vivens supposes that, whatever it is, it must be something akin to the element of fire, a point that he supports by drawing attention to the familiar hot-bloodedness of birds. He suggests, finally, that all living creatures are endowed with this substance to some degree, and that in the long run it must be thought to have a close affinity with the principle of life itself. Flight is thus intimately related to the idea of vitality.

Refining his ideas further by the introduction of more mechanical principles, Vivens supposes that the control of a bird's specific gravity is assisted by adherence of large quantities of air, helping to lighten it. When its body is in equilibrium, the wings are able to act like true oars, with the additional advantage that their movement may set free from the body a number of bubbles of the light elastic fluid, whose tendency to rise will again help to sustain the bird in the air.

Georges Buffon

In the introductory essay to one of the most celebrated and sumptuous of eighteenth century works on ornithology, the nine-volume *Histoire naturelle des oiseaux,*[44] Georges Leclerc, Count of Buffon (1707–88), makes only a few desultory remarks about flight, which interests him much less than does descriptive taxonomy. Celebrating the birds' speed and facility of movement, he makes the obvious points that they are light and that their muscles are strong. Without attempting to explain why it should be significant, he also takes up John Ray's point about the curved shape of the wing: 'the form of the wings, convex above and concave below.'[45] Despite the detail and the loving care of his book, Buffon shows himself to be one of the many ornithologists whose delight in birds paradoxically coexists with an almost complete disregard for the beauty of their flight. Writing briefly about how one may experiment with the flight patterns of the eagle, he demonstrates this disregard by concerning himself with the bird's responses to temperature rather than with the function of its wings:

> When one wants to prevent an eagle from rising too high and being lost to our sight, it is sufficient to pluck his belly; he then becomes too sensitive to the cold to be able to rise to great heights. (I.44)

Avoiding Fabricius's attribution of an almost mystical power to the feathers, Buffon simultaneously discards any responsiveness to the strange phenomenon of lift.

As a natural consequence of his scant interest in aerodynamic matters, Buffon makes the traditional error of believing that a bird would avoid flying with a following wind because otherwise its feathers would be ruffled. For this reason, he says, the bird of paradise is especially careful to fly only when the winds are light. In conditions of strong wind, the fluttering of its long decorative feathers would make navigation very difficult (III.154).

Silberschlag

For the great German churchman, scientist, and ornithologist Johann Esaias Silberschlag (1721–91), the principal force enabling a bird to fly derives from the power of compressed air. In common with Leonardo, he believes that the bird's wing squeezes the air beneath it and sustains itself on the region of high pressure so created. As a consequence of the curvature of the wings and the flexibility of the feathers, the downstroke creates much more compression than does the upstroke. Although fundamentally mistaken, his discussion is developed with intelligence.

Considering the process by which the compressed air again spreads out beneath the bird, he supposes that the air finds more difficulty in escaping the more the bird pushes itself forward, so that the faster it flies the better it is able to sustain itself:

> But this compression is in direct proportion to the speed; and in the same way the force of the forward movement and the resistance of the air increase in proportion. If that were not so, the wind would meet the bird at the same speed as that with which the bird flies towards it, and its resistance would be the same as that of a wind moving at the same speed

> as the bird. That this is not so one may observe from the sails of windmills which often move faster than the wind which drives them.[46]

In an attempt to show that the concavity of the wing holds the compressed air for long enough to enable the bird to remain in level flight (p. 228), Silberschlag suggests an experiment:

> If one spreads chaff on a table and moves an artificial wing towards it as fast as possible, using a movement similar to that of a bird, one will find that the chaff first begins to move when the stroke is almost completed; also, that the air, which is met by the oar of the wing, spreads out very little, mostly blowing forwards under the wing at the point where the primary feathers end. (p. 223)

Proving his point in a different way, Silberschlag describes how birds flap two or three times before rising from the ground, a movement that he attributes to the need to compress the air before lift becomes possible (pp. 223, 226).

Although he includes only a few brief comments about the bird's centre of gravity, Silberschlag is careful to consider the relevance of its position, pointing out that the body must be so arranged and angled that in flight the centre of gravity is between the wings, while on the ground it is vertically above the feet (pp. 241–43).

In keeping with the generally high quality of his thought, Silberschlag is among the first to show true awareness of the relative motions of bird and wind:

> No bird can remain still while soaring, and if it seems, when there is a strong head wind, as if the bird is hanging motionless in the air, the reason is that the bird is soaring as fast against the wind as the wind is carrying the bird along. (p. 249)

In the same context he makes the accurate observation that a bird can use its airspeed to gain more lift while leaving its ground speed unchanged. Although Silberschlag may not have appreciated all the variables of speed, angle of attack, and displayed wing area, his description shows insight:

> If a gust of wind should blow, the bird half pulls its wings together, or often more than half does so, and by means of its tail gives itself a diagonally downwards direction. It would then sink, but the strong wind raises it and it remains quietly soaring in the place where it was. In times of calm, on the other hand, it is completely impossible for the bird to practise this trick. (p. 250)

Later, Silberschlag analyses how birds can occasionally appear to rise vertically upwards by means of a 'zoom climb.' He is careful, nevertheless, to point out that this is always executed directly into wind:

> This spring into the air occurs only in connexion with rapid flight. When birds use their tails to direct themselves straight upwards, their flight is immediately into the wind which then not only stops their forward movement, but also lifts them several feet into the air, and indeed often forces them back a little. After this they are able to place themselves once again horizontally and to set themselves softly down by folding their wings. This whole trick is dependent upon a rapid forward motion of the wings and is quite impossible if the bird approaches slowly. (pp. 252–53)

Mauduit

The long dictionary of ornithology contributed to the massive *Encyclopédie méthodique* (1782, 1784) by Jean-Pierre-Etienne Mauduit de la Varenne[47] begins with a series of prolix and comparatively unoriginal 'discourses.' Despite the inordinate length of the work, Mauduit has, like Buffon, very little to say about the mechanism of flight itself. While he shows some interest in the *alula* and the tail, he makes only a few conventional remarks about the process of flight. The most salient of his comments is symptomatic of the almost universal misunderstanding of relative motion in the air, an error that persisted throughout the eighteenth century. Noticing the convexity of the wings, he suggests that a bird may benefit from a following wind, which, he says, will strike against the dipping underside of the leading edge and therefore help to push it along over the ground.[48] With almost unlimited space at his disposal, Mauduit severely disappoints any reader who might hope to find an article living up to the scientific aspirations of the encyclopaedia as a whole.

Huber

Better things are found in smaller compass in an attractive little book on the flight patterns of birds of prey, published in 1784 by the Swiss naturalist François Huber (1750–1831).[49] In view of the care and precision of his observations, undertaken with the assistance of a loving wife and willing servant, it is particularly poignant that he gradually became totally blind as the consequence of an infection contracted when he was fifteen. The charm of his book lies principally in Huber's enthusiasm for a subject that he treats with delighted sensitivity. Making some attempt to classify birds according to their modes of flight, Huber distinguishes two types of wing, 'rowing' and 'sailing' (*rameuse* and *voilière*). The rowing wing is stiff, of high aspect ratio, has pointed tips, and lacks any readily visible wingtip slots. The sailing wing he describes as broader and softer with slots formed by separated primary feathers.

Describing the action of the two kinds of wing, and attempting to formulate a mechanical explanation of flight, Huber invokes the idea of the inclined plane associated with the elasticity of the air. Noting, as many others had done, that birds need to vary their mode of flight in order to control their groundspeed, Huber assumes that a steady wind has aerodynamic effects.

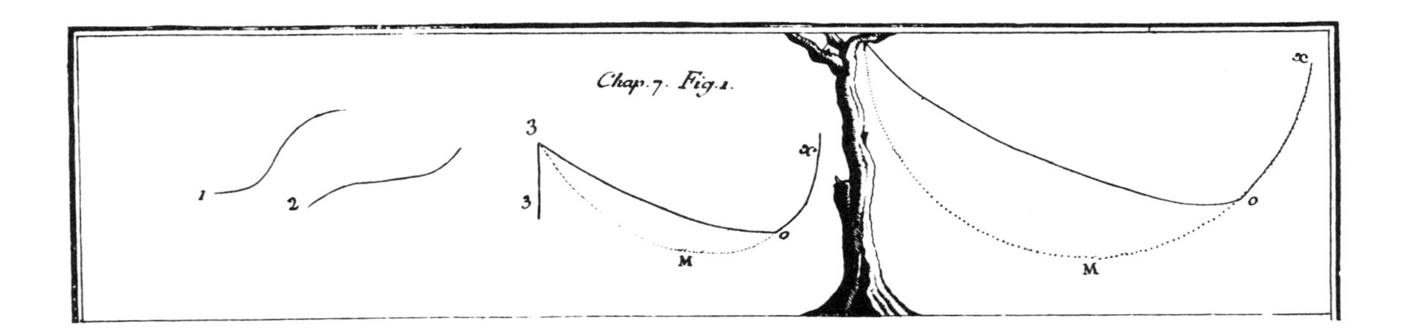

16. (a, b) Rapid climbing and dives followed by zoom climbs. François Huber, *Observations sur le vol des oiseaux de proie* (Genève 1784) plate 5. (Reproduced by permission of the British Library)

Accordingly he attempts to introduce into his argument a number of subtleties based on that proposition:

> The effect of the rowing wing is to overcome the resistance of the elastic fluid on which it acts. The air, elastic in the calmest conditions, becomes more so in one direction when the wind is following, and less so in the other direction.
>
> The rowing wing, striking against the wind, meets resistance that raises the bird as it moves forward. But it advances both because of its specific weight and because of the capacity of a sharp wing to cut the wind.
>
> When the rowing wing is working in a following wind, it meets no resistance capable of raising the bird; it meets even less than it does when the wind is calm; its effect, in this case, is to sustain the bird in the horizontal direction and to maintain its speed, either by its own forces, or by the help of the wind that acts on it insofar as the individual bird knows how to control it. (pp. 7–8)

Huber goes on to attempt an explanation of how the action of the wing causes the bird to move forward. Adopting a post-Borellian attitude to the direction of the wing's movement, he is not inclined to believe that the beating action is downwards and backwards. In explaining how the downstroke sustains flight, he tries to find a mechanism both more realistic and subtler than Borelli's:

> There is this difference between the flying oar and the ship's oar, that the one strikes immediately beneath itself while the other strikes down and back.
>
> In order to explain how it is possible that in striking straight downwards, vertically under itself, the rowing wing carries the bird forward, one must observe that the underpart of the wing is formed into a kind of arch whose most curved part is situated at the front of the wing where the primaries and secondaries begin.

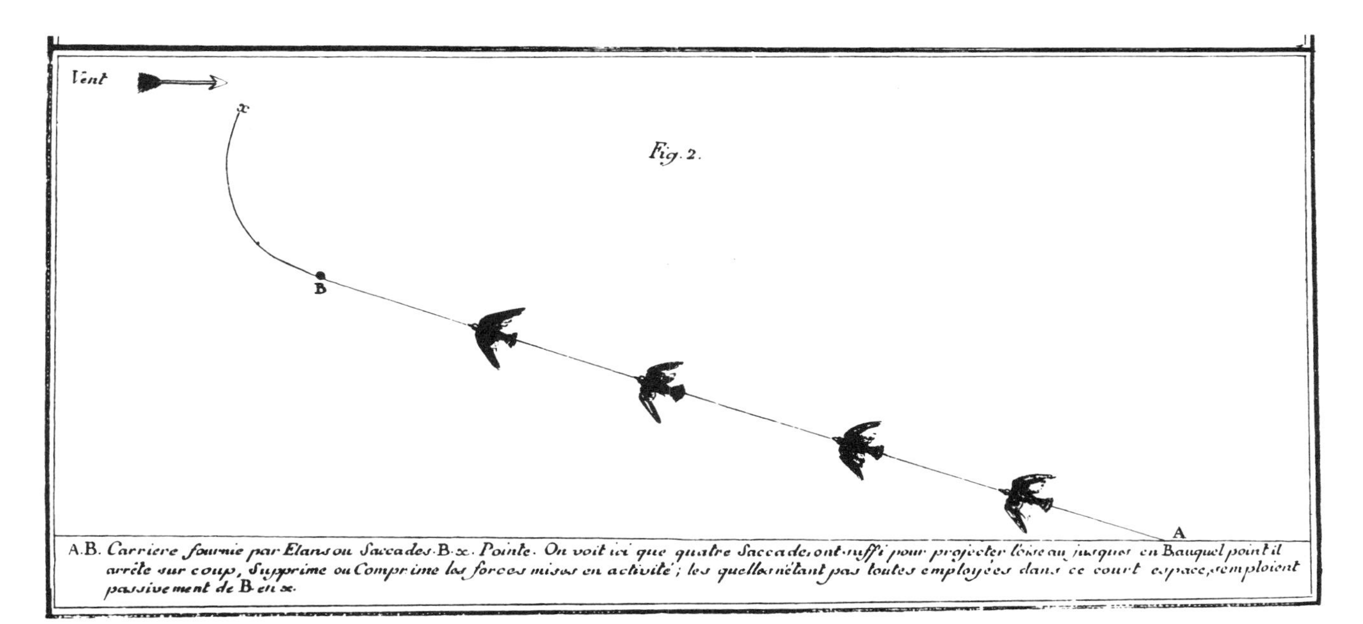

A.B. Carriere fournie par Elans ou Saccades. B.x. Pointe. On voit ici que quatre Saccades ont suffi pour projecter l'oiseau jusques en B auquel point il arrête sur coup, Supprime ou Comprime les forces mises en activité; les quelles n'étant pas toutes employées dans ce court espace, s'emploient passivement de B en x.

However little noticeable this arch may be when the wing is stationary, it becomes noticeable when the wing strikes the air with force. Then the solid part that borders the front of the wing cuts the air while the rest gives way because of the strength of the stroke, and this very small inclined surface, forcing the air backwards more than downwards, is the cause of the forward movement that would not occur if the wing were perfectly flat and equally firm over its whole surface. (Butterflies fly in a tumbling fashion because their wings are flat.)

All parts of the wing contribute to the forward movement, and the elastic primaries, giving way and immediately springing back, necessarily carry forward the body that they accompany. The springiness of the air, reacting with still more force, doubles the means of projection; for, giving way with resistance during the strongest period of the blow that strikes it, it acts in its turn during the moments between beats with the force of a spring that is released after having been forced in the opposite direction.

The sailing wing also forms an arch that is necessary for forward movement for the same reasons. But the movement is slower in proportion as the beats are less strong, less frequent, and the primaries softer.

It cannot even project the bird forwards unless the wind is from the rear. The wings are the bird's rudder; to turn to the right, the left wing beats strongly, the right moves less according as the turn is smaller or more complete; it remains almost immobile when the bird turns on itself.

It is easy to understand the gradations from a pirouetting movement to movement in a straight line, the latter being the straighter the more the movement of the two wings is equal.

When the bird glides, it turns without making any noticeable movement of its wings; in this case it is by lowering a little the side to which it is turning and by raising the opposite side that it moves in a circle and in a more or less flat spiral. (pp. 8–10)

Huber is most interested, however, in the flight paths of the various birds of prey and their quarry, which he illustrates with delicate line drawings. Without saying anything explicit about the relationship of potential and kinetic energy, he nevertheless accurately describes how a diving bird can, in aeronautical parlance, 'trade speed for height':

> He does this by suddenly reopening his wings, which he was keeping folded during the descent. This movement is sufficient not only to arrest the descent, but also to carry him, without any effort, back to the level from which he dived . . . The dive and the climb together are called a *passade,* which looks much like the to-and-fro movement of a child's swing. (pp. 29–30)

Barthez

As the eighteenth century drew to a close, the lack of a cogent explanation of flight seemed increasingly anomalous in comparison with the general advance in other branches of mechanics. Among those most clearly aware that the phenomenon posed a real problem for which a solution must be found was Paul-Joseph Barthez (1734–1806). Having absorbed much of the earlier literature, Barthez includes, at the end of his book on animal movement, an attempt at a fully rational analysis.[50] Despite the observations of Borelli, whom he several times quotes, he begins with the assumption that the wing is beaten backwards. Unlike Pliny and Leonardo, he believes that during the downstroke the wing is expanded rather than contracted, thus providing a more extended surface to press against the air. For Barthez, however, these remarks constitute only the beginning of the discussion. Rightly sensing that the phenomenon of lift still requires explanation, he prefaces a statement of his own views with a refutation of some earlier opinions:

> Bird flight, which depends on the action of the wings, has always been explained hitherto by saying that the wings act like oars against the air which offers resistance to them; or that their movements are reflected in the opposite direction by the reaction of the elastic fluid. I shall show in detail how erroneous these explanations are. (p. 194)

To speak of the rowing action of the wings, says Barthez, is to point to the phenomenon without explaining it. Unless some resistance can be found to oppose the movement of the wings, these will merely move in an angular way about the body without causing displacement of the whole bird. Mentioning Borelli and Parent, Barthez perceives a tendency to avoid explaining how it is possible for the wings to be opposed by what must necessarily be a very great reactive force (pp. 195, 196).

After eight pages of refutation, Barthez comes to the point and tries to find an explanation of his own. His opening statement is nevertheless disappointingly unoriginal:

> I first of all remark in general that the movements of the bird's wings for flight are singularly analogous to those executed by the arms of a man when he is using them to swim. (p. 198)

Admitting that this does not get us very far, he goes on to propose that the bird's movement is caused by reciprocal movements of muscles within its body:

> The resistance which the air opposes to the movements impressed on the bird's wings by the pectoral muscles makes the action of these muscles function reciprocally (to the extent of this resistance) so as to move the sternum and the ribs (where they have their origin) and consequently the trunk of the bird's body, towards their points of attachment at the humerus. It is thus that these muscles pull the bird's body in a direction opposite to that of the movement of the wing, that is to say upwards and forwards. (p. 199)

Although it would surely have surprised Barthez to be told so, his theory is distantly analogous to Galen's 'tonic flight' and might even be thought to offer a modernised equivalent of Aristotle's argument about the bipartite basis of animate 'voluntary motion.'

In a number of succeeding paragraphs, Barthez attempts to develop his theory in further detail. Far from clarifying his ideas, his description of the interrelated movements of muscles and bones is confused and does nothing to improve on the general idea of a bird's pushing itself forward against the resistance of the air. Summarizing his argument thus far, he can do no better than to return to the familiar analogy with swimming:

> The resistance which the air opposes to the movements of the feathers and vanes provides the support which is necessary for the action of the wing muscles. In this respect the fan of the wing that is deployed laterally in relation to the bones of the wing so as to give the greatest possible extent to the mass of resisting air, can be compared to the hand that a man who is swimming with his arms spreads out flat so as to seek the greatest resistance in the water. (p. 201)

Wishing to write as full an account as possible, Barthez considers contributory factors that assist the process of flight. First he returns to the idea of elasticity, discussing the springiness not only of air but also of the feathers. Although once again the argument is confused, he clearly believes that the flexibility of the feathers provides a further means of harnessing propulsive forces:

While the wing is being lowered, the primaries and secondaries which are flexible and elastic prolong the effort of the air by their flexibility, and despite their elasticity add a new resistance to that of the air. From this it results that with the movement of the wing being developed for longer and greater effort, the reciprocal action of its flight muscles impresses on the body of the bird still greater projective movements.

When the lowering of the wing is finished, and when the body of the bird begins to respond to the projective movements that have been impressed upon it, the principal feathers of the wing, returning because of their springiness, give to the wing bones an upward and forward impulse. (pp. 201–02)

Barthez finds a second contributory force in the wind, the energy of which can be gathered below the concave undersurface of the wing. In addition to his acceptance of the false idea that a bird is able to extract energy from a steady horizontal wind, Barthez introduces a comment showing that he has some insight into the effects produced by varying the relative positions of the centre of gravity and the centre of pressure:

If the wings and the tail are spread and curved underneath, the wind, held and reflected below these curves, pushes and raises them at the same time. If all of these surfaces, spread at the same time and carried by the air, do not equally sustain the front and back of the body, one of these parts will descend or be raised relative to the other around the bird's centre of gravity. (p. 207)

Understanding that lift is increased by increasing its angle of attack, Barthez is also aware of how a bird can increase lift by increasing the effective area of its wing. After paraphrasing Huber's description of the *passades* of birds of prey (see earlier), he goes on to explain how lift may be increased:

The bird gives its body a raised position, lifting the front part with force. Thus the bird's body is pushed against a new layer of air, the surface of which, rising from below to above, makes a more or less large angle with the surface of the layer of air on which it was formerly carried. The momentum with which it is endowed is expended in relation to this new layer and the bird is projected against it obliquely. (p. 209)

Barthez's last suggestion about supplementary forms of lift is the most fanciful of his ideas. Understanding the falsity of many early comments about the way air in the cavities of bones and feathers might reduce a bird's specific gravity, but supposing, as with some of his predecessors, that the introduction of air aids the movement of the flight muscles, Barthez believes that it may also contribute to stability:

> The principal use that I attribute to such a compression of internal air in the bones of the wings and body of the bird is to prolong and to augment the efforts of the wing muscles, insofar as they bring about the movements of flight. But in addition this air, when it is thus compressed in the bones, cannot but make birds more stable in the uncertain movement of the air, whether they are flying or remaining still as if suspended: it cannot but contribute to the resistance with which they oppose the wind, which would otherwise carry them off in an entirely passive manner. (p. 227)

Despite his lengthy analyses, Barthez adds little to the eighteenth century's understanding of the fundamental mechanics of flight. He nevertheless shows that he has many intuitive insights into matters of balance and control that he lacked the scientific skill to explain (see later). What is most interesting about his book is the urgent need he felt to resolve puzzles of long standing by an appeal to all the physical knowledge available to him.

The mechanics of stability and control

For centuries, the function of a bird's tail was misunderstood, and even today few people appear to be aware of its primary uses. The *alula,* which delays the onset of turbulence over the wing and hence helps a bird to fly slowly without stalling, was still less well understood, and indeed went altogether unnoticed by many early bird watchers. I summarize here a few representative views from Aristotle to Barthez.

The misapprehension that the tail is primarily a steering device is at least as old as Aristotle:

> In winged creatures the tail serves, like a ship's rudder, to keep the flying thing in its course. The tail then must like other limbs be able to bend at the point of attachment. And so flying insects, and birds (Schizoptera) whose tails are ill-adapted for the use in question, for example peacocks, and domestic cocks, and generally birds that hardly fly, cannot steer a straight course. Flying insects have absolutely no tail, and so drift along like a rudderless vessel, and beat against anything they happen upon; and this applies equally to sharded insects, like the scarab-beetle and the chafer, and to unsharded, like bees and wasps. Further, birds that are not made for flight have a tail that is of no use; for instance the purple coot and the heron and all water-fowl. These fly stretching out their feet as a substitute for a tail, and use their legs instead of a tail to direct their flight.[51]

Perhaps because of our familiarity with aeroplanes and with the elementary principles of mechanics, most people today would doubtless object immediately that as a bird's tail is not arranged vertically, the analogy with a ship's rudder is false. The tail seems more obviously suited to act as an elevator, controlling the bird's movements up and down. In fact, the truth is a good deal more complex than either explanation. The tail serves mainly as a flap but is also used to assist with pitch control and often creates supplementary turning forces when it is twisted vigorously to one side. Accurate observations of the twisting effect abound in Leonardo's manuscripts, which remained generally unknown, and a few perfunctory comments, correcting and supplementing Aristotle, were offered by Fabricius in his *De alarum actione* (1618). Before Borelli, however, the authority of Aristotle was dominant. The generally accepted analogy with a ship's rudder powerfully sustained the tendency to treat movement through air according to the same principles as apply to movement through water.

A more serious misapprehension was introduced by Borelli, who adopted a position totally opposed to that of Aristotle. Insisting that the tail could have no part in lateral control, Borelli asserted that it served merely to vary the bird's pitch angle. His theories, frequently quoted, exerted a widespread influence on the design of many flying machines that, in emulation of birds, omitted any form of yaw control. While builders of ornithopters sometimes envisaged a combination of differential flapping and banking to simulate the basic control techniques used by birds, the absence of vertical surfaces in many fixed wing designs made controlled flight virtually impossible.

Frederick II

Avoiding oversimplified explanations, Frederick II and his son Manfred base their brief passages about the tail and its functions on careful observations of birds in flight. In his usual matter-of-fact style, Frederick avoids the use of any term such as 'rudder,' preferring to speak simply of *cauda*. Unlike most other early writers, he makes no overt analogy with the steering of a boat but implies that a bird creates turning forces primarily by the use of its wings. His own paragraph on the tail includes a remarkable insight into its function as a contributor to the overall lift, which is varied according to need:

> The tail has a manifold function in birds. When expanded it partly sustains the weight of the body and diminishes the labor of the wings as they propel the avian body in any of the four directions—upward, downward, right, or left.[52]

While in his supplementary comment King Manfred first reverts to the old idea of the analogy with a ship's rudder, he goes on to describe how the bird controls its posture while resting, referring to the way the tail is moved to keep the centre of pressure in line with the centre of gravity and the feet:

> In the first place, just consider the help a long tail provides when the bird rests not only on high cliffs but on lofty trees and then flies off with the tail as a (rudder) steering apparatus, or when simply roosting the tail certainly acts as a shield against gusts of wind and other insults. (p. 91a)

Frederick II was probably the first to describe the movement of the *alula* as a bird comes in to land. He offers an explanation that, while not entirely accurate, is a commendably intelligent attempt to account for the way it makes sustained flight possible at high angles of attack:

> As the bird makes a descent it draws in and closes its other wing feathers but extends the *empiniones* (the bastard wing). Were all the wing feathers extended, the on-rushing air would lift the bird and hinder its descent. Were they all closed (both quill feathers and the bastard wing), the bird would fall, heavily in fact, and be without power to direct or control its landing. But with the *empiniones* alone expanded, descent is not obstructed but is controlled and directed to whatever point the bird desires. (p. 89b)

Fabricius

Although perfunctory compared to those of Leonardo, the analyses of control techniques by Hieronymus Fabricius show an intuitive grasp of the processes at work. Developing his argument about the air-gathering capacity of the concave wings, he shows that he understands the tail's contribution to lift, pointing out that it also assumes a concave shape to gather the air. Improving on Aristotle's observations, he adds that the tail can move in four directions: up, down, left, and right. Although he does not attempt to describe in detail how the tail is twisted for this purpose, he makes the interesting remark that although they are set vertically the tails of fishes are able to move in the same four directions. It is nevertheless clear to Fabricius that the tail of a bird provides only a supplementary force to assist in turning, the main turning action being contributed by a differential beating of the wings. If it beats one wing hard while moving the other slowly, a bird can, he says, make so sharp a turn as to reverse its direction almost immediately. In summary he says that the tail has two functions: first, to gather air, making the body lighter and thus contributing to lift; second, to help the bird to steer its course.[53]

Willughby

Nearly a century after Fabricius, Francis Willughby (1635–72) made a similar assessment of the role of the tail. Playing down the idea of directional control, he not only writes of the tail's contribution to overall lift, but also develops Fabricius's implicit comment about its function in providing pitch control:

> The Tail doth not only serve for directing and governing the flight, but likewise for supporting and keeping even the body. Hence the *Colymbi,* which have no Tails, fly very inconveniently, as it were erect in the Air, with their heads straight upward, and their Tail [rear parts] almost perpendicularly downward.[54]

In this passage Willughby is evidently describing grebes in slow flight, as at takeoff, when the comparatively short tail is well depressed and the wings are held at a high angle of attack. According to his understanding of the way the bird is supported in the air, the absence of a tail inevitably makes it impossible for the bird to maintain its body in a normal level flight attitude.

Borelli

Early in his discussion of the tail Borelli offers a crisp refutation of Aristotle's ideas:

> The mistaken nature of this opinion may be demonstrated through reason and experience. If the rudder of a ship were fixed in the same fashion as that of a bird, i.e. so attached that the axis lay not vertically but horizontally, so that the surface of the rudder could be moved only upwards and downwards against the surface of the water, one would be able to observe that the ship could not be manoeuvred to right or left.[55]

Since, according to Borelli, the tail could be twisted sideways to only a slight degree and since it was clearly unable to create the forces needed for the tight turns of which many birds are capable, another mechanism had to be found to account for lateral control. The difficulty was increased by the known ability of pigeons to carry out most ordinary manoeuvres even when deprived of their tails (I.312). Borelli offers, in fact, virtually the same explanation as Fabricius, describing differential beating of the wings and some movement of the outer wing in a rearwards direction, drawing analogies both with the use of oars and with the differential rowing motion of a swimmer executing a turn. Appealing to experience, Borelli says that one may observe how pigeons turn by raising one wing higher than the other to produce a stronger beat (I.311–12).

Stressing the function of the tail as an elevator and as a producer of additional lift, Borelli observes that birds spread the tail when climbing or descending, and especially when counteracting the effects of reduced airspeed on landing (I.315). He is especially concerned about the interesting problem of how a bird makes a soft landing. As the bird approaches the ground, it expands both its wings and its tail, making concave surfaces like those of a ship's sail. Just before touchdown, it beats its wings forwards to produce a force towards the rear, counteracting its forward speed, while at the moment of touchdown it allows its legs to flex, thus absorbing some of the energy of descent. Without, of course, understanding in detail the aerodynamics of this process, Borelli provides here a generally true description of the landing procedure of many birds.

Writing as though he wished to refute a widely held belief, Borelli includes a chapter (I.316–18) in which he discusses whether or not birds execute turns by steering with their heads and necks rather than with their tails. Analogous to Leonardo's ideas about the manipulation of the centre of gravity for control purposes, the theory suggests that birds turn by moving their heads sideways (see Figure 11). According to Borelli, a triangle of velocities (or of 'impetus') may be drawn to show that the resultant of the bird's forward velocity and of the velocity at which the centre of gravity is moved sideways will be a new forward motion at an angle to the original. Although such a triangle is inapplicable, no sideways force being applied to the bird from outside, Borelli does not dispute this aspect of the mechanics. His objection is based, instead, on observation. Many small birds that have short necks and a small mass are yet able, contrary to what one might expect, to turn quite sharply; heavy birds with long necks, on the other hand, make only slow and ponderous turns. This interesting theory of manoeuvre, appealing strongly to the seventeenth century predisposition to treat problems of motion by geometrical representation, was later quoted by many commentators with the false implication that Borelli himself held it in favour.

Silberschlag

While treating the tail essentially as a rudder that can move in all directions—up, down, right, and left—Johann Silberschlag points out that many birds fly quite well even if totally deprived of their tails. Some, he says, have such small and ineffective tails that their manoeuvres must be controlled by the wings alone, while others, as Aristotle had thought, use their rearward-pointing feet as a substitute rudder.[56]

When he attempts to explain the contribution of the tail to directional control, Silberschlag is puzzled to find that the effects of moving the tail up and down are contrary to what he expected:

> As soon as the bird lifts its tail, it sinks forwards; if it presses the tail down, it thereby raises its breast. At first I tried to clarify this perception from the function of the oncoming air, but found too many difficulties to enable me to arrive at a convincing explanation. Finally I discovered the reason for the mechanism within the bird itself . . . (viz., the leverage of the muscles). (p. 238)

Prior to the development of aerodynamics it was natural for a careful and mechanically minded observer such as Silberschlag to think in terms of levers and therefore to have difficulty in finding a rational explanation of the bird's movements. What Silberschlag had in fact observed was the use of the tail as a flap to increase lift and drag and so make slow flight possible. In conjunction with the depressing of the tail, the body is raised to increase the angle of attack of the wings, a configuration that may often be seen when a bird is approaching to land on a branch or perch. Although Silberschlag had observed this flight pattern, he did not understand the relationship between slow flight and pitch change:

> Occasionally birds use their tails in order to lessen the rapidity of their movement and to arrest themselves in forward flight. They then spread this rudder out, hang it steeply downwards, and use it as if it were an anchor, which one may observe when a bird wants to land. (p. 240)

Elsewhere he adds to his error by associating the use of the tail as an airbrake with the effects of wind. He does not, as so many others had done, speak of the difficulty of flying in a head wind, but suggests, more interestingly, that the presence of a head wind is necessary to produce enough lift, or 'resistance':

> The bird can even increase the resistance of the air by a spreading out of its tail, which happens, and must happen, especially when it wants to fly slowly, or is not flying against the wind, because then the resistance of the air is less. (p. 239)

In view of this, it is perhaps surprising that Silberschlag thinks birds fly for preference into wind, since their feathers might otherwise be ruffled (p. 239).

When Silberschlag describes the use of the tail for lateral control, he again attempts to base his explanation on the theory of levers. This time, however, his description is less easy to reconcile with the most common movements:

> Observe the way a kite turns, how hard it works its tail to bring the movement of its wings to the correct angle and then to hold it there. If it wants to establish its flight at an angle to the right, it turns its tail at an angle to the left, and vice-versa. (p. 240)

Although birds can sometimes be observed using the tail to control a turn by 'holding off bank,' it is not systematically applied as Silberschlag believes. Thwarted in his attempts to make sense of his observations by recourse to aerodynamics, he has to rely instead on the idea that the bird levers itself around: by pushing the tail to the left, the whole of the bird's body is caused, he thinks, to follow in the same clockwise direction, so that the head will be turned to the right.

The dominance of the lever in Silberschlag's mechanical theories is again revealed in his discussion of the *alula,* which he calls the *Lenkfittig,* or 'steering wing.' It has, he says, two main functions:

> This little afterwing has two functions. In the first place it broadens the wing tip where the wing's strength is greatest, and helps the bird to rise. In the second it enables the bird to steer itself quickly. For when the wind is streaming powerfully against him and he finds it necessary to turn quickly, he places the steering wing about which he wants to rotate in a position perpendicular to the wind; it then becomes a pivot about which the other wing, whose steering wing has remained folded, is turned like the sail of a windmill. (pp. 229–30)

As an afterthought Silberschlag adds to this a comment that arises from close observation and that, while attributing the effect to the wrong cause, accurately describes the true function of the *alula* as an antistall and high-lift device:

> But this appendage also decreases the speed of the forward movement, which can be perceived in the case of ravens flying too fast and wanting

to land on a roof. Both steering wings immediately become visible, and not only does the forward movement lessen, but the ravens lift themselves as if springing into the air, draw their wings a little together, and so gain the advantage of being able to lower themselves slowly on to the roof without damaging their feet by a sudden jerk. (p. 230)

Paraphrased later by Barthez,[57] Silberschlag's description of how ravens intentionally 'balloon' their landings so as to touch down at minimum forward speed in a stalled condition is as vivid an account as one could wish of the conditions in which the *alula* serves its main purpose. It is difficult to see what more could have been said about the matter before the beginnings of true aerodynamic science a century later.

Mauduit

Silberschlag's erroneous supposition about the way the tail functions to afford pitch control was repeated in a different form in Mauduit de la Varenne's tedious monograph.[58] Failing, in the first place, to understand the principles of relative velocity in a fluid medium, Mauduit says that if the bird lowers its tail when it is flying with a following wind, the tail will act as a sail to help the bird along. He then goes on to say that if the bird lowers its tail when the wind is blowing from in front, the effect will be to raise the bird. Mauduit does not seem to be referring to an increase in lift resulting from flaplike action. He proceeds, rather, to speak of the way this process can be used to help steer a course, presumably implying that by lowering the tail the bird causes the wind to raise the angle of its head.

Reflecting the attitudes of the experimental age in which he lived, Mauduit begins his account of the *alula* by discussing the results of amputation. Mutilated in this way by birdcatchers, the birds rapidly recover from their wounds and show no subsequent sign of ill health. Thereafter they are nevertheless unable to fly except in short hops, an effect that Mauduit does not think attributable to the small reduction in wing area. In his view, the removal of the *alula* changes the aerodynamic characteristics of the wing's leading edge in such a way as to vitiate the downstroke:

> When the *alula* has been cut away, the wing, when rising, pushes the air aside instead of cutting through it, and, when being lowered so as to find something against which to push, it beats into the vacuum which it has just made by rising and by displacing the layer of air.[59]

Of no scientific value, Mauduit's comments on flight control are nevertheless of historical interest. Included in a large and prestigious work, they masquerade as well-informed and up-to-date pronouncements by a spokesman of the Enlightenment. It is ironic that such worthless and totally confused statements should have appeared just as the era of manned flight was getting under way with the experiments of the Montgolfiers.

Barthez

A much better understanding of the function of the tail is shown by Barthez, who is more responsive than Mauduit to the organic flexibility of birds and whose thinking is less dominated than that of Silberschlag by analogies with machines. Denying Borelli's unduly simplified mechanical account, he points out that the tail can be twisted so as to create turning forces,[60] and he suggests that, especially for takeoff, it may be used as a pair of ailerons to help maintain lateral balance:

> It can, in certain cases, facilitate and maintain the equilibrium of the motive forces of the wings, being moved more to the side of the weaker wing, so that the bird is not turned over by that wing.
>
> This effect is most marked when the flight commences, a time when the wings may be unequally deployed for various reasons (even in quite still air) and be differently inclined to the line of flight that the bird wants to follow. (p. 205)

Departing still further from the common assumption that the tail is primarily used for directional control, Barthez moves very close to an understanding of its high-lift, high-drag function:

> When the tail, moderately extended, is widened and becomes concave on the underside, it collects a great deal of air, which is useful for the support of the bird, for allowing it to rise, or for slowing its descent. Such a state of the tail is most commonly associated with an expansion of the wings. (pp. 206–07)

When he comes to consider birds that soar in circles, he emulates his classical forebears by paying special attention to the kite. Unable to account for the continued creation of lift, he has recourse to Galen's idea of invisible trepidation or palpitation of the wings. Passing from the matter of lift in general to the control of the turn while soaring, Barthez once again invokes differential drag rather than the angling inwards of the direction of lift. Differential drag is a concept of importance to Barthez, who returns to it many times, even using it to explain the stabilising effects of the fletching of an arrow (p. 206). He explains changes of direction in the soaring patterns of birds by saying that when the outside wing is raised it presents a smaller surface to the air and has its primary feathers more widely separated than the inside wing. As a result, the inner wing has greater resistance (i.e., drag), which causes the bird to turn (p. 210). A similar suggestion about turning by means of differential drag is included in some remarks about the flight of insects. Puzzling over the function of the antennae, he says:

> One of the functions of the mobile horns on the head of insects may be that the insect extends or places sideways the antenna on the side towards which it wants to turn its body. This antenna being thus more firmly fixed

> in place by the resistance of the air provides a relative point of support that can only facilitate the turning of the body to the side. (pp. 216–17)

Although based on the common misapprehension about the effects of a steady horizontal wind, his account of the way the tail acts as an elevator shows real insight arising from close observation. Aware that variations in trim would require compensatory pitch control if the bird were to remain in level flight, he attributes that role to the tail:

> The agitations of the bird's tail also serve to assure and modify the lateral movements which result from the circles which it describes. They must in any case be varied because of the different resistances that the wind or noticeable currents of air can make to the movements that the bird makes in the different parts of these circles. (p. 211)

Barthez's ideas about stability and control are a good deal more impressive than his theories of flight in general. Had his thinking been less dominated by drag and more concerned with lift, he might well have been able to clarify some control techniques that were not adequately explained until well into the nineteenth century.

Leonardo's theory of bird flight and his last ornithopters 4

Before the late nineteenth century, by far the most voluminous notes on the theory of bird flight were made by Leonardo da Vinci. Although frequently mentioned in histories of aviation and discussed in detail in two Italian books,[1] they have rarely been given serious analytical attention, nor have they been properly related to the mechanisms of Leonardo's last and most ingenious ornithopters.[2]

Beginning in his early thirties, Leonardo's interest in bird flight grew ever stronger, finally becoming even more obsessive than that of his great mediaeval predecessor Frederick II of Hohenstaufen.[3] His copious manuscripts show an increasing concern not only with flight in general but also with detailed analyses of how birds are able to manoeuvre. In 1505 he set down a summary of his ideas, filling most of a short notebook known as the *Codice sul volo degli uccelli.* Many further notes are found in the manuscripts now lettered K (1504–09) and E (1513–15).

Incomparably more impressive than anything else from the period, the analyses are nevertheless strikingly repetitive, the same points often being made as many as six or eight times in virtually identical language. Many of the notes were evidently intended to form the basis of a definitive and properly organised treatise, the overall strategy of which offered various possibilities:

> I have divided the Treatise on Birds into four books; of which the first treats of their flight by beating their wings; the second of flight without beating the wings and with the help of the wind; the third of flight in general, such as that of birds, bats, fishes, animals and insects; the last of the mechanism of this movement.[4]

Another version of this plan, drafted at about the same time, appears to incorporate a number of improvements, especially in its emphasis on matters of fundamental principle:

> To speak of this subject you must needs in the first book explain the nature of the resistance of the air, in the second the anatomy of the bird and of its

> wings, in the third the method of working of the wings in their various movements, in the fourth the power of the wings and of the tail, at such time as the wings are not being moved and the wind is favourable, to serve as a guide in different movements.[5]

Making a further memorandum to himself a dozen pages later, Leonardo mentions another topic for inclusion:

> Before you write about creatures which can fly make a book about the insensible things which descend in the air without the wind and another [on those] which descend with the wind.[6]

While, as one would expect, Leonardo's ideas about bird flight developed over the years in many matters of detail, he rarely deviated from three basic principles. Interested, as were many experimenters, in the compressibility of the air, he suggested, first, that the force that sustains a bird in flight is created by a region of air compressed under the wings and between the wings and the body. Second, while accepting that in flapping flight a bird uses its muscles to establish the region of compression, he also believed that by continuous cyclic changes of the angle of attack of its wings, a bird could extract energy from a steady horizontal wind and so remain airborne with very little effort. The third principle, of lesser importance, had to do with a bird's capacity to vary its effective weight.

Although Leonardo's elaborations of his three principles are of interest in relation to the development of his thought in general, their prolixity is out of proportion to the significance of the underlying theoretical arguments. Contrary to the assertions of some of his adulators, Leonardo's contribution to the history of flight lies much more in the care with which he set down his observations of the movements of birds, and in the methodical design of practical artificial wings, than in any true insights into aerodynamic processes. The account given here is therefore comparatively brief.

The first principle

A bird is able to stay airborne, according to Leonardo, because it rests on a layer of air dense enough to support its weight. The density can be maintained provided the air is continually struck downwards by an object moving faster than the rate at which the air can escape:

> As the atmosphere is a body capable of being itself compressed when it is struck by something which is moving at a greater rate of speed than that of its flight, it is compressed into itself and becomes like a cloud within the rest of the air.[7]
>
> Unless the movement of the wing which presses the air is swifter than the flight of the air when pressed, the air will not become condensed beneath the wing, and in consequence the bird will not support itself above the air.[8]

Although this theory was fully formulated only in the 1500s, Leonardo implicitly alludes to it as early as 1485. Making the usual Aristotelian assumption that the air was very attenuated at high altitude, he points to the remarkable power of the eagle to sustain itself in such surroundings:

> Observe how the beating of its wings against the air suffices to bear up the weight of the eagle in the highly rarefied air which borders on the fiery element![9]

While sharing, as I discuss later, many contemporary misconceptions about relative velocity, Leonardo was nevertheless aware that the same process of compression would result whether the object were moving through still air or a wind were impinging on a stationary object:

> The movement of the wing against the air is as great as that of the air against the wing.[10]
>
> There is as much to move the air against the immovable thing as to move the thing against the immovable air.[11]
>
> The movement of the air against a fixed thing is as great as the movement of the movable thing against the air which is immovable.[12]

Reasoning that if a region of high pressure is created, a corresponding area of relatively low pressure must exist elsewhere, Leonardo concluded that low pressure would be found above the wing, thus assisting the creation of a firm support:

> It is necessary to say . . . that the air is compressed beneath that which strikes it and it becomes rarefied above in order to fill up the void left by that which has struck it.[13]
>
> The air which surrounds the birds is as much lighter above than the ordinary lightness of the other air as it is heavier below, and as much lighter behind than above as the bird's movement is swifter in its transverse course than is that of its wings towards the ground, and similarly the heaviness of the air is greater in front of the contact of the bird than below it in proportion to the two above mentioned degrees of lightness of the air.[14]

If the bird is to remain in level flight, the amount of 'resistance' created by the compressed air will have to be equal to the bird's weight. Leonardo suggests a simple relationship between the effect of the resistance, the rate of descent of a gliding bird, and the rate of downward movement of a flapping wing:

> Unless the bird beats its wings downwards with more rapidity than there would be in its natural descent with its wings extended in the same posi-

tion, its movement will be downwards. But if the movement of the wings is swifter than the aforesaid natural descent then this movement will be upwards, with so much greater velocity in proportion as the downward stroke of the wings is more rapid.[15]

The way the bird causes the condensation beneath itself is to flap its wings not only downwards but also in towards the centre line of the body, thus creating high pressure between the squeezing wingtips and its sides:

> When the bird desires to rise by beating its wings it raises its shoulders and beats the tips of the wings towards itself, and comes to condense the air which is interposed between the points of the wings and the breast of the bird, and the pressure from this air raises up the bird.[16]

Observing the upturned primary feathers of birds such as rooks, Leonardo makes an ingenious suggestion for an additional source of support:

> When the bird descends with a great slant without beating its wings all the extremities of the wings and tail bend upwards, and this movement is slow, for the bird is not only supported by the air beneath it but by the lateral air towards which the convex surface of the bent feathers spreads itself at equal angles.[17]
>
> Lateral power checks the descent of heavy things; as may be seen in the case of a man pressing with his feet and back against two sides of a wall as one sees chimney sweepers do. Even so in great measure the bird does by the lateral twistings of the tips of its wings against the air where they find support, and bend.[18]

The feathers of the wing are so arranged, Leonardo believed, as to lie together more closely on the downstroke than on the upstroke. The feathers thus form a kind of compound valve, allowing the wing to function as a pump to compress the air:

> The feathers spread out one from another in the wings of birds when these wings are raised up, and this happens because the wing rises and penetrates the thickness of the air with greater ease when it is perforated than when it is united.
>
> The spaces between the feathers in the wings of birds contract as the wings are lowered, in order that these wings by becoming united may prevent the air from penetrating between these feathers, and that with their percussion they may have a more powerful stroke to press and condense the air that is struck by these wings.[19]

More important than this lessening of resistance to the flow of air is Leonardo's conception of the cyclic motion of the wing. Assuming that a bird beats its wing downwards and backwards, thus creating lift and propulsion, he believed that after the downstroke the wing is turned edge on to the air for the upstroke, just as an oar is feathered in water.[20] While the angle of incidence of a wing does indeed undergo constant changes during the flapping cycle, the sequence is quite different from what Leonardo supposed.[21]

The second principle

Leonardo's second principle is concerned with techniques used by birds to make the winds do the work of creating the resistance under them. He correctly noted that birds often benefit from upcurrents caused when winds pass over hills and over waves at sea.[22] In addition, however, he constantly adhered to the idea that a freely flying bird is subject to wind forces in the same way as a captive kite. Failing to understand that in the case of a bird flying in a steady horizontal wind the relative motion of the bird and the air is exactly the same as that which is created when a bird is flying in calm conditions, Leonardo believed that it would feel the wind in its face if it turned upwind, and that it would avoid flying downwind so as to prevent its feathers being ruffled:

> It very seldom happens that the flight of birds is made in the direction of the current of the wind, and this is due to the fact that this current envelops them and separates the feathers from the back and in addition to this chills the bared flesh.[23]

Whether the bird were flapping or gliding, Leonardo thought that it could take advantage of this wind force. If it faced the wind and held its wings with the leading edge raised, the wind would strike the undersurface and lift the bird, as in the case of a captive paper kite, by a wedgelike action. In such a position the bird is, in Leonardo's phrase, 'above the wind.' When descending, the bird would normally lower the leading edge, so placing its wings 'beneath the wind.' By alternating these angles of attack, birds could maintain both altitude and forward flight against the wind, the impetus acquired during a descent being more than sufficient to keep them airborne when flying 'above the wind.'

Leonardo uses the term 'reflex movement' to refer to flight with the leading edge raised:

> With a slight beating of their wings they enter the wind with a slanting movement, this movement being below the wind. After this impetuous movement they place themselves slantwise upon the course of the wind.
>
> This wind after entering under the slant of the bird after the manner of a wedge raises it up during such time as the acquired impetus takes to consume itself, after which it descends afresh under the wind and again acquires speed; then repeating the above-mentioned reflex movement upon

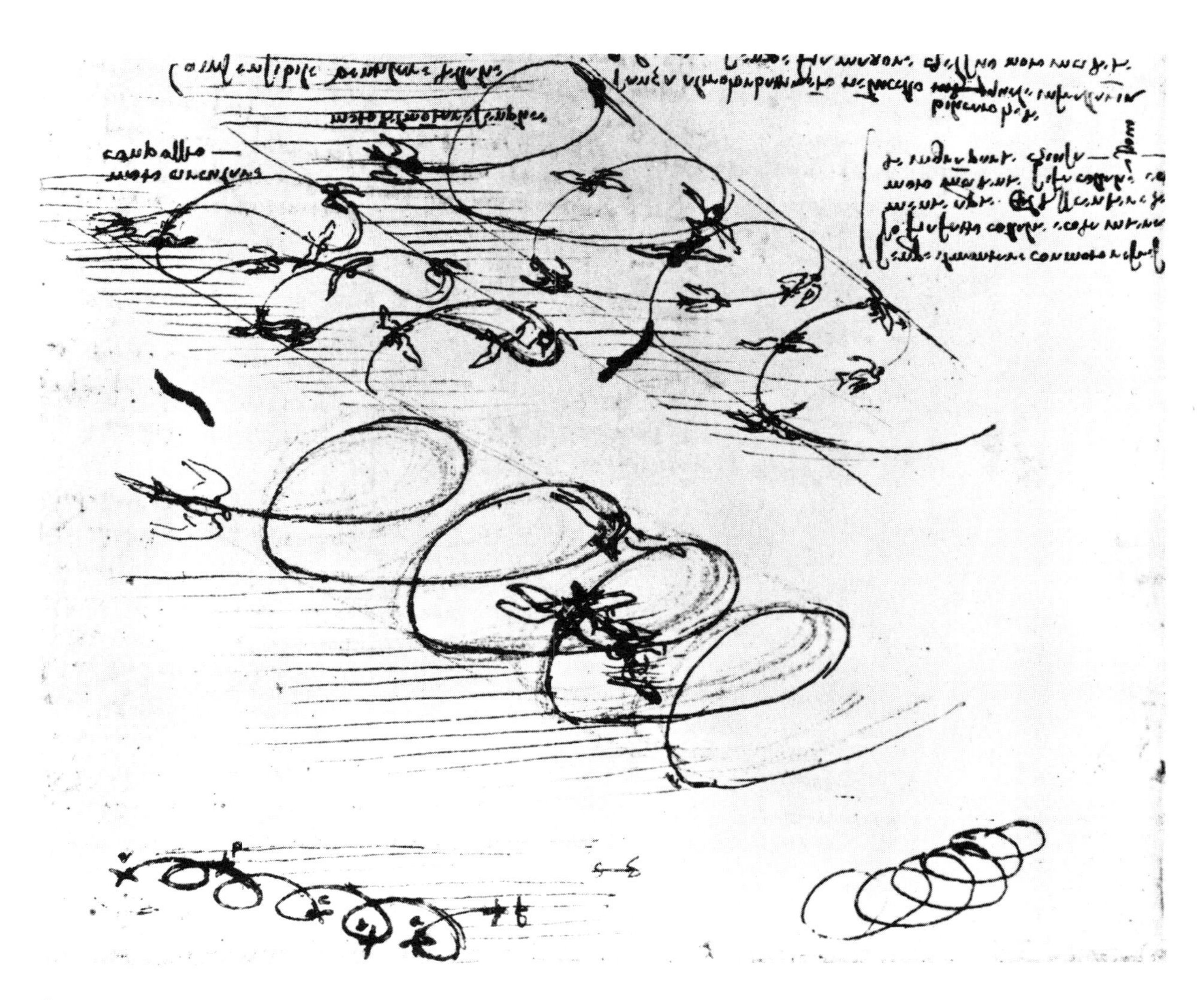

17. Leonardo's theory of the circling climb, showing (top and lower left) 'simple circular movement' and (lower right) 'complex circular movement.' *Codice Atlantico* f. 308^{r-b}, c. 1505. (Biblioteca Ambrosiana)

the wind until it has regained the elevation that it lost, and so continuing in succession.[24]

The bird moves against the wind without beating its wings and this is done beneath the wind as it descends, and then it makes a reflex action above this wind until it has consumed the impetus already acquired, and here it is necessary that the descent should be so much swifter than the wind that the death of the acquired impetus at the end of the reflex movement may be equal to the speed of the wind that strikes it below.[25]

Several times Leonardo describes a more complex version of this cyclic alternation of angles of attack by which he explains the circling flight of soaring birds.[26] He appears to have been observing not only the birds' use of updrafts but also the more delicate operation of 'dynamic soaring' in a wind gradient.[27] As he lacked, of course, the necessary concepts, he could account for this beautiful form of flight only by developing an elaborate theory based on the inadequate mechanics with which he was familiar.

Distinguishing two different kinds of circling, associated respectively with strong and light winds, Leonardo defines them as 'simple circular movement' and 'complex circular movement' (see Figure 17). In the simple case the bird holds its wings 'above the wind' while gliding upwind. At the end of the turn to the downwind direction, it lowers the leading edge so that the wings are angled downwards relative to the horizontal. This causes the wind, now blowing from behind the bird, to continue pressing on the lower surfaces of the wings and hence to continue causing the bird to rise. In this way Leonardo is able to explain how birds can continuously spiral upwards without flapping.

The complex movement, associated with light winds, is merely a curved variant of the way birds gather and release impetus by alterations of the angle of attack in forward flight. When soaring by this means, the bird executes the same half circle in the upwind direction as for the 'simple movement,' but it begins the downwind half by again raising the leading edge. This is necessary because a lowered leading edge would cause the bird to descend rather than rise if its forward movement were greater than that of the following wind. At the end of the turn to face upwind again, it lowers the leading edge once more, diving to acquire impetus that it then consumes with a raised leading edge, flying 'above' whatever light wind may be blowing.

Examining many of the consequences of this theory of flight, Leonardo occasionally encounters difficulties. Among the most troubling is the apparent impossibility of a bird's using these techniques in the downwind direction if the wind speed exceeds the bird's airspeed:

> If the bird descends to the south while the wind is blowing from the north it will make this descent upon the wind, and its reflex movement will be below the wind; but this is a vexed question which shall be discussed in its proper place for here it would seem that it could not make the reflex movement.[28]

A further comment, perhaps based on observations of birds stalling when flying in very gusty conditions, suggests that they need to be careful to control the effects of a headwind:

> When it is at the end of its ascent it will have used up its impetus and therefore will depend upon the help of the wind, which as it strikes it on the breast would throw it over if it were not that it lowers the right or left wing, for this will cause it to turn to the right or left dropping down in a half circle.[29]

Some years before Leonardo made these notes, he had shown signs of understanding the principles of relative velocity in the air:

> The movement of things that fly is much swifter than that of the wind. For if it were not so no bird would move against the wind. But its movement against the wind is as much less than its natural course within the still air as the degrees of movement of the wind are less than that of the bird.
>
> Let us say the bird moves in the still air at a speed of six degrees and the wind of itself moves at a speed of two degrees, then this wind following its natural course takes away two degrees of speed from this bird and consequently of the six degrees there remain four.[30]

The same idea recurs among the passages on bird flight that he set down towards the end of his life. Contradicting what he said earlier about gliding flight, he now seems to understand that the only forces involved are gravity and the horizontal propulsion of the wind:

> Of the things that are moved by others with simple movement the thing moved is as swift as the swiftness of its mover: therefore the bird carried by the wind in the same direction as this wind will have a speed equal to it.
>
> But if the objects carried by the wind slant more towards the ground than the direction of the course of this wind the thing moved will be swifter than its mover. And if the slant of the objects carried by the wind turns towards the sky this is a clear sign that the movement of the thing moved is slower than that of the wind. The reason is that when the slant is turned towards the ground it produces this movement by reason of its gravity and not by the help of the wind. But when the slant of the movement made by the thing moved is towards the sky, this slant is caused merely by the shape of the thing moved, for it bears itself like a leaf which caught by the wind in its breadth raises itself up merely by the help of the wind, and moves as much as its mover.[31]

The third principle

Combining the forces of flapping and of the wind, birds are able to fly with less expenditure of energy than at first seems likely, a point that encourages Leonardo to think that human flight will indeed prove possible.[32] In addition, it seems that he shares with many others the belief that the form of a mass, such as a bird, affects its weight and hence the ease with which it can sustain itself in flight. Writing of birds that fly with 'bounds' (that is, repeatedly closing their wings), Leonardo says: 'The bird which flies with a bound acquires impetus in its descent, because in the course of this by closing its wings it acquires weight, and consequently velocity.'[33] This point, more fully explored a generation later by Pierre Belon (see Chapter 3), is related to Leonardo's careful consideration of the bird's centre of gravity, the manipulation of which is given great importance in his notes on flight manoeuvres.

Flight control

Leonardo attempts as far as possible to base his understanding of the mechanics of flight control on direct observation of birds in the air. In contrast to the efforts of many others who attempted to design practical flying machines, it is greatly to his credit that by far the highest proportion of his notes on bird flight are directly concerned with stability and control. Constantly pondering the problem of how a man might one day fly, Leonardo was almost the only thinker before the Wrights to see that success would depend on a thorough prior analysis of methods of control.

The ideas about flight control that underlie many of his highly detailed observations may be distinguished under the following headings: (1) effects of wind, (2) differential flapping, (3) control of inertia, especially the inertia of the wings, (4) use of the tail, (5) control of the position of the centre of gravity, (6) use of the feet as airbrakes, (7) steering by use of the *alula*.

1. Effects of wind

Leonardo explores various ways in which a bird might manoeuvre by controlling the kitelike action that resulted, he thought, from the relative velocity of the bird and the wind. Not only may a bird climb by circling in a wind, varying its angle of attack so as to gain more height when flying upwind than when flying downwind, but turns may be made by allowing the wind to impinge on the two wings in differing degrees. When gliding at 90° to the direction of the wind, a bird will be in danger of executing an unintentional turn, 'weathercocking' so as to face into wind. Leonardo assumes that, as is often the case, the bird will glide with its wings held at an anhedral angle. The wing on the windward side, which will therefore be 'under the wind,' will be pushed down by pressure on its upper surface, while the wing on the lee side will be raised, causing the bird to bank and turn:

> If this falling movement is made to the east when the wind is blowing from the north then the north wing will remain under the wind and it will do the same in the reflex movement, wherefore at the end of this reflex movement the bird will find itself with its front to the north.[34]

The manuscripts are filled with comments about the way birds adjust the attitude of their wings to control sudden effects of this kind produced by gusts:

> When the bird wishes to avoid being turned over by the wind it has two expedients, one of which is to move the wing that was above the wind and place it suddenly below the wind, that is to say the one that was turned to the wind; the other is to lower the opposite wing so that the wind that strikes it on the inside is more powerful than on the wing that faces the wind.[35]

As Leonardo's asides often indicate, he is clearly aware of the possible applications of such control techniques to manned flight:

> A bird as it rises always sets its wings above the wind and without beating them, and it always moves in a circular movement.
>
> And if you wish to go to the west without beating your wings, when the north wind is blowing, make the falling movement straight and beneath the wind to the west, and the reflex movement above the wind to the north.[36]

Most of Leonardo's explorations of control techniques are connected with the effects of wind, the power of which he greatly respected and which, if properly harnessed, appeared to offer ample supplementation to the limited strength of human muscles.

2. Differential flapping

Making the rational assumption that birds often turn by thrusting more vigorously with one wing than with the other, Leonardo allows his description of lateral manoeuvres to be coloured by his observations of swimming and rowing techniques:

> A bird makes the same use of wings and tail in the air as a swimmer does of his arms and legs in the water.
>
> If a man is swimming with his arms equally towards the east, and his body exactly faces the east, the swimmer's movement will be towards the east. But if the northern arm is making a longer stroke than the southern arm then the movement of his body will be to the north-east [*sic*].[37]
>
> The bird often beats twice with one wing and once with the other and it does this when it has got too far over to that side.
>
> It also does the same when it wishes to turn on one side; it takes two strokes with one wing backwards, keeping the opposite wing almost stationary pointing towards the spot to which it ought to turn.[38]
>
> The bird beats its wing repeatedly on one side only when it wishes to turn round while one wing is held stationary; and this it does by taking a stroke with the wing in the direction of the tail, like a man rowing in a boat with two oars, who takes many strokes on that side from which he wishes to escape, and keeps the other oar fixed.[39]

Although he does not fully understand the forces at work, Leonardo observes that birds bank when making a turn:

> The bird which rises with a circular movement stays in a slanting position with the breadth of its wings, and the circle in which it revolves will be so much greater in proportion as its position is more slanting; and this circle will be so much smaller as its position is less slanting.[40]

When the bird turns sharply, the wings are held more nearly straight, with less anhedral droop:

> The bird which makes the shortest revolving movement prepares the extreme extension of its wings with less slant, and for this reason the circle of its revolving movement is so much more curved as the revolving movement is shorter.[41]

3. Control of inertia

One of Leonardo's most interesting ideas about flight control concerns the use of what we now call inertia. By moving one wing quickly backwards, the bird will cause itself to turn towards that side because the inertia of the wing will pull it around:

> When the bird wishes suddenly to turn on one of its sides it pushes out swiftly towards its tail, the point of the wing on that side, and since *every movement tends to maintain itself,* or rather *every body that is moved continues to move so long as the impression of the force of its mover is retained in it,* therefore the movement of this wing with violence, in the direction of the tail, keeping still at its termination a portion of the said impression, not being able of itself to follow the movement which has already been commenced, will come to move the whole bird with it until the impetus of the moved air has been consumed.[42]

The same principle can be applied if the bird wishes to slip to one side:

> When the bird wishes to let itself fall on one of its sides it throws its wing down rapidly on the side on which it wishes to descend, and the impetus of this movement causes the bird to drop on this side.[43]

4. The use of the tail

Having observed the many changes of position of the tail, Leonardo had no doubt that it was used for a variety of control purposes. Noting that it is often twisted from the horizontal, he understands that it can assist in turning:

> If the tail is struck above, the bird is raised in its front part. And if the tail is somewhat twisted, and shows its front slanting under the right wing, the front part of the bird will be turned towards the right side.[44]

Not only does Leonardo frequently allude to such passive control functions, but he interprets the occasional rapid twisting movement of the tail as an active beating of the air, comparable to that of the wing:

> There are many times when the bird beats the corners of its tail in order to steer itself, and in this action the wings are used sometimes very little, sometimes not at all.[45]

Even a single repositioning of the tail may be a percussive stroke, the tail surface acting to pivot the bird around a point of resistance:

> When the tail is thrust forward with its face and it strikes the wind, this makes the bird move suddenly in the opposite direction.[46]

5. Control of centre of gravity

With true engineering insight, Leonardo explored many aspects of the relationship between centre of gravity and centre of pressure, and the consequent effects on the flight path. He was especially clear sighted about the effect that variation in the relative position of the two centres would have on pitch control:

> When the bird wishes to go down it throws its wings backward in such a way that the centre of their gravity comes away from the middle of the resistance of the wings and so it comes to fall forward.[47]

Reiterating several times a commonplace of late mediaeval mechanics, Leonardo takes it as axiomatic that any falling object, including a freely gliding bird, always moves with the centre of gravity to the front:

> The descent of the bird will always be by that extremity which is nearest to its centre of gravity.
>
> The heaviest part of the bird which descends will always be in front of the centre of its bulk.[48]

These assumptions about the importance of the centre of gravity lead Leonardo to formulate a rational explanation of how a bird might recover from a tail slide:

> If the bird falls with its tail downwards by throwing its tail backwards it will regain a position of equilibrium, and if it throws it forwards it will come to turn right over.[49]

The position of the centre of gravity will have comparable effects in relation to lateral balance:

> The equal power of resistance of a bird's wings is always due to the fact of their being equally remote in their extremities from the bird's centre of gravity.
>
> But when one of the extremities of the wings is nearer the bird's centre of gravity than the other the bird will then descend on the side on which the extremity of the wing is nearer to the centre of gravity.[50]

On the other hand, a bird with its wings fully outstretched on both sides cannot fall in a continuous sideslip. Once initiated, any such movement would rapidly become a forward dive as the centre of gravity took the lead.[51]

Applying these principles, Leonardo several times describes manoeuvres controlled by the bird's drawing in a wing on one side. The resulting asymmetrical position of the centre of gravity then causes a turn:

> The wing extended on one side and drawn up on the other show the bird dropping with a circular movement round the wing that is drawn up.[52]

More interesting, especially in relation to the later tradition about 'head-steering,' discussed by Borelli and others (see Chapter 3), is Leonardo's comment that the same effect can be created by moving the head rather than the wings:

> If the head . . . is bent down towards one of the wings open equally, the slanting descent will then proceed between the head and the wing which the head is near.[53]

6. Use of the feet as airbrakes

While only partially understanding the role of the feet as flaps and airbrakes, Leonardo was aware that birds sometimes use their feet as well as, or instead of, their tails to control descent:

> The bird in descending against the wind lowers its feet as the wind strikes them, and this it does in order not to disarrange the tail from the direction of the whole body when it wishes to lower itself.[54]

7. Use of the alula

Although always convinced that the *alula* was in some way concerned with directional control, Leonardo suggested a number of distinct functions. In notes made about 1500, he proposed that the only role of the *alula* was to help govern climbing and descending:

> There are two helms on the humerus of the wings of each bird, and these without making any change of wings have power to cause the birds various movements between ascending and descending; it is only in the transversal movements that the helm of the tail takes part.[55]

Further notes made a few years later clarify how Leonardo thought the *alula* was used for pitch control of this kind:

> The rudders of the wings of birds are the parts which immediately place the bird above or below the coming of the wind, and with their tiny movement cleave the air in whatever line along the opening of which the bird can then penetrate with ease.[56]

In another manuscript he repeats this idea, but develops it further to include the possibility of making a turn by the use of only one *alula:*

> If the bird wishes to rise it spreads the helm in the opposite direction to the way the wind strikes it; and if to descend it spreads the top part of the helm slanting to the course of the wind. If it turns to the right it spreads the right helm to the wind, and if it turns to the left it spreads the left helm to the wind.[57]

On the succeeding page of the same manuscript, Leonardo approaches a little more closely to the truth when he begins to think that the *alula* may act as an airbrake:

> These helms or fingers show themselves fronting the air down which the slant of the bird is gliding, and by thus striking upon it with these helms it resists it as it glides.[58]

Returning repeatedly to the subject in notes made towards the end of his life, Leonardo attempts to improve on these explanations. Having asked himself why the 'helms' should be used for turning when control techniques involving the whole wing seem readily available, he answers by saying that in many manoeuvres a movement of the whole wing would be inconvenient or would cause problems of instability.[59] In one of the longest passages devoted to the *alula,* where he shows an explicit awareness that it is analogous to the thumb of the human hand, he asks the additional question why these steering devices should be placed in front of the bird, while the rudder of a ship is behind:

> This was not done with ships because the waves of the water are thrown up in the air to such a great height when smitten by the impetuous blow of the moving ship as would render the movement of the rudder very difficult from the gravity it would have acquired, and moreover it would often get broken. But since air within air has no weight but has condensation which is very useful the rudders (or helms) of the wings have a better use in a thick substance than in a thin one, the thick offering more resistance than the thin.[60]

Leonardo thinks of the *alula* as a highly active device, requiring positive and sensitive control by the bird. He writes of its strong feathers and sinews and of the 'marvellous power' of the whole structure. Neither he nor any other observer before the nineteenth century realised that the function of the *alula* is in large part automatic, its feathers being lifted away from the leading edge of the wing by the changing position of the centre of pressure at high angles of attack. Lacking the aerodynamic knowledge necessary to solve the problem, Leonardo was obliged to offer a series of incomplete answers which plainly left him feeling that something more remained to be said.

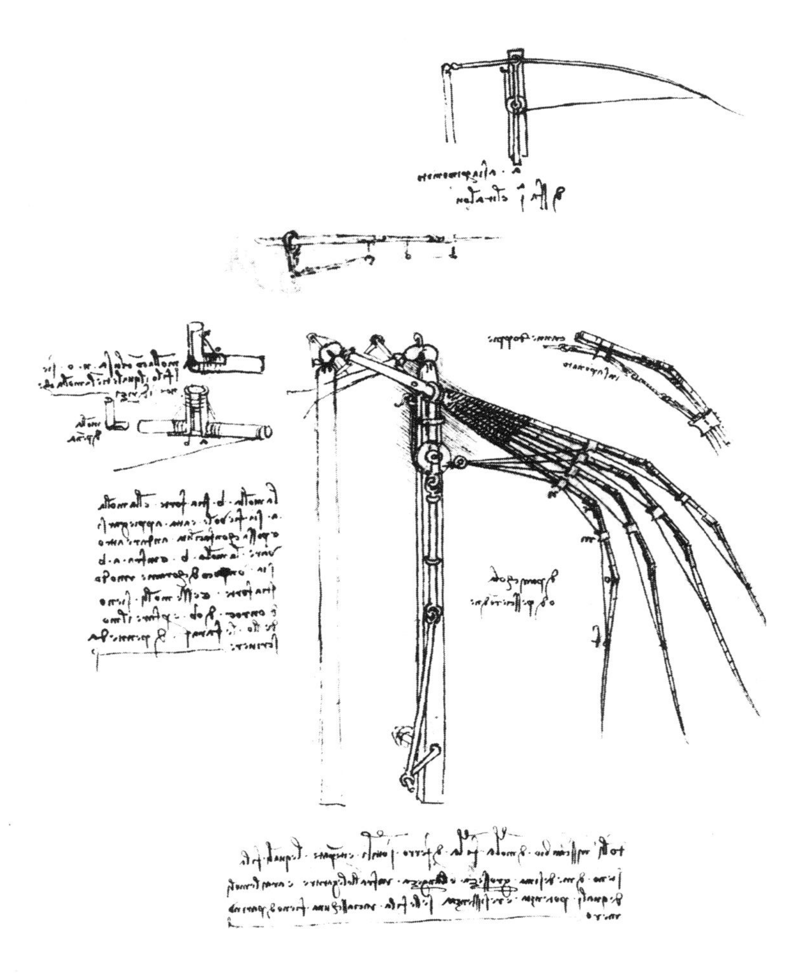

18. Mechanism for flapping, twisting, and 'squeezing' the wing of an ornithopter. Leonardo, *Codice Atlantico* f. 308$^{r\text{-}a}$, 1495–97. (Biblioteca Ambrosiana)

The last ornithopters

Having formulated, to his own satisfaction, theories of lift, propulsion, and attitude control, Leonardo believed himself to be in possession of all the principles that a man would need to apply in order to manage a pair of artificial wings. The notes on bird flight are accordingly filled with semiveiled comments suggesting that he envisages a return, now better armed with theoretical guidance, to the enthusiastic design of flying machines with which he had filled many pages a decade or two earlier.

The ornithopter mechanisms have been analysed too often for a complete rehearsal of Leonardo's ideas to be needed here.[61] Some points of detail have nevertheless been passed over too lightly, and at least one crucial aspect of the movement of the later wings has been misrepresented, especially by the late Charles H. Gibbs-Smith, whose views have gained wide currency.[62]

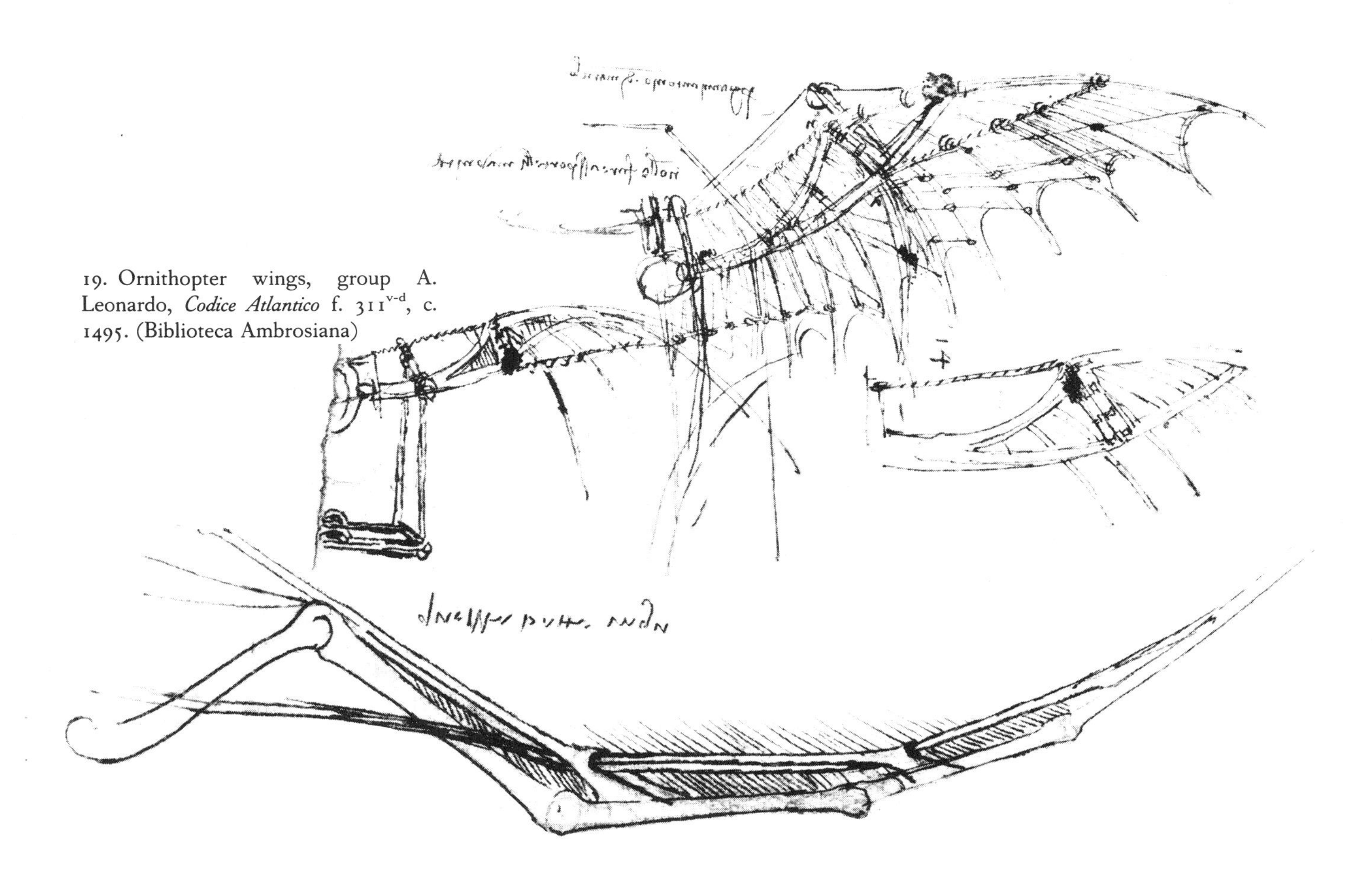

19. Ornithopter wings, group A. Leonardo, *Codice Atlantico* f. 311^{v-d}, c. 1495. (Biblioteca Ambrosiana)

In the first place it should be noted that although almost all of Leonardo's notes on flight control refer to birds, the structural details of his last ornithopters are modelled more closely on the wings of bats, which in the fifteenth and sixteenth centuries were still usually considered to be birds of a special or abnormal kind. The flight patterns envisaged for the ornithopters were essentially those of the birds that Leonardo had so frequently observed.

I shall be concerned with five main designs, all of them closely related in concept and basic function (Figures 19–22). For convenience I have lettered them A to E. Wing A (Figure 19) is shown in a set of fairly hurried sketches setting out ideas which appear to be more fully developed later. Wings B and C (Figure 20) form a pair, C being a more fully developed variant of B. Wing D (Figure 21) appears to be a preliminary sketch for wing E (Figure 22).

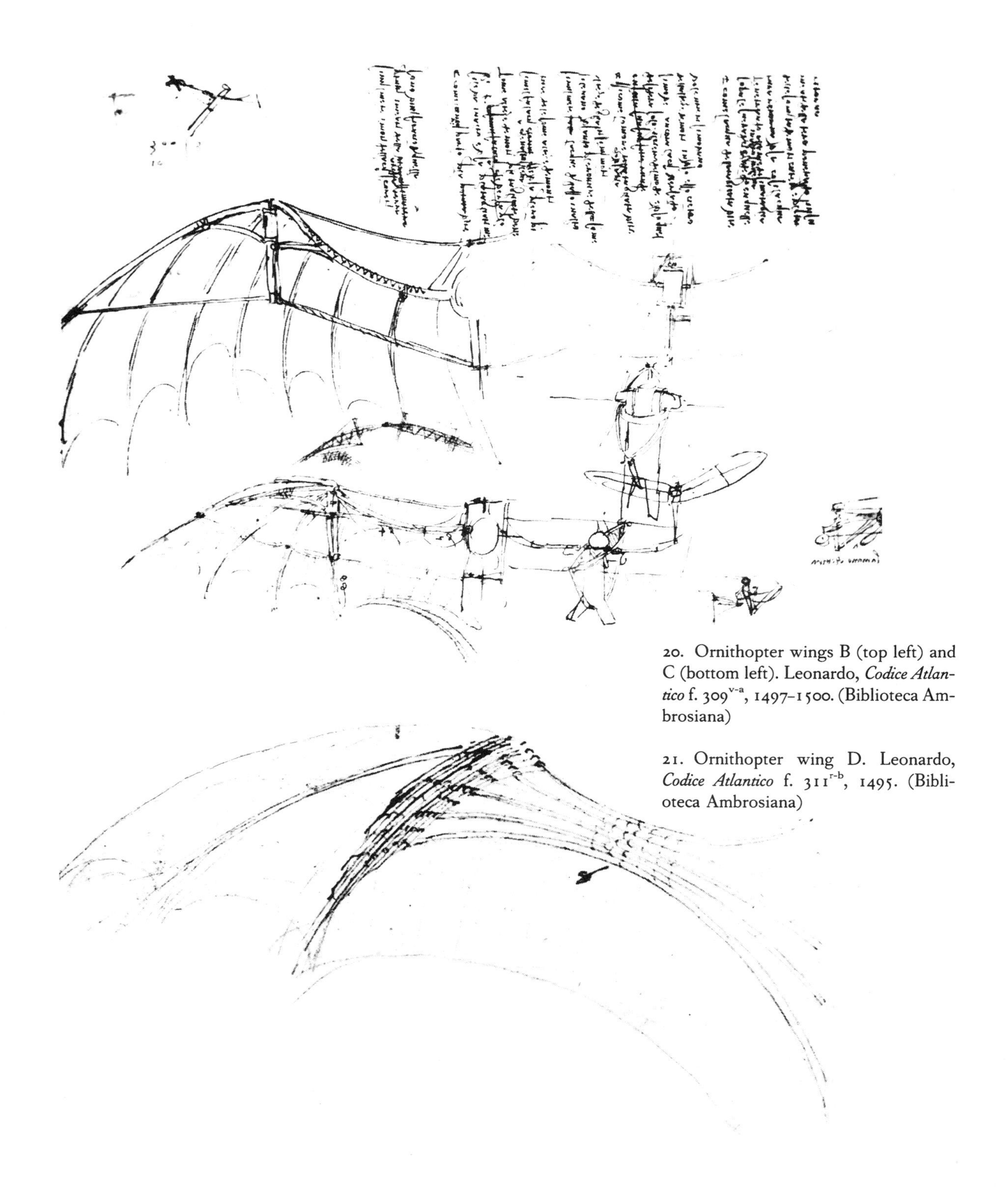

20. Ornithopter wings B (top left) and C (bottom left). Leonardo, *Codice Atlantico* f. 309[v-a], 1497–1500. (Biblioteca Ambrosiana)

21. Ornithopter wing D. Leonardo, *Codice Atlantico* f. 311[r-b], 1495. (Biblioteca Ambrosiana)

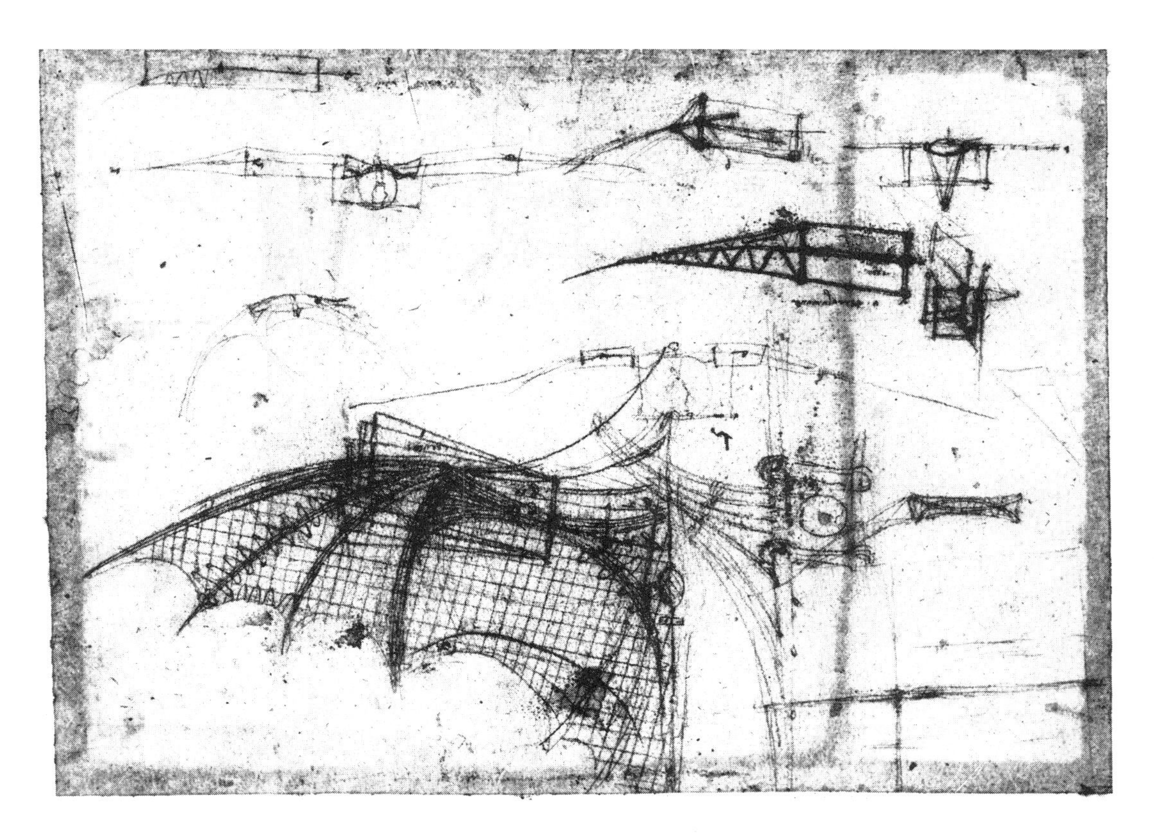

22. Ornithopter wing E. Leonardo, *Codice Atlantico* f. 22^{v-b}, 1497. (Biblioteca Ambrosiana)

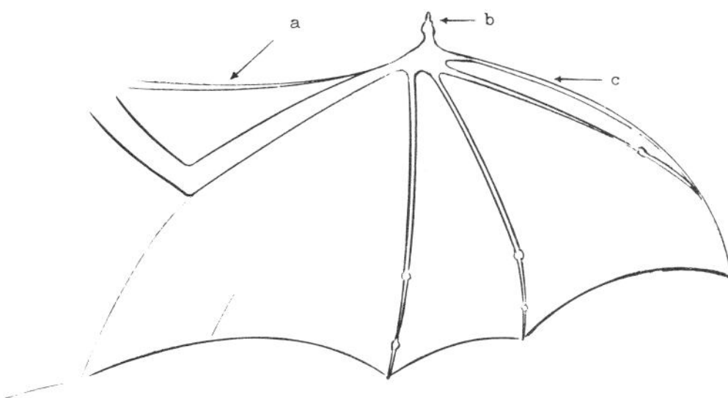

23. Schematic representation of a bat's wing, showing (a) the occipito-pollicalis muscle; (b) digit 1, the 'thumb'; (c) digit 2, held close to digit 3.

As commentators have noted, all five wings are articulated at a point near their half-span, enabling the outboard area to be flapped downwards relative to the inboard. Designed to simulate the movement of a bird's or bat's wing, articulated at the wrist, this arrangement is intended, in the cases of C and E, to work in conjunction with wingtip structures shaped like a five-fingered hand. In the case of a bat wing, digits 3, 4, and 5 serve as rigid members for the wingtip, with digit 2 lying close in front of digit 3. Digit 1, the 'thumb,' projects forward and is more or less clear of the membrane (Figure 23). Leonardo's design uses digits 2–5 to spread the wingtip, with the 'thumb' taking the place of the bat's digit 2. Referred to the structure of a bird's wing, this leading-edge digit can be interpreted as a permanently closed *alula.*

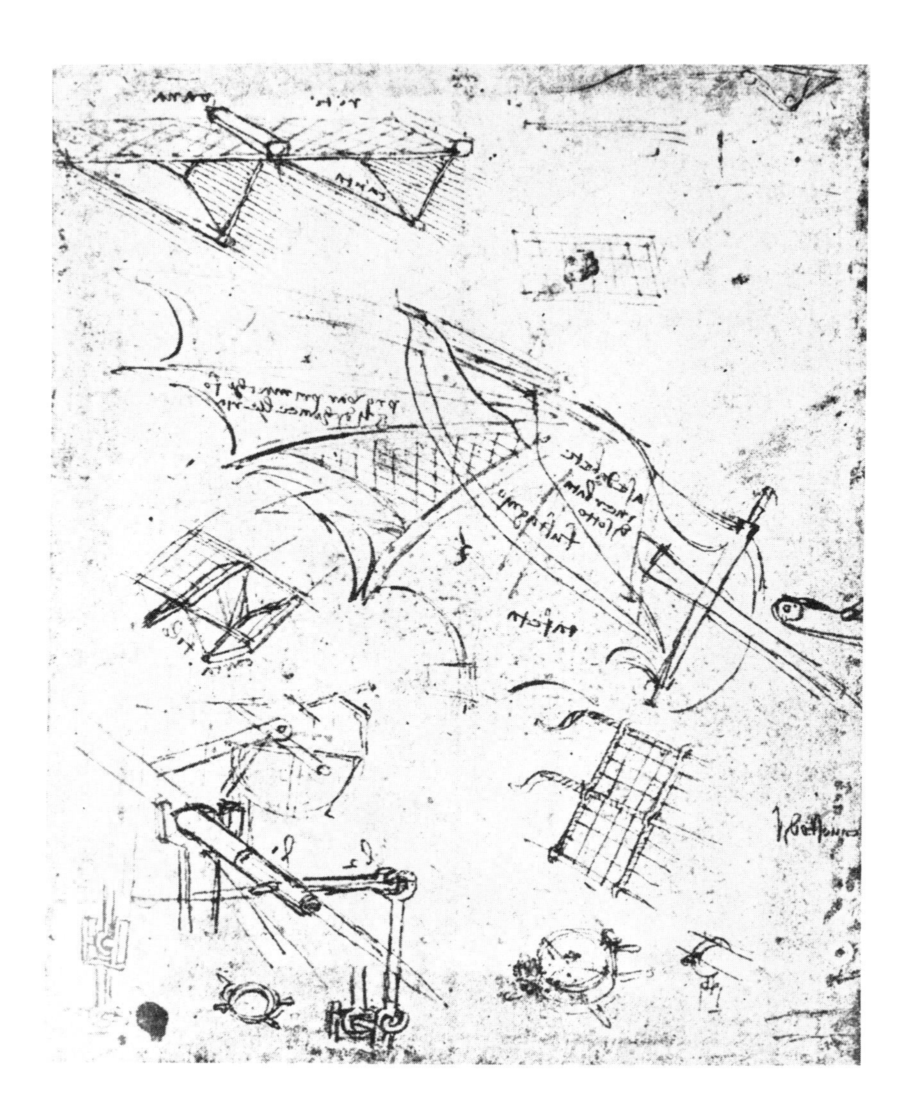

24. Ornithopter wings with flap valves (see top and middle left) operating in conjunction with a network covering. Leonardo, *Codice Atlantico* f. $309^{\text{v-b}}$, 1487–90. (Biblioteca Ambrosiana)

Starting with the sketches for A, Leonardo appears to have abandoned his earlier, cumbersome ideas for including flap valves to reduce air pressure on the upstroke (Figure 24)[63] and to have returned to a more practical design of wings with a uniform fabric covering. A note to A reads 'Do not make it with panels, but covered uniformly' (*nolla fare assportelli ma unita*).

All of the wings have two or more wooden members on the inboard section, serving the function of ulna and radius in fixed combination. (Throughout these designs Leonardo regularly omits the complication of a wing section representing an independent humerus.) In addition, all of them appear to have inboard leading edges formed by a stout cord serving the function of the occipito-pollicalis muscle (see especially the lower sketches in A). In every case some kind of supportive netting is presumably to be used for the cover. The most fully realised drawing of the net appears in E, for which Leonardo has used a ruler to show a single network stretched across the whole frame. As can be seen on the third-finger rib, on the inboard ulna, and on the trailing edge, the netting is intended to be sewn to the frame. The double lines used to draw the trailing edge of the main wing in A, and of the area between the fourth and fifth fingers in E, may indicate that curved cane members ensure the separation of the ribs, but indications elsewhere suggest that a simple fabric trailing edge may be intended, the tension on the cane ribs being sufficient to keep them apart.

Taken in conjunction with details from the other sketches, the most fully realised drawing, wing E, gives the clearest indication of how in his maturity Leonardo aspired to simulate the main movements of a living wing. The fifth finger of the hand is formed of two parallel members hinged to each other along their whole length. The inboard part, around which the outboard member turns along with the rest of the wingtip, is rigidly fixed to the ulna-radius combination. Interpreting this mechanism in relation to B, Mr Gibbs-Smith stated:

> The most significant feature in this type of aircraft is the abandonment of the pure ornithopter, in which the whole of each wing is flapped, and the adoption of a partially fixed-wing configuration in which only the outer portion of the wing is flapped. . . . This shows, I believe, that Leonardo had been thinking of the structure of a bird's wing, and had realised that the inner part moves far slower than the outer, and therefore produces mainly lift rather than thrust: so he wisely economises on the amount of wing-surface that has to be flapped in relation to the muscular resources of the pilot, and thus concentrates the movement and effort where it can best be utilised.[64]

As reference to other details of the drawings shows, this is true only in part. As Lilienthal did centuries later, Leonardo's flier was to adopt an upright position with low centre of gravity. The feet were to rest on a horizontal bar, only the hands and arms being used for motive power (see later). In the fully developed version of this late machine (see B, lower right, and E), only the flier's head was to project above the surface of the wing. The mechanism is operated from below, as in Miller's familiar Degen-derived machine of 1843 (see Figure 25, and cf. the central front-elevation sketch in E). The inboard half of the wing is not, however, intended to remain rigidly horizontal. As several of the sketches show, Leonardo wished to use a double ring or rectangular framework to surround the pilot's head: The inner ring was rigid, and to it, on either side, were hinged the inboard ends of the main spars. The details of structure and exact function of the outer ring or frame seem never to have been fully worked out, but in C it is clear that Leonardo intended it

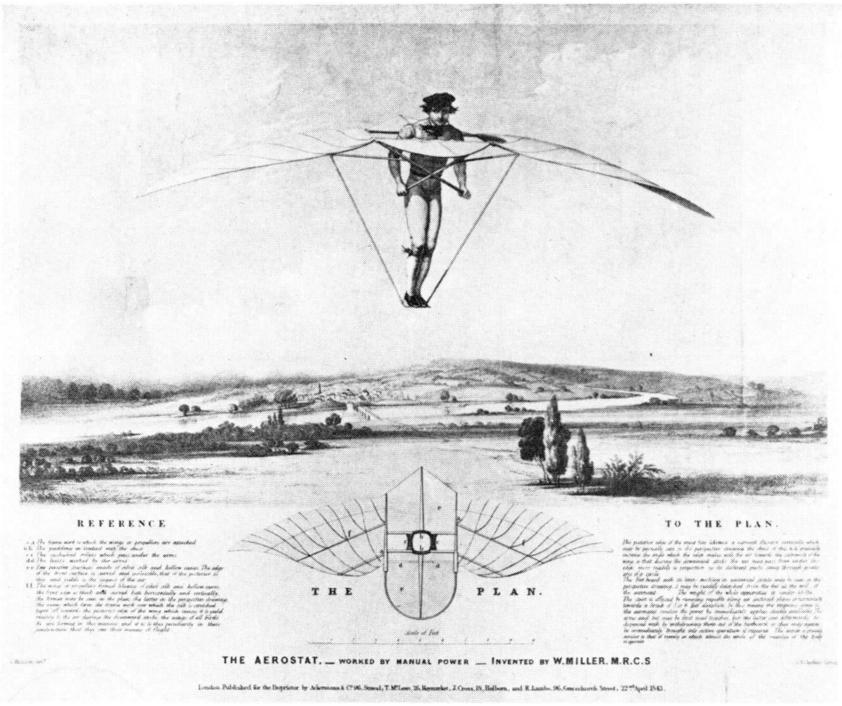

25. W. Miller's ornithopter of 1843, derived from Degen's machines of the early 1800s. Cf. the position of the aeronaut in Leonardo's last ornithopters. (Reproduced by permission of the Royal Aeronautical Society)

to act, among other things, as the inboard point of attachment for the whole wing. The diagram to the right of E seems to indicate that the fore-and-aft members of the outer frame should rotate in sockets at the ends of the lateral members. Whatever arrangement he might finally have chosen, Leonardo certainly wanted the whole wing to be flappable about its centre, while the wingtip would be free to descend still further about its hinges.

In addition to these two main freedoms of movement about hinges, the cane spars and ribs could also be flexed (see later) to create a more continuously curved shape on the downstroke. The curvature, quickly sketched in some of the smaller diagrams of E, serves to create the squeezing action that compresses the air under the bird to provide support.

Two principal mechanisms are proposed for allowing the wingtip to move backwards and so provide the propulsion needed for forward movement. The first idea, shown in A and D, is to set the wingtip hinges at an angle to the centreline, so that the flapping stroke is automatically down and back. Leonardo seems, however, to have preferred a system that would allow the flier to exercise full control over the degree of backward motion. In A he continued to use the systems of ropes and pulleys that proliferated in the complex ornithopters of the 1480s. The details are far from fully worked out, but ahead of the leading edge of the largest wing appears to be a set of pulleys on a post that is presumably to be thought of as projecting vertically down from the plane of the wing. One cord passing over the pulley and attached to the leading edge of the wingtip can be used to pull the wingtip down. The function of the second cord is unclear. Another cord-pulley mechanism is shown on the wing at bottom left: an equal pull on the handles would produce a simple downstroke; a stronger pull on the forward handle would flex the wing down about its main spar and so create a backward movement. The diagram does not indicate how the pilot would control two such pairs of handles simultaneously.

The final solution, roughly sketched at the bottom of B, is elegantly developed in E. Although they are not shown in the main diagram of B, sets of king and queen posts are fixed vertically through the wing, attached to the main spars and to the ribs. Five posts are shown on the main wing in E, while others are indicated in the small diagrams of B and E. Cords are attached from the two inboard posts to the outboard rib posts, and thence to the wingtips. Leonardo also speculates on the idea that these cords might be made to function better if laced to the ribs, above and below, by further triangulated cordwork (see the small diagram in E, upper right).

26. Ornithopter pilot. Leonardo, *Sul volo degli uccelli* f. 6(5)^r, 1505. (Turin, Biblioteca reale)

Leonardo now has a well-shaped, flexible, highly manoeuvrable pair of wings that can be flapped straight down, or down and back, with varying degrees of curvature. They can be differentially operated and can be disposed at will to enable the flier to harness the forces of the wind. Having approached more closely than ever before to a simulation of natural flight, Leonardo feels free to abandon the painful search for maximum muscle power that had earlier led him to design machines using a combination of leg and arm movements. He now relies wholly on the hand and arm muscles of an erect flier. As is indicated in the sketch in the middle of E, the flier holds the bottom ends of the two inboard vertical posts. By pulling these towards himself, he lowers the whole wing and swings the wingtips down about their hinges. By pulling both inwards and slightly backwards, he twists the wing and creates the backstroke. Other sketches in E, together with the elevation diagram at the bottom of B, show the effects of these actions. Nothing more detailed is needed, but at the top right of E Leonardo explores, as usual, one or two more complex and ingenious mechanisms for achieving similar results.

At the same period Leonardo was sketching movable control surfaces that would simulate the actions of a bird's tail.[65] A combination of the last ornithopter wings and appropriate control surfaces would have put Leonardo in a position to imitate virtually all of the flight techniques that he attributed to birds: circular soaring flight would be possible with the fully controllable, flexible wings; differential flapping would initiate turns, which could also be produced by rapid asymmetric movements to control inertia; the tail could be used as elevator and rudder, and the low-slung flier could exercise further control by moving his body within the comparatively spacious cockpit provided.[66] The only significant omissions are the relatively unimportant matter of the use of the feet as airbrakes and the directional control that Leonardo attributed to the *alula*. Given Leonardo's tendency to oversophistication of design, it was doubtless wise of him not to graft a bird's *alula* on to a bat's wing and so impair the elegance of these last fine ornithopters.

Will man fly? 5

Is flight to be considered a peaceful or a violent activity, a liberation, or a disguised step towards further bondage? In premodern times, psychological unease was increased by associations of flight with the idea of inescapable judgement and vengeance. In his commentary on the mysterious flying roll, or flying sickle of Zechariah 5:1ff, Saint John Chrysostom (?345–407) had powerfully expressed his awareness of the frightening power of wings and of their potential as a source of human suffering:

> What, forsooth, is this which is here spoken? and for what reason is it in the form of a 'sickle,' and that a 'flying sickle,' that vengeance is seen to pursue the swearers? In order that thou mayest see that the judgment is inevitable, and the punishment not to be eluded. For from a flying sword some one might perchance be able to escape, but from a sickle, falling upon the neck, and acting in the place of a cord, no one can escape. And when wings too are added, what further hope is there of safety?[1]

Whether man would ever be able to fly like a bird was as much a moral and spiritual as a practical matter, the technical difficulties clearly expressing the presumptuousness of any intention to force one's way towards the heavens. Ecclesiastical writers of later centuries grew increasingly sympathetic, but even the scientifically minded Bishop John Wilkins (1614–72) was well aware that 'it may seeme a terrible and impossible thing ever to passe through the vaste spaces of the aire.'[2] As the growth of the mechanical sciences nevertheless made it more likely that man would indeed one day learn to fly, fear of the consequences of so immodest an aim began increasingly to be expressed as concern about the likely practical misuse of an invention that society might find difficulty in controlling. From the seventeenth century, indeed, admonitory comment on the evils that might follow became a commonplace trope. Among the most pungent, most imaginative, and most prophetic is a passage in the introductory pages of Johann Daniel Major's *See=Farth nach der neüen Welt, ohne Schiff und Segel,* a book that aims to take the reader on an improving armchair tour of the world. Writing about man's general tendency to abuse his talents, Major considers the possibility, which he considers quite real, that we might eventually be able to fly:

> What a completely new and dangerous appearance the world would have, how much more hazardous, indeed how much more abominable the world would seem to have become for all posterity. What treachery, robbery, and assassination, what other sins and shamefulness would be heaped upon one another! Towns and castles, whole provinces and kingdoms, would presumably soon be obliged to fill the air either by means of the frequent firing of canon or by stirring up rising smoke; or else to protect themselves thoroughly with large iron gratings, used as nets, and to arm themselves if not against total invasion, at least against the frequent throwing of fire and stones by the flying army which, like Lucianic birds of prey, darting down from the world of the moon, would otherwise raze everything to the ground. From this activity alone the world would seem a thousandfold hateful and more ruinous than from the . . . misuse of the compass, of gunpowder, and of printing.[3]

Major contrasts this scene of destruction with an idyllic vision of the contemplation of nature and the pursuit of knowledge. Passing from ideas about practical flying, he transforms the aerial perspective into a metaphysical view of truth:

> But let us set before the eyes of the mind a much more lovely spectacle: a beautiful basis for present rapture; a splendid and more than royal palace of perfection, situated far across the sea, in an as yet unknown land, in which people, who while still alive become earthly gods rising on a golden chain towards Heaven, and who, benefiting from the continual health-giving air of spring, perceive with glowing lynx- and Argus-eyes all that lies between Heaven and Earth, all the revealed and concealed treasures of Nature, with her innermost crevices, people who know the history of her early times, who know all human arts and sciences, all discoveries of genius that have ever been made, and to the glory of God can tell her measures and laws.[4]

Although most writers were content to offer only warnings, some took the opportunity to point the moral explicitly. In his *Mathesis biceps,* published in the same year as *See=Farth,* Johann Caramuel Lobkovitz wrote:

> God denied to men the faculty of flight so that they might lead a quiet and tranquil life, for if they knew how to fly they would always be in perpetual danger. Whose life would then be free from danger? What house would be safe from robbers? What city would be safe against the enemy? In truth no care, no foresight would be sufficient to protect men, especially at night,

from the flying foe. And so mankind was treated mercifully when he was denied the wings which he would have misused.[5]

A more highly coloured version, lacking the explicit moral, is offered by l'Abbé Pluche, whose catalogue of ills ends with a scene of barbarous isolation:

> *Prior* : . . . The Art of Flying would be the greatest Calamity that could happen to Society.
>
> *Chevalier* : On the contrary, Sir, I should think this Invention would save us Abundance of Labour. We should be sooner acquainted with what we are desirous of knowing; and if we had once found out a small Machine, we could soon build a larger. We should not only traverse the Air our selves, but might likewise convey Cargoes of Merchandize through that Element. By this Means Commerce ———
>
> *Prior* : You have a charming Penetration, *Chevalier,* and are the best in the World, at guessing the Advantages we might receive from this Invention; but these Advantages would not countervail the Disorders that would be introduced.
>
> *Count* : This is certain, that were Men capable of Flying, no Avenue could be inaccessible to Vengeance and inordinate Desires. The Habitations of Mankind would be so many Theatres of Murder and Robbery. What Precautions could we take against an Enemy who would have it in his Power to surprise us both by Day and Night? How should we preserve our Money, our Furniture, and our Fruits from the Avidity of a Sett of Plunderers, furnished with good Arms to force open our Houses, and with good Wings to carry off their Booty, and elude our Pursuit? This Sort of Trade would be the Refuge of every indigent and impious Person.
>
> *Prior* : I may add to this, that the Art we are speaking of, would intirely change the Face of Nature; we should be compelled to abandon our Cities and the Country, and to bury ourselves in subterraneous Caves, or to imitate Eagles and other Birds of Prey; we should retire, like them, to inaccessible Rocks and craggy Mountains, from whence we should from Time to Time sally down upon the Fruits and Animals that accommodate our Necessities; and from the Plain, we should immediately soar up to our Dens and Charnel Rooms.[6]

The close association of the ideas of flying and of isolation appears several times in commentaries on the progress of Christian society. Although less gloomy in his vision than l'Abbé Pluche, Nehemiah Grew suggests that the capacity to fly would have inhibited the development of urban life:

> Had [Man] been a Bird, he had been less Sociable. For upon every true or false ground of fear, or discontent, and other occasions, he would have

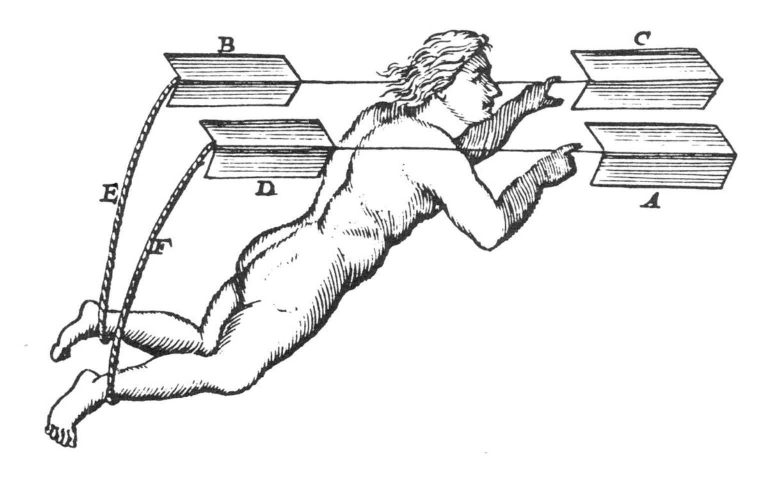

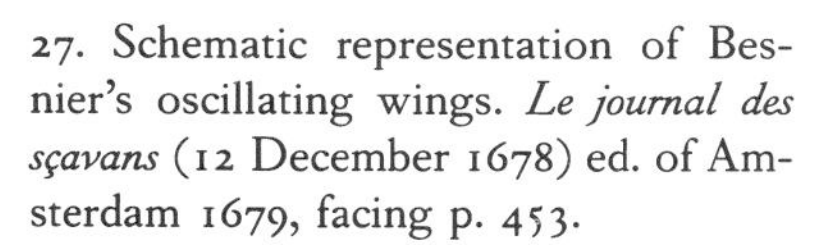
27. Schematic representation of Besnier's oscillating wings. *Le journal des sçavans* (12 December 1678) ed. of Amsterdam 1679, facing p. 453.

> been fluttering away to some other place: And Mankind, instead of cohabiting in Cities, would like the Eagle, have built their Nests upon Rocks.[7]

The same point is made in William Derham's pious *Physico-theology.* Introducing a quotation from Grew, he suggests that flight is a source of dangerous knowledge:

> The Art of Flying . . . in some Cases might be of good use, as to the Geographer, and Philosopher; but in other Respects might prove of dangerous and fatal Consequence; as for Instance, by putting it in Man's Power to discover the Secrets of Nations and Families, more than is consistent with the Peace of the World for Man to know; by giving ill Men greater Opportunities to do Mischief, which it would not lie in the Power of others to prevent.[8]

Moral questions are more indirectly raised by Joannes Ludovicus Hannemann, acting as *Praeses* for a dissertation by Georgius Matthias Hirsch. If man could fly, Hannemann asks, would he be better to do so naked? No: nakedness for us is unacceptable, however suitable it may be for Indians who remain undepraved by it nor have their decorum violated by it. In any case we should need protection from heat and cold and from the attacks of flies and fleas. A flying outfit would therefore be necessary, made perhaps from waxed linen or parchment or lambskin. The whole should be covered with feathers and made to lie smoothly on the skin without wrinkles or folds. It should in particular fit tightly over the buttocks and loins, though not so tightly that it might split and let in the inclement air.[9] These comments may perhaps have been prompted at least in part by the classical nudity of the famous illustration, in the *Journal des sçavans,* of Besnier and his flying apparatus (see Figure 27).

The early eighteenth century tendency to surround the notion of flight with images of restraint, with clear indications of limit, finds emphatic expression in Louis-Bertrand Castel's curious *Traité de physique* of 1724. Developing a modernised version of the Aristotelian theory that terrestrial objects transported to the sphere of the moon would still tend to fall back to earth, Castel speaks in passing of the possibility of flight. Even if one could learn to fly, he says, the last layer of air (*la dernière voûte de l'Atmosphére*) would form a barrier beyond which we could not pass. Our substance has no affinity with that of the moon, but belongs here below.[10]

The first connotation of flight is nevertheless the idea of escape. The point is taken up, if with irony, in Johnson's *History of Rasselas.* Designed as a means to avoid oppressive containment, the wings that Rasselas has fashioned for him, and that fail him at the first attempt, are a potential source of misery. While Rasselas wants to fly towards excitement and new experience, Johnson, in common with many of his predecessors, suggests that the general availability of wings would lead to rapine, warfare, and invasion. The skilful mechanical 'artist' of the Happy Valley warns Rasselas:

> If men were all virtuous . . . I should with great alacrity teach them all to fly. But what would be the security of the good, if the bad could at pleasure invade them from the sky? Against an army sailing through the clouds neither walls, nor mountains, nor seas, could afford any security. A flight of northern savages might hover in the wind, and light at once with irresistible violence upon the capital of a fruitful region that was rolling under them. Even this valley, the retreat of princes, the abode of happiness, might be violated by the sudden descent of some of the naked nations that swarm on the coast of the southern sea.[11]

When, in the late eighteenth century, it became clear that flight was a real possibility, frightened voices were heard calling for protective legislation. Among the most totalitarian were seven suggestions for state control of flying machines proposed by Laurent Gaspard Gérard:

1. All flying machines might be owned by the State;
2. If individuals were allowed to own them, the flying machines should be built by artisans specially selected for their probity; a magistrate's permission would be needed before a machine could be built;
3. An individual would be permitted to use his flying machine only for the benefit of himself, his wife, and his family; children could fly the machine only when accompanied, a restriction which would avoid *la débauche des jeunes gens;*
4. Magistrates might allow the construction of a flying machine only after having received a detailed statement of the use to which it would be put;
5. If private ownership were nevertheless prohibited, a depôt might be set up from which flying machines could be hired;

6. It might be a legal requirement that on each flight the hirer be accompanied by a government-nominated copilot—a 'strong, brave, and honest' man—who would ensure that the hirer did not deviate from his stipulated route; if the copilot could not prevent the making of such a deviation, he would file a report on the delinquent's behaviour;
7. Given the difficulty and danger of their job, copilots should receive a generous State salary.[12]

Although more than a century of further patient effort was required before there was any genuine need for control, predictions of disaster continued to be made. Even so keen an enthusiast as Frederick Marriott included a modernised version of the theme in an article that he published in 1880:

> Like the invention of gunpowder, it is doubtful whether it would be a boon or a bane to mankind. Many leading interests would be paralyzed. Railroads and ocean steamers would become utterly useless. Hundreds of millions of invested money would be lost. War, as at present carried on, would become a farce when 'strategical positions' could be overhung by a cloud of torpedo-bearing air-ships. Engineers and geographers would lose their vocation, for the trackless air requires no embankments or bridges, and needs no survey but that which the polar-needle can supply. The heart of Africa would be as accessible as London, Paris or New York.[13]

Comments in the same spirit were sometimes included even in the context of scientific discussions. Following in the tradition of Borelli and his disciples, Silberschlag offers a few simple calculations designed to demonstrate that a man can never aspire to fly by the power of his unaided muscles. Just let him try the effect of unsupported wings, says Silberschlag, and he will realise how impossible the idea is. And in any case, he adds, following the comments of Major, if flight were possible, whose house could be adequately protected from flying thieves?[14]

Not only the general populace but also the foolhardy aviators themselves would of course be in danger. That the hazards were real was emphasised by a number of well-authenticated accounts in early annals showing that in some sense men had already been known to fly, though involuntarily, out of control, and in frightening circumstances. Powerful gusts of wind passing over curved and irregular surfaces can sometimes generate substantial pressure differences resulting in lifting forces. People caught in such systems have been raised from the ground and killed when dashed against trees and buildings. Occasionally men have survived such terrifying experiences when by good luck they were gently set down again on the earth. In one of many mediaeval and Renaissance rimed chronicles of marvellous events, Jean Molinet, writing in about 1500, mentions 'what does not happen often: a man snatched up to the clouds by a violent wind at Bruges in Zeeland. After a good hour he was deposited, alive, right in front of Ceriché.'[15]

Such happy accidents occasionally formed the basis of moral tales or amusing and risqué stories. A generation or two after Molinet, a sixteenth century Norman French writer who used the anagramming pseudonym Philippe d'Alcripe but whose real name appears to have been Philippe le Picard, compiled a small collection of allegedly historical marvels. In the early pages of his book is a scabrous account of the experiences of two tarts in Rouen. Nicknamed Jeanne-Yellow-Arse and Marion-the-Can, they were walking through the town one Friday in search of custom when a violent wind filled their skirts, raised them from the ground, and carried them as high as to the second gallery over the gateway to the Prioress's house. There they held on to the cornice work with their arms, remaining suspended for two hours with their skirts folded over their heads and their charms on show. Several young workmen, clearly in no hurry, took time in getting them down. Le Picard ends with a short, platitudinous moral:

Ceux là sont en peril souvent
Qui sont en la merci du vent.[16]

Fears were increased by accounts of intended flight ending in disaster. Some of these point the moral through more or less coarse burlesque, as in le Picard's story of a French labourer, a solid, square-set man who was reputed to be a great swearer of oaths and drinker of curdled milk. One day when, as Picard says, the fumes of the curdled milk had gone to his head, he hit on the idea of making flying apparatus for himself. Without saying anything to his wife he went to his barn, where he cut his winnowing basket in two and used the halves to make a pair of wings. Fixing them to his back, he passed his arms through handles that he had attached to them and then tried to fly. When he failed to lift himself from the ground, he realised that he had forgotten about the need for a tail, 'which is a great help to birds in flight.' After thinking about the matter for some time, he fixed a coal shovel between his legs, tying the handle along his belly. When he was ready, he climbed to the top of a pear tree so as to 'gather the wind.' 'Having finally cast himself into the air, it happened that either because he had not fully established his equilibrium or because some branch of the pear tree impeded his tail, he fell head first into the drain of his dunghill where he broke his shoulder.' The wound never healed sufficiently to allow him to perfect his invention.[17]

Among stories that probably have more historical foundation is one concerning a sixteenth century clockmaker called Denis Bolori. Evidently an artisan of skill and imagination, Bolori lived between 1515 and 1536 in Troyes, about 100 miles southeast of Paris. He was sufficiently respected to be commissioned to make the town clock of Rigny-le-Ferron, which was named after him following its installation in 1530. According to the last of his descendants, a secondhand dealer who told the story to Grosley, a dedicated historian of Troyes, Bolori made wings that could be flapped by manipulating a system of springs. Having jumped from the tower of the Cathedral of Saints Peter and

28. The tower of the Cathedral of Saints Peter and Paul, Troyes, from which the clockmaker Bolori is said to have flown in 1536.

Paul, he is said to have managed to glide in an easterly direction. While he was still airborne, one of the springs broke, causing him to crash into the fields surrounding the priory at Foissy, some 2 or 3 kilometres east-southeast of the cathedral. As a result of the fall, he died of his wounds in 1536.[18]

Whether true or legendary, Bolori's fall into the priory fields has associations with marauding invaders, flying lovers, and the sexual symbolism of flight. Originally housing both monks and nuns, the Priory of Foissy became an entirely female institution in about 1484. By 1535, there were 46 nuns and 13 lay sisters housed in a convent by the Seine, surrounded by spacious woods, gardens, and vineyards. Bolori's descent out of the skies reintroduced a male element after more than fifty years of female rule in this small Arcadia. However little intended, his visit was an invasion, an affront quickly avenged by fate.

Junonian punishment is still more painfully symbolised by an apparently true story concerning the childhood of John Williams, Archbishop of York (1582–1650):

> So far was he a stranger to wanton Lusts, that his Acquaintance marvelled, that the more the Sin came near to him, so comely a Feature wanting not Enticement, the further he ran from it. *Arthur Wilson,* in his History of King *James,* by some secret Whisper came nigh to the Discovery of the Reason. Not that he was an Eunuch *ab utero,* as he bluntly delivers it; but he had suffer'd an adventitious Mischance, being about 7 Years old, which compell'd him to actual Chastity. He took a leap, being then in long Coats, from the Walls of *Conway* Town to the Sea-shore, looking that the Wind, which was then very strong, would fill his Coats like a Sail, and bear him up, as it did with his Play-fellows: But he found it otherwise, for he did light with his Belly upon a big ragged Stone, which caused a secret Infirmity, fitter to be understood, then further describ'd; and want of timely Remedy, the Skill of good Chirurgery being little known in that Climate, continued it to his dying day. They that traduc'd him when he came to be Lord-Keeper, not only to be Amorous, but to be Incontinent with a great Lady, and taught common Fidlers to sing it, may blush at this Discovery, if they be alive; but if they died without Repentence, it may be they want the tip of his Finger dipp'd in Water to cool their Tongues.[19]

There is something archetypal about this account. Underlying adult experiments with flapping wings may be the conviction of many children that they can fly, a conviction often felt so strongly that they cannot resist throwing themselves from walls and windows.

Long before most grown men seriously believed that they might be able to fly by the use of artificial wings, the urge to do so was expressed, with less risk of injury, through theatrical simulation. Adding an unusual vertical dimension to the more natural horizontal movement on stage or in ceremonial processions, such spectacles readily captured the attention and imagination of

29. Earliest printed illustration of a kite in an English book. Robert Fludd, *Utriusque cosmi maioris scilicet et minoris metaphysica, physica atque technica historia* III (Francofurti 1619) opposite p. 138.

audiences. Although they did not normally involve flight in any real sense, they were sometimes performed with the help of equipment—especially wings and kites—having some ancillary aerodynamic function. There is no doubt that their growing familiarity in the sixteenth and seventeenth centuries influenced the visual imagination of many who speculated about the possibility of building a true flying machine.

Cords, wires, pulleys, moving platforms, and sliding panels were in common use for assisting the creation of lifelike spectacles at late mediaeval and Renaissance banquets and other ceremonial occasions. Angels, classical gods and goddesses, winged dragons, and a variety of other fantastic creatures were caused to pass overhead or to make startling descents into the centre of the arena. Arranged both in and out of doors, such ceremonial entertainments, often costing huge sums of money, were set up with meticulous care and great ingenuity. Sometimes a *trompe-l'oeil* effect was aimed at, the machinery and wires being concealed as far as practicable from an impressionable audience; on other occasions a frankly theatrical artifice was evident.

An early account of a simulated flying angel, impersonated by a man, is included in Juvénal des Ursin's description of the entry of Queen Isabeau of Bavaria into Paris in 1389:

> The bridge through which she passed was completely decked in blue taffeta with golden fleurs de lys. And there was a lightly built man in the guise of an angel who, by means of well-made machinery, came from one of the towers of Notre Dame de Paris to the place of the bridge, and entered through a slit in the covering at the time when the Queen was passing, and placed a fine crown on her head. And then, by means of the arrangements

> that had been made, was drawn up through the slit, as if he was returning to heaven by his own forces.[20]

A comparable indoor spectacle, though this time in a purely secular context, was arranged as a part of the entertainment at a famous banquet known as the 'Feast of the Pheasant,' given by Philip the Good, Duke of Burgundy, on 17 February $14\frac{53}{54}$. A contemporary description is given by Mathieu d'Escouchy (c. 1420–c. 1485):

> From one end of the room, at the highest point, a fiery dragon moved off, flying in the air along most of the length of the room, and passing beyond the guests so that no one knew what became of it.[21]

Later in the banquet, a further aerial spectacle was arranged, although on this occasion even an earthbound animal was made to fly. D'Escouchy's account suggests that wires were probably stretched from corner to corner of the banquetting hall and made to cross in the centre:

> From one of the ends of the hall, high up, there departed a hare that, when it was noticed, was shouted at by several men in guise of falconers and huntsmen who cried 'Tally-ho, tally-ho!' And immediately there departed from another corner a falcon that flew down to attack the hare. From another small corner a second falcon departed that approached so precipitately and attacked the hare so hard that it was slain in the middle of the hall. After the noise of the hunt was over, the hare was taken and presented at the Duke's table. (pp. 149–50)

In the elaborate court masques of seventeenth century England simulated flight was an especially popular spectacle. James Shirley's *Cupid and Death,* performed on 26 March 1653, contains a typical stage direction, the matter-of-fact tone of which indicates how normal it had become to ask the producer to arrange such effects:

> The scene is changed into a pleasant Garden, a fountain in the midst of it; walks and arbours delightfully expressed; in divers places, Ladies lamenting over their Lovers slain by Cupid, who is discovered flying in the air.[22]

An occasional variant was the flying of large decorative kites at festivals and ceremonial processions. Given the form of dragons, eagles, or other figures of symbolic power and importance, the kites greatly impressed both the populace and the chroniclers. The written accounts frequently suggest that magical arts or unusual mechanical skills were used to enable the figures to fly freely over the heads of the observers.

30. Funambulist 'flying' down a wire. Johann Caramuel Lobkovitz, *Mathesis biceps* I (Campaniae 1670) plate 18.

Among the most celebrated of the ceremonial kites was one in the form of an eagle flown by Regiomontanus (1436–76) to greet the Holy Roman Emperor Frederick III on his entry into the city of Nuremberg in 1471. Another, whose creator was anonymous, was flown in 1530 in honour of Charles V of Spain, on the occasion of his visit to Munich.[23] That the habit of arranging such spectacles was still alive towards the end of the sixteenth century is indicated by a passage in Scipion Dupleix, who describes the entry of Henry, Duke of Anjou and King of Poland, into Krakow on 12 February $15\frac{73}{74}$:

> What seemed most rare at this ceremony was that a white eagle appeared, flying continuously above the head of the King all along the streets and arranged with such artifice that one could not see whence proceeded its movement.[24]

A similar account in the three-volume *Dictionnaire des origines, découvertes, inventions et établissemens* adds that the eagle flapped its wings, but this is probably no more than an unauthentic late embellishment.[25]

Flight simulation requiring great skill in balance and involving the control of small amounts of effective lift has been practised by funambulists from mediaeval times to the present day. At the end of his passage on flight, Caramuel Lobkovitz describes one such performer whom he saw in Madrid. A rope was stretched across the marketplace, at an angle of about 30°, from a high window on one side to a column on the other. The funambulist climbed out through the window, made the sign of the cross, addressed the crowd that had been invited to watch him fly, and then, balanced on the rope near the window, performed a number of dextrous tricks that included jumping and dressing and undressing the upper part of his body. After having entertained the crowd in this way for about half an hour, he spread grease on his chest, lay on the rope with his arms and legs outstretched, and allowed himself to slide down across the marketplace. At the bottom end he was saved from injury by a pile of cushions arranged against the pillar. That such displays were often daring and spectacular is frequently attested. Pierre Boaistuau (1500–66), for example, writes of 'those actors we have seen in our time flying through the air on a rope, with such dexterity and danger that the Princes and Lords who were present could not bear to watch.'[26]

Flight simulated on a rope was sometimes made to seem more lifelike by the addition of wings to the funambulist's arms. Bishop Wilkins describes the technique:

> It is a usuall practise in these times, for our *Funambulones,* or Dancers on the Rope, to attempt somewhat like to flying, when they will with their heads forwards slide downe a long cord extended; being fastned at one end on the top of some high Tower, and the other at some distance on the ground, with wings fixed to their shoulders, by the shaking of which they will break the force of their descent.[27]

While flapping could hardly have been effective, carefully made wings attached to the arms can create significant amounts of lift and drag that a funambulist may use to control his descent. Although he is neither flying nor truly gliding, a man manipulating such wings is using techniques similar to those of free-fall parachutists or of 'birdmen' such as Clem Sohn and Léo Valentin.

Together with the minor contribution to technological thinking made by these practitioners of flight simulation, one may perhaps notice their traditional moral sensibility. According to Hugues le Roux and Jules Garnier, historians of circus aerobatics, it is proverbial in the profession that chastity is essential to success on the rope: 'Love destroys the centre of gravity in tight-rope dancers, and as a rule equilibrists . . . might rank with the Roman vestals.'[28]

Despite the proliferation of moral and social doubt, and despite the stories of well-deserved failure, the seventeenth and eighteenth centuries saw an increasing tendency among scientists and natural philosophers to deal seriously with the question of artificial flight. Joseph Glanvill, asking his readers to keep open minds, suggests that 'to them, that come after us, it may be as ordinary to buy a *pair* of *wings* to fly into remotest *Regions;* as now a *pair* of *Boots* to ride a *Journey.*'[29] Some respected men of science were even reputed to have achieved a modest success. Giovanni Francesco Sagredo (1571–1620), an influential Venetian gentleman and amateur of science who was a friend and collaborator of Galileo, prepared himself for a practical flying attempt by studying the structure of birds.[30] According to Tito Livio Burattini, Sagredo began by catching a falcon, whose feathers he removed. He then 'diligently observed the proportion of these to the body, in respect both of weight and of size, and made wings for his own body according to the same proportion.'[31]

31. Winged trapeze artist. Hugues le Roux and Jules Garnier, *Acrobats and Mountebanks,* trans. A. P. Morton (London 1890) 240.

Equipped with these, the materials of which Burattini does not specify, Sagredo threw himself from a height, arriving unharmed on the ground many yards distant from his starting point.

The lively and imaginative Flemish alchemist and physician Jan van Helmont (?1577–?1644) appears to have been seized, late in his life, with a bout of enthusiasm for flight. Although van Helmont left no writings on the subject, Caramuel Lobkovitz reports having been present at an address that he gave in Brussels, in 1640, before Prince Emanuel of Portugal. According to Caramuel, a book called *Ars volandi,* published in Belgium in that year, attempted to persuade its readers that flight was not only possible but easy. Helmont undertook to defend the book's thesis, speaking 'so ingeniously, with so much erudition, and with such eloquence that, if you had heard him, you would have believed that you could immediately spread your wings and fly.'[32]

The great Robert Hooke (1635–1703) was more cautiously optimistic and also, it seems, more practical. Although he returned to the subject again and again throughout his life, on one occasion reporting the trials of a Mr Gascoyn in about 1640,[33] he remained constantly aware of the inadequacy of human muscle power. Writing of his work on the subject in about 1658 or 1659, he said that he had applied his 'Mind to contrive a way to make artificial Muscles; divers designs whereof I shew'd also at the same time to Dr. *Wilkins,* but was in many of my Trials frustrated of my expectations.'[34] Hooke's early biographer Richard Waller saw some of these designs, which have not survived. According to Waller, they included a sketch of a primitive helicopter and 'some contrivances for fastening succedaneous Wings, not unlike those of Bats, to the Arms and Legs of a Man.'[35]

In the mid-sixties, Hooke again showed his openness to new ideas by making what was, for his time, a most unusual suggestion: ''Tis not impossible to fit a Pair of Wings for a Man to fly with, which may be contriv'd somewhat after the manner of the long Fins of . . . flying Fish.'[36] The strangely equipped aeronaut would still need an external power plant, which Hooke described intriguingly, if also obscurely, as 'an Artificial Repository or Magazine of Strength, which for Weight and Bulk would not be too cumbersome.'[37] Only a little later, in *Micrographia,* he returned to the need for an additional source of energy:

> To me there seems nothing wanting to make a man able to fly, but what may be easily enough supply'd from the Mechanicks hitherto known, save onely the want of strength, which the Muscles of a man seem utterly uncapable of, by reason of their smallness and texture, but how even strength also may be mechanically made, an artificial Muscle so contriv'd, that thereby a man shall be able to exert what strength he pleases, and to regulate it also to his own mind, I may elsewhere endeavour to manifest.[38]

By February 1675, Hooke felt sufficiently confident to mention to the Royal Society that 'there was a way, which he knew, to produce strength, so as to give to one man the strength of ten or twenty men or more, and to contrive muscles for him of an equivalent strength to those in birds.' In order to apply this additional force in the most efficient way, 'a contrivance might be made of something more proper for the feet of man to tread the air, than for his arms to beat the air.'[39] A leader among the century's rational thinkers, Hooke felt, as these passages reveal, little excitement at the idea of flight itself, and no sense of awe or wonder. Nor does he appear to have shared Leonardo's interest in aerodynamic theory. Almost his only expressed concern, indeed, was to solve the mechanical problem with the application of ingenuity and technological skill.

An influential scientist of little more than a generation earlier had shown a greater capacity for personal involvement: among the most enthusiastic of the early theoreticians of flight, especially in his younger days, was the much-respected French Minim, Marin Mersenne (1588–1648). A copious writer and an active correspondent, Mersenne frequently alluded to the subject. The degree to which artificial flight intrigued him is indicated by his having devoted to it the first chapter of one of his most widely read books, *Questions inouyes, ou recreation des scavans* (1634):

> *Whether the art of flying is possible, and whether men can fly as high, as far and as fast as birds.*
> Many people believe that the art of flying is impossible because the air is too rare and too weak to bear the body of a man, no matter what wings he might have; but if one considers the character and the beating of wings used by geese and other large birds when they want to raise themselves so as to fly, and the weight of the pieces of paper and wood that one can make fly by pulling them on the ends of a thread, and which are called flying birds, one may perhaps change one's opinion. For a man can raise himself into the air provided he has big enough and strong enough wings, and

> enough industry to beat the air as is necessary. This can be achieved with certain springs which will make the wings move and beat as fast as is required in order to fly. But it would be necessary to accustom children to this exercise and to make them start flying over water so that they would not be in any way hurt if they happened to fall; and I do not doubt that they would learn this exercise as easily as they learn to dance on a rope, turn somersaults, and a thousand other kinds of dangerous jumps which are as difficult for those who have not learned to make them as is the art of flying.
>
> Now it is still more difficult to tell whether men can fly as far, as high and as fast as other birds because that depends on many experiments, much knowledge, and very long practice, which we lack and which will perhaps be known to posterity. That is why I simply add that it is perhaps hardly more difficult to fly than to swim; and that as we find the art of swimming very easy when we have learned it, although we should have held it to be impossible, so we shall consider the art of flying to be very easy when we have achieved it.
>
> *Corollary*
>
> There are a number of experiments that one can make that will serve to establish the art of flying: for example, one can find what weight can be lifted by artificial birds, for if one makes little huts of wood in which a child can be housed, the child will feel assured and will lose any kind of fear, and will be able to try his artificial wings without risk, because he will only be able to fall with the wooden rooms.[40]

As he grew older, Mersenne began to be less hopeful of success. His doubts may have started at the time of his correspondence with Christophe de Villiers in 1634. On 3 March de Villiers wrote to Mersenne about the flight of Bolori in Troyes. Prepared to accept only the possibility of a glide, de Villiers asserted: 'Manned flight is possible from above to below, but not from below to above, because of the great weight which cannot be raised by wings.'[41] In July and August of 1640 Mersenne received letters from Descartes in which the subject of flight was raised. In the first of these Descartes responded to Mersenne's evident desire to believe that some kind of manned flight might be possible:

> I should like to see the results before believing the propositions. It is indeed possible to make an object remain in the air for a time, but not to make it stay there unless it is held from below; such as the piece of iron which will remain suspended in the presence of a single magnet without flying to it, but will doubtless be retained by a silken thread so slender and placed so well out of sight that it cannot be seen, which is nothing more than a puerile trick; but as regards birds, they beat the air more or less according to whether they need to stop or advance, which cannot be imitated by any machine made by man.[42]

In a letter to Mersenne of 30 August 1640, Descartes returned to his objections:

One can indeed make a machine, metaphysically speaking, which will sustain itself in the air like a bird; for birds themselves, at least in my view, are such machines; but it is not possible physically or morally speaking, because one would need springs so subtle and at the same time so strong, that they could not be made by man.[43]

Mersenne's correspondents continued to throw doubt on the matter. A year or two after receiving the letters from Descartes, he was again discussing flight, this time with Théodore Deschamps. On 26 March 1642 Deschamps reinforced Christophe de Villiers's point, asserting that the human muscles were insufficiently strong to allow a man to raise himself with wings. Gliding he nevertheless thought possible, though he believed that this could be achieved only when a wind was blowing.[44]

When writing to Theodore Haak on 13 December 1640, Mersenne himself had spoken about the difficulties of raising heavy weights, alluding this time to shadowy and highly exaggerated accounts of a flight by an inhabitant of Reims variously known as Lisson, Desson, Du Son, d'Egmond, and other names usually beginning with 'D.'[45] The real person behind these stories may have been Nicolas Desson, Sieur d'Aigmont, a landscape painter, inventor, and entrepreneur born in 1604. A mediocre artist, Nicolas Desson lived for a time in Paris[46] and Brussels.[47] In 1635 he settled in Holland, where he became famous for the invention of a spring-operated ship. In 1664 he visited England, where he caught the attention of members of the Royal Society. Notes of his work were included in the *Philosophical Transactions,* which called him 'that excellent Mechanician.'[48] He was known in particular as an inventor of watches, lenses, and a new style of carriage.[49] In *Närrische Weiszheit und weise Narrheit,* Johann Joachim Becher discusses him with a mixture of respect and scorn. Although, Becher says, the ship sank like lead, giving rise to a new Dutch proverb, 'It goes forth like Desson's ship,' he had devised a new and efficient technique for desalinating seawater and had invented a remarkable silent gun.[50] If this is indeed the Desson whose flying machine was so often talked about, he must have spent the 1640s in Paris, from which most of the reports were sent. Desson's earliest experiments, however, seem to have been made in Brussels. According to the somewhat coarsely humorous *Arliquiniana,* a sequence of sociopolitical dialogues and monologues published in 1694, it was there that Desson made an early and unsuccessful attempt to fly from the top storey of a house:

Haven't you heard, I said to him, of the imagination of a certain man called du ..., who wanted to travel from Brussels to Saint Germain in a machine that he said he could drive through the air. That adventure, said Harlequin to me, was told to me once, but I have forgotten it. Here it is, I said: You know that C... was rather hot-blooded; he had been guilty of some indiscretions that the king had pardoned, but he did not want to pardon him the violence that he did to a coachman once on returning from a voyage. He killed this coachman and escaped, to Brussels I believe. There he found a man who became his friend and who promised to obtain his pardon for

> him, basing his claim on the idea that he would go to find the king in Saint Germain in a machine that he had conceived and that the king, seeing him arrive through the air in so extraordinary an equipage, would not fail to grant his request. With this idea he shut himself up at the top of a house in a large attic, and he remained there until he had completed his machine to which he attached sails and a kind of rudder. He asserted that the air would support it and that by means of the wind and this rudder he could make it travel wherever he wished. When the machine was finished this man took leave of all his friends; he had made a large opening in the attic in order to get it out. As he was on the point of sending it out into the open air, he realised that it might be too light, and that the wind might carry it too high. So as to avoid this accident he filled two sacks, each weighing fifty pounds, and attached them one to each foot, so that he could travel in the middle of the air without fearing that he might be carried off too precipitately. Hardly had they pushed the machine into the air than it fell twenty paces away on to a small house, the roof of which it stove in. Du ... broke his legs. Afterwards he wanted to sue the owner whom he said to be responsible for the harm which the roof of the house had done him. As the owner was a simple fellow who feared the law, he was on the point of reaching an agreement that would have cost him quite a bit if one of his friends had not prevented the abuse of his simplicity.[51]

In expressing his scepticism about Desson's flying machine, Mersenne casts doubt on the practicability of his earlier ideas about the wooden huts for children. Mersenne's report is more matter of fact than the story in *Arliquiniana* and, although the scale is more grandiose, the details may well be closer to the truth of what Desson attempted:

> Here they are talking about a man from Reims who was once in your England and who has a machine of 32 square feet that he says he can make fly through the air anywhere he wants, with 8 or 10 men able to accompany him; time will tell the truth of this, and what will happen as a result. I hold it to be very difficult, if not impossible, to make a dwelling fly, or to fly a machine furnished with rafters, etc.[52]

Another version of the story appears three years later in a letter to Mersenne from Théodore Deschamps (c. 28 December 1643):

> I am reminded of this new Icarus who wanted to leave Paris to travel in his machine through the air to dine in Constantinople and return to sup the same day in Paris, and who suggested that he would make very small machines for a single man costing only one pistole each. What is to be said about it at present? I don't believe you should entrust yourself to his machine for your own voyage.[53]

By 1648 Desson's machine had grown still more impressive, and the details more precise. In a letter to Hevelius (14 March 1648) Mersenne responded to comments about Burattini[54] by referring once again to the claims being made in Paris:

> Here we have an unusual inventor who says that he can make a flying machine that could fly from here to Constantinople in a single day. Its wings are 32 feet long and 12 feet wide; and he even believes that he can carry in it, along with 6 men, 4 or 5 large siege machines, so that nothing could possibly resist it. And I fear that these things may indeed at some time see the light; the old times are overwhelmed and obliterated these days by the ingenuity of our inventors.[55]

In 1647 Mersenne published *Novarum observationum physico-mathematicarum,* a book conceived in a more serious vein than the *Questions inouyes.* By the time he came to write it, his doubts about the possibility of manned flight had grown substantially. While still showing clear signs of finding the idea emotionally attractive, he was now prepared to say that all experiments would probably be fruitless:

> You may determine the length of the wings that a man must use if he wants to fly; since the length of a swallow's wing is about twice the length of its whole body, the length of a wing for a man of 5 feet must be 10 feet. The rest can readily be deduced. Nevertheless many other kinds of birds should be examined, for example geese, partridge, doves, hens, vultures, eagles, before you commit yourself to the winds. I fear that we may lack the muscles that could impart to the necessary wings the motion necessary for flight, and that the future efforts of men made for this purpose may be vain.[56]

In August 1647 Mersenne fell ill. Pleurisy was wrongly diagnosed, he was incompetently bled, and he remained weak until his death on 1 September 1648. Although he continued to correspond during the last year of his life, he lacked some of his former zest. Had he continued in full vigour, he might have been inspired to develop ideas of his own in response to the excitement generated in 1647 and 1648 by the most innovative and tenacious practical experimenter of the century: Tito Livio Burattini.

Part Two Practice

Burattini's flying dragon 6

Largely as a result of frequent written reports by Pierre Des Noyers, secretary to the queen of Poland, mid-seventeenth century scientists took a great deal of interest in the aeronautical work of an Italian engineer, Tito Livio Burattini. Born on 8 March 1617 in Agordo, north of Venice, Burattini early became a travelling scholar.[1] By 1637 he had reached Egypt, where he helped John Greaves with his famous work on the pyramids.[2] Despite his youth, he appears greatly to have impressed Greaves, who referred to him, with slight inaccuracy, as 'Titus Livius Burretinus, a Venetian, an ingenious young man.' In 1641 Burattini returned to Europe, spending a little time in Germany and travelling on to Poland, where he lived until 1645. In 1647, after a visit to his native Italy and a second period in Egypt, he settled permanently in Poland, where he attempted, soon after his arrival, to establish himself at the court of Władysław IV (30 May 1595–20 May 1648). Although poor and not altogether well received to begin with, Burattini fairly soon received a measure of royal patronage, partly owing to the presence in the court of Władysław's Italian wife Maria Louisa, daughter of Charles I, Duke of Mantua. Despite some setbacks after Władysław's death, which was followed by a period of bitter warfare in Poland, Burattini became a courtier of distinction and power, married into an important Polish family, and grew wealthy. After the death of Maria Louisa in 1667 his fortunes again diminished for a time, but by the early 1670s he was once more in favour. The date of his death is unknown, but he was no longer alive in 1682.

From at least as early as his expedition with John Greaves, Burattini appears to have taken a serious interest in scientific and mechanical matters. In Poland Stanisław Pudłowski, a celebrated scientist and professor in Krakow University, knew and befriended him. He followed the latest news of scientific achievements, much of it communicated through correspondence, and established a reputation as a skilled engineer, being especially talented as a maker of telescope lenses. He also invented an improvement to a form of balance designed by Galileo, proposed a universal linear measure (the 'metro,' the length of a 'seconds' pendulum), and perfected a water clock.

Correspondence among Des Noyers, Roberval, Mersenne, Thévenot, and Huygens contains many references to a flying machine on which Burattini was

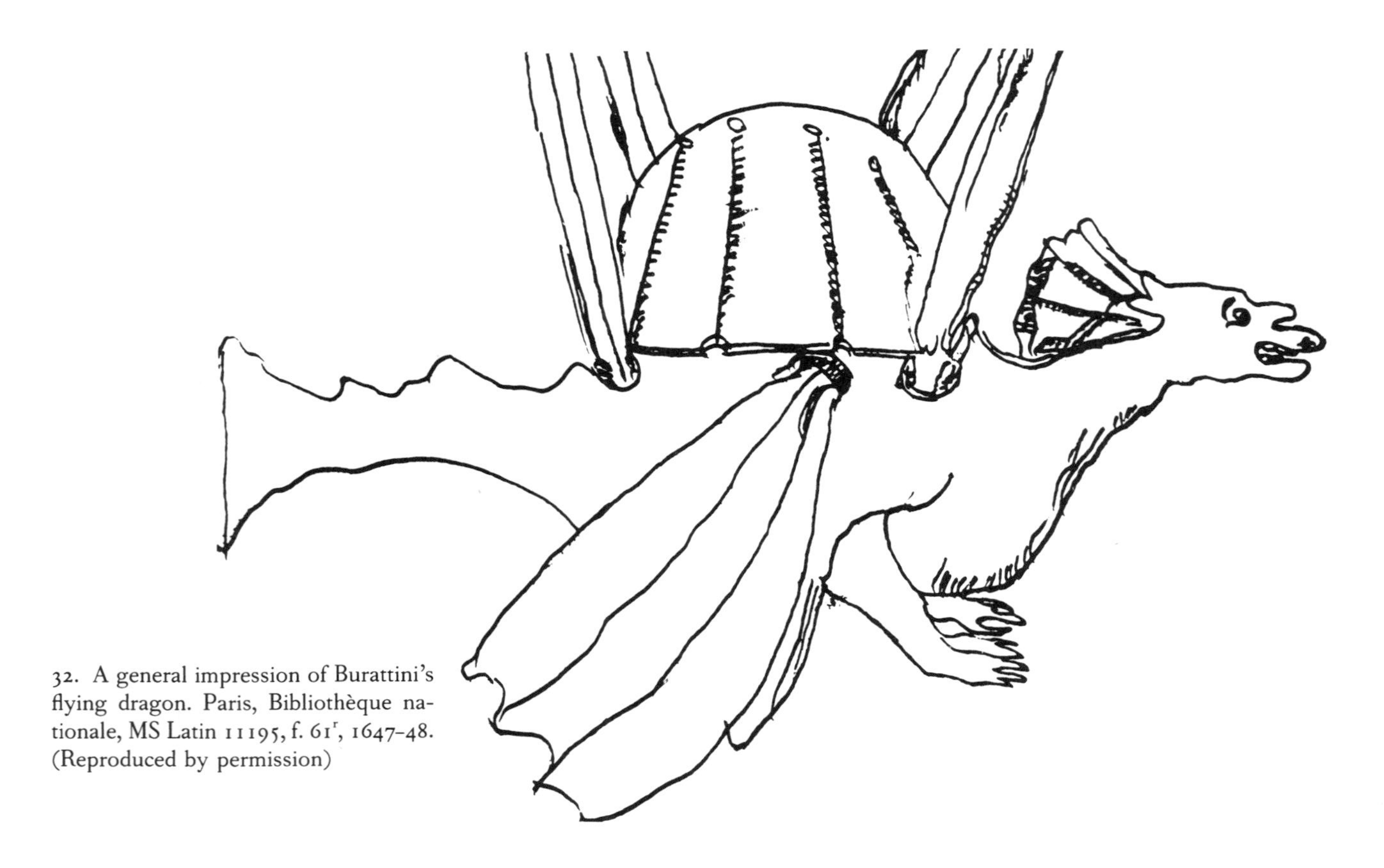

32. A general impression of Burattini's flying dragon. Paris, Bibliothèque nationale, MS Latin 11195, f. 61r, 1647–48. (Reproduced by permission)

working in 1647 and 1648. Having apparently induced Władysław, a patron of science, to take an interest in the project, Burattini asked for sufficient support to cover the cost of materials, but he offered to carry out the work for nothing, putting his invention at the disposal of his royal employers and hoping for a just reward should it prove to be a success. Presenting his ideas to the court, he wrote a short treatise setting out his aims and the principles on which he was working. One of a number of copies of this was sent to Gilles Personne de Roberval by Des Noyers, accompanying a letter dated 4 December 1647. Unconvinced at the time, Des Noyers wrote in a sarcastic tone:

> A little while ago there arrived here a mathematician who tells us that he has just come from Arabia, and because he comes from far away he thought he could tell lies. The manuscript that I am sending you, together with the figure that accompanies it, will tell you the rest and make it unnecessary for me to go into further detail. He put it into Italian to show it to the Queen. The figures are badly copied.
>
> The Queen commands me to write to you to say that if this machine is a success she wants me to come and fetch you so that you may visit her in her Realm. And I believe that if I did this I would be hardly less well rewarded than its creator.[3]

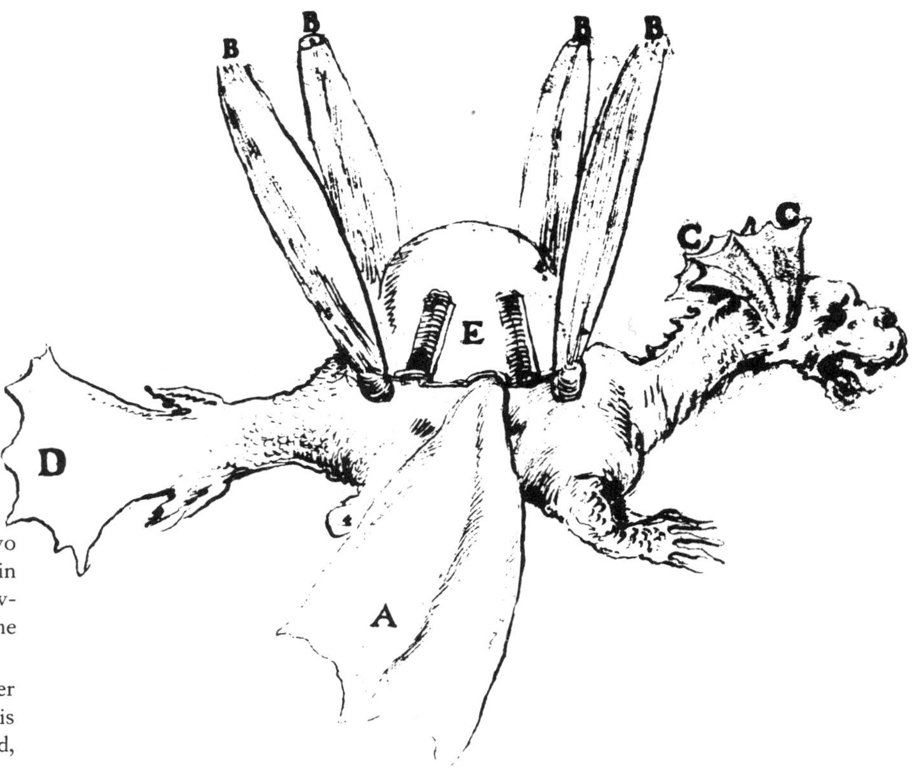

33. The more detailed of the two sketches of Burattini's dragon, MS Latin 11195, f. 56^r^. On 55^v^, facing, the following notes, in French, are keyed to the letters:

A. Master wing to which, on the other side of the dragon, a second wing is opposed. Both serve to push it forward, and also to keep it up.

B, B, B, B. Four wings that move up and down, and do nothing but lift the dragon up.

C, C. Two little wings that do nothing else than push forward with the two master wings.

D. The tail that causes the machine to turn to the desired side, and to raise and lower it.

E. Movable cover that can, when one wishes, be expanded in a circular fashion over the machine, holding it up sufficiently to prevent its falling precipitately.

A single movement governs all of these things, even though they move at different times.

All the wings fold up on being raised, and when they are lowered they spread out, and everything, as I have just said, is governed by a single movement. (Reproduced by permission)

Soon afterwards, having grown convinced of Burattini's seriousness and mechanical skill, Des Noyers changed his opinion. Although the manuscript is not in Burattini's hand, it is clearly authentic.[4] The treatise is written in black ink on eight sides of a single quarto gathering of unlined paper, measuring 228 mm × 173 mm. Enclosing these as a cover is a further single folded sheet, bearing the title *Ars volandi* on the front. On the inside back cover is a drawing of the dragon (Figure 32). This is the figure that Des Noyers rightly describes as 'badly copied.'

The treatise is at present bound at the back of a mathematical dissertation, *De recognitione aequationum,* written in the same hand. After the title page, *Ars volandi* (f. 50^r^), an additional quarto gathering of smaller sheets has been bound in (ff. 51–54), containing notes on flying drawn from Flayder's *De arte volandi.*[5] Between this gathering and the first page of Burattini's text (f. 57^r^) a further single folded sheet (ff. 55–56) has been tipped in. This contains a more delicate sketch of the dragon (f. 56^r^) drawn in a lighter ink, together with notes in French (f. 55^v^). Both the sketch and the notes may have been prepared especially for Roberval, to accompany the Italian treatise (Figure 33).

Burattini not only reveals his obsession with the problem of flight but also shows a keen interest in the growing science of mechanics. 'Flying is not impossible as has hitherto been universally believed': heading his discussion with this correct prediction, he begins with an expression of concern that, despite its many achievements, human ingenuity has yet to solve certain recalcitrant problems, among which is the art of flying. So difficult has such a thing always seemed to be that flight has become a type of the impossible act: 'When ordinary people want to say that something is impossible, they say that it is as impossible as to fly.'[6] From his earliest years Burattini had nevertheless found it strange that human inventiveness had failed to achieve skill in a kind of movement granted to animals inferior to man, especially since man is able to swim under water, an element antithetical to breathing. He accordingly made many attempts to fly that he chooses not to discuss in detail. These experiments, which all ended in failure, were made about ten years before the writing of the manuscript and may thus be dated to about 1637, when he was twenty.

By 1647, being then more experienced in mechanical matters, Burattini believed that he could solve the problem by the application of new principles. He therefore proceeds to give a brief survey of three different sets of ideas, only two of which, in the event, appear to have any direct application to the machine he draws and describes. The first has to do with buoyancy and specific gravity. Having stated Archimedes' principle, he says that it will form the whole basis of the discussion to follow, leading the reader to expect that his flying machine will be lighter than air. In fact, although lightness of construction is implied, the Burattini dragon is heavier than air. Developing the Archimedean train of thought, Burattini enunciates the incorrect but conventional Aristotelian idea that the speed at which metals and stones sink through water is proportional to their specific gravities, and that the same is true when they are dropped through air, although in that case the velocities will be much greater because of the lesser density of the medium.[7]

Using this false principle to arrive at a correct conclusion, Burattini next says that a substance as light as air would neither rise nor sink, while one lighter than air would naturally move upwards towards the region of fire. Aware that no such substance appears to exist, he diverts his attention towards the known structure of birds, pointing out that although they are more lightly constructed than men, even their smallest feathers are denser than air and therefore sink when dropped.[8] Not only is the beating of a bird's wings necessary to enable it to rise, but even if a bird were to throw itself from a height with its wings outspread and motionless, it would not be able to sustain itself but would fall to the ground with significant impact. Burattini here ignores the familiar problem of continuous soaring and may even imply, incorrectly, that a bird is incapable of landing safely from the glide.[9]

In contrast to traditional ideas about velocities of descent, the remaining statements of principle in Burattini's manuscript are based on the most up-to-date scientific thinking available to him. He was particularly influenced by Galileo, having evidently read not only the published works but also many of the manuscripts that were circulating freely at the time. Burattini appears to have acquired at least a part of this material at the time of his second visit to Italy in the mid-1640s.

His next set of ideas, concerning impact or percussion, is a rather lame development of physical principles that Galileo had explored much more thoroughly in a passage written in about 1635 though not published until after his death.[10] Although he makes no attempt to offer an explanation of flapping flight, Burattini gives some thought to the way in which the air resists the beating of a wing. Leaning heavily on Galileo for his premises, he first divides impact, or percussion, into three categories: 'firm,' the stroke being made against a stationary solid object; 'contrary,' the stroke being made against an approaching solid object; and 'fleeting' or 'escaping,' the stroke being made against a solid object that is retreating from the striker. Considering next the idea of percussion against fluids, Burattini points out that fluids can offer surprisingly strong resistance: a sword blade or stout piece of wood can be shattered if it is beaten against the surface of water, while a bullet will bounce if fired obliquely into a river. The air also can exert strong resistance against a 'firm' stroke, being able, for example, to break a piece of card that is held in the hand and whirled fast through it. Similarly, a reed held perpendicularly by one end and moved through the air will, by inclining away from the direction of motion, demonstrate the power of fluid resistance.[11]

Continuing with his categories, Burattini defines a 'contrary' stroke in the air as one made against the wind, while an 'escaping' stroke is made in the same direction as the wind. In commenting on the 'escaping' stroke, Burattini shows that he has some grasp of relative motion:

> The escaping stroke is that which is made with the wind, and it will be greater or lesser according to the relative speed of the wind to that of the stroke. If the motion of the wind is equal to the motion of the stroke, the latter will be of no value and will be like two horsemen one of whom is attacking the other with a lance: if the speed of the one is equal to that of the other, the striker will not be able to wound the man who is fleeing even if he has the point of his lance against the enemy's body.[12]

Not pursuing this matter further, Burattini moves on to his third set of ideas, with an account of the principles of mechanical advantage as applied to the functions of the lever. Once again, these ideas are directly derived from Galileo, whose treatise on mechanics was first published in a French translation in 1634. Chapters V and VI of the book provide Burattini with his basic concepts.[13] Obscurely understanding something of scale effect and aware of the effort that would be needed to flap the wings of a large flying machine, Burattini says that his invention must be no bigger than would allow it to carry two men. The levers for flapping the wings must be worked both powerfully and fast, so that the smaller the machine the more practicable it will be for men to manage. Having set out the main principles on which his invention is based, Burattini might now have been expected to describe his dragon in detail. Instead, however, the manuscript ends quite abruptly, with a cursory final paragraph giving only a general idea of the machine's configuration:

> I do not think it wise to divulge the internal design of the machine, which would in any case be difficult to explain. Let it suffice for the present to show the exterior, which, as may be seen, is made in the form of a dragon in which up to two men can stand, one of whom works while the other is resting, as is done on board a ship. It can travel by night also, with the help of a compass, and has provision for food and drink for a few days; and, what is more important, it is built in such a fashion that if any of the wings should break, it will not crash to earth but will sink gently downwards so that those inside will suffer only the most minor injuries.[14]

Despite his assurances about the fundamental importance of specific gravity and the principles of impact, Burattini makes no attempt to invoke those ideas in discussing the nature of his invention. All we may presume, from the general tenor of the brief and ill-organized manuscript, is that he intended to build the dragon of light materials, to keep its overall scale appropriate to the strength of a man, and to use lever mechanisms offering an appropriate mechanical advantage.

How much practical work, if any, Burattini ever undertook on the full-scale flying machine is uncertain, but there is no doubt that he completed at least three working models, described in several letters by Des Noyers. In the first of these, written to Roberval on 14 January 1648, he reports that 'our engineer from Arabia has made here a little model of his flying machine.'[15] Exhibited at court in the hope of promoting the main project, it worked sufficiently well to impress a critical observer. Writing to Thévenot a few weeks after the event, Des Noyers included many details of the construction in his account of the demonstration:

> We have here a gentleman and mathematician called Burattini who has shown the King of Poland the design and a small model of a machine for lifting a man and allowing him to fly . . . as far as the model goes I can tell you that the one he presented to the King was four or five feet in length, including the tail, which by means of a cord which he has arranged to come out of the tail can rise into the air, the cord causing springs and wheels to move inside. This model lifts a cat that is put into it and is sustained in the air for as long as the wheels are made to move by the cord, and to do that one must be at the same height as the machine. It was evident that if the cat had the judgement to apply himself, he could, since his strength is sufficient, raise himself into the air . . . seeing the effect produced by his model he has no doubts about the large machine which remains to be built. For that he is requesting five hundred crowns, not having the means to pay for it himself. I do not know whether our princes will be curious enough to want to do that; he asks eight months in which to finish it. He makes the wheels partly of wood, partly of whalebone; all the ribs of the wings are of whalebone.
>
> The covering on top of the dragon spreads out very wide when a spring is released . . . The tail turns in any direction to serve as a rudder, and if the dragon were to fall into water it would serve as a boat. He believes that

> it could not be made to fly against the wind. Finally, he does not speak in any way like a charlatan but gives evidence of being skilled in mechanics. With all that he is poor.[16]

Writing to Mersenne a month later, Des Noyers repeats these details but varies his earlier statement about the manuscript's having been written in Italian so that it might be shown to the queen. It was intended, he now says, for the king's eyes.[17] While this seems less convincing, it is clear that Burattini was hoping for royal patronage in general, the particular source being a matter of small importance.

On 18 March 1648 Des Noyers once again wrote to Roberval. Along with many other matters, he returned to their discussion of the flying dragon. Both Roberval and Mersenne having written to him about Desson's claims, Des Noyers made implicit comparisons with Burattini's machine:

> In answer to your letter of 29 January I may say that since you show such care for my person I shall not entrust myself to the air as I had resolved to do. I shall wait here for the machine that is being made in France which Father Mersenne, who writes to me about it with admiration, describes as much more powerful than you do. He says that it will carry much more sail than you say, and because the French have many enemies in Germany, which has to be crossed to get to France, I shall wait at Camp Volant so as to arrive there with greater safety.[18]

During his discussions with Burattini about the dragon, Des Noyers suggested that Roberval's advice might be sought. Burattini readily agreed to this and undertook also to write further notes for Roberval about the external and internal construction of the machine. It is probably at this point that Burattini made the more detailed drawing, with the annotations in French, that now accompanies the Italian manuscript. As his ideas were to be considered by a distinguished scientist, Burattini was more guarded about the likelihood of success with the full scale project:

> If he had the means he would, he says, make the experiment himself, although he admits that the forces and proportions of a [small] machine may not suffice for a larger one, which may well be the case. . . . He even agrees to send you a model so made that it can be demounted and reassembled and carried anywhere. . . . He insists that he will be obliged if you will give him reasons that will disabuse him of the opinion he holds. (f. 453^{r-v})

Burattini promised that the new model would soon be ready for despatch. On 21 May 1648, writing this time to Mersenne, Des Noyers reports that it is finished and speaks of a possible confrontation between Burattini and Desson:

> Mr Burattini writes to say that his machine is ready and that I shall see its effect in Warsaw. He says that if he were in the presence of Mr Desson of Reims he would let him see the said machine and its effect and that he would then take it to pieces and hand it to him to reassemble, because he well knows, he says, that he would never manage it and that he could thereby judge that he is not in possession of this secret. Also he does not promise so great an effect.[19]

Thirteen years after the event, Thévenot discovered among his papers a copy of Des Noyers's description, which he had intended to send to Huygens. On 7 May 1661 he accordingly included it in a letter to Huygens, who replied on 21 July 1661. Finding it hard to imagine how the dragon functioned, Huygens objected that it could not properly be thought of as an automaton:

> In order to make the wings beat one must pull on a cord, which is something I do not understand, and yet the machine rises into the air. From this it seems that it was not an automaton of the kind that can rise by its own power.[20]

Huygens was not alone in his puzzlement over the cord. As Roberval had also failed to understand its function, Des Noyers returned to the matter in a letter dated 21 October 1648:

> I do not recall having said that the cord served to hold the machine up when it is in the air. I must have expressed myself badly at that point since this machine can only rise into the air while one moves the wings. Now to make them move while it is in the air the man would have to be there with it. And for that one would need something other than this little model. As regards your belief that the cord by means of which the wings were moved also served to keep the machine either straight or turned more towards one side than the other, it does nothing of the kind: the tail that moves like that of a bird, to the right, to the left, up, down, or at an angle, causes that effect so as to make it rise, descend, and turn.[21]

In this letter Des Noyers used the past tense in writing about the cord because Burattini had changed his technique for flapping the wings:

> Instead of that cord that I told you about for moving the wings, he has included a rod, which when pushed for example into the body makes the wings rise and when pulled makes them lower, in such a way that the wings flap as fast as one pushes and pulls the rod. And only a very little force is needed to make this movement, which is quite vigorous. (f. 456^{v})

It seems unlikely that the demountable model was ever sent to Paris. Indeed, Des Noyers was even rather tardy in forwarding the additional notes that Burattini had prepared. They were finally included in a letter dated 2 December 1648, but only after Des Noyers had been chided by Roberval for his omission (f. 460^{r}).

The full-scale dragon almost certainly remained unbuilt. On 3 July 1648 Theodore Haak wrote from London to Mersenne, saying 'If M. Burattini had only finished his flying chariot we could meet and converse more easily.'[22] On 21 October 1648 Des Noyers was still quoting Burattini as saying that he needed eight months to carry out the work,[23] while on 5 May 1649 Des Noyers wrote to Roberval about the siege of Paris, 'which I have been able to see in imagination better even than if I had been there. If M. Burattini's dragon had been built we could have come to see you, but that will have to be for another time' (f. 465^{r}). Burattini's hopes of royal patronage had been thwarted by the death of Władysław IV, since which time the Poles had been occupied with the more important matter of a major war. Although Burattini appears to have remained in favour with the queen, there was little hope of his being given the necessary subsidy. In his letter of 2 December 1648, Des Noyers reports the depressing news that 'our princes are not curious enough to lay out five hundred crowns to make the experiment' (f. 460^{v}).

Burattini's notes about the flapping mechanism, appended to the more detailed drawing, appear to indicate that it was to be operated by a single lever, cr possibly by a pair of levers arranged like oars and no doubt worked by hand. A system of pulleys, cords, and linkages would have been needed to transmit the power to the wings, the wheels of wood and whalebone mentioned by Des Noyers being presumably a part of this mechanism. Since the cat placed in the model used for the royal demonstration could not work the levers, it seems that Burattini incorporated a minor modification allowing them to be worked first by the cord and later by the rod emerging from the tail. At no stage did Des Noyers explain how the operation of cord or rod could be compatible with anything like free flight, and it seems likely, in fact, that the model could merely be made to jump into the air by every jerk of the operator's hand. Assuming Burattini to have been as good an engineer as his reputation suggests, he could certainly have arranged the wings so that they spread on the downstroke and folded on the upstroke. Although such a system for producing lift would have been very inefficient, it might, in conjunction with the probable action of the 'master wings' discussed later, have been just sufficient to allow a model with a small cat or kitten inside to be raised briefly from the ground. It is highly improbable that a machine of this design with an overall length of 4 or 5 feet could have risen into the air carrying a fully grown domestic cat.

Although none of the internal machinery is shown, some further indications of the dragon's structure are given by the two sketches. While the materials used for the body can only be guessed at, the general shape suggests a framework covered in waterproofed fabric. The hemispherical canopy, evidently also of cloth, is held down by either four or eight retaining members, presumably the spring release mechanism mentioned in Des Noyers's letter to Thévenot. Despite Burattini's theories of percussion, which lead the reader to suppose that the dragon was to be pushed along through the air, the large central wings do not seem to be arranged to move backwards to any significant extent. Shown at or near the end of the downstroke, they seem to be built so as to flex about the leading edge, which in both illustrations is lower than the trailing edge. If they were so made, such wings might produce usable thrust and lift. The general structure of the wings is partly shown on the undersurface of the forward starboard wing, B, where two central spars of whalebone are shown emerging from a more solid member of circular section. How the spars open and close like a fan is not clear. As the members at the wing roots are drawn in much the same style as the springs holding the canopy in place, it may be that strong helical springs form some part of the wing-flapping mechanism. Although the two small fins just forward of the vertical tail surface look like dorsal and ventral surfaces, it is possible that the illustration misrepresents a horizontal tail, although even without such a surface a sufficiently flexible fin could be made to exert forces in any desired direction and so have the effect indicated in Burattini's notes.

Burattini's flying machine rapidly became famous and was several times alluded to in social chat of the day. Quite soon after the demonstration of the model, Maria Louisa's regular correspondent Mère Marie Angélique Arnaud reported an exaggerated account of it:

> I should let Your Majesty know, in order to divert you a little, that they are saying here that in your realm there is a man who has invented a way to fly and who can cover eighty leagues in a day. This caused Sister Catherine and myself to say that if we could do as much, we could still have the chance of seeing Your Majesty at least once more, even though when flying one cannot see the world. But despairing of any such success, we put all our hope in going, when it shall please God to be merciful, to Paradise.[24]

The aeronautical events in Poland had almost as rapid an effect on imaginative literature. In his *Histoire comique,*[25] written in 1648–49, Cyrano de Bergerac included an open allusion to Burattini. Cyrano had undertaken exciting adventures with a fine imaginary flying machine, the operative part of which consisted of a crystalline regular icosahedron. Describing its subsequent fate, he says that it fell into the hands of the famous Polish engineer, whose reputation was therefore entirely unearned:

> I doubt whether it lost its transparency, for I saw it later in Poland in the same condition as when I entered it for the first time. Now, I knew that it had fallen below the line of the Tropic in the kingdom of Borneo; that a Portuguese merchant had bought it from the islander who found it, and that, after passing from hand to hand, it finally came into the possession of that Polish engineer who is now using it to fly.[26]

While little reliance should be placed on Cyrano's ironic remarks, included in a fictional context, there may here be a faint hint that Burattini undertook some work on the testing of a full-scale machine. Some thirty-five years after the demonstration to Władysław, Johann Joachim Becher offered much more than a hint when he asserted that Burattini had managed to lift himself three times from the ground, though without achieving sustained flight.[27] Whether Becher's account is in any way trustworthy is, however, doubtful. Reporting at second or third hand, he writes in a sardonic tone, describing the dragon as a 'chariot' made of straw.

After the end of the seventeenth century very little more is heard of the flying dragon, and it seems probable that as a result of the upheavals of 1648 Burattini ceased aeronautical work. The general neglect he has suffered among prehistorians of aviation is nevertheless unjust. However inadequate the principles that he applied, Burattini worked with real seriousness and completed a rational design for what he believed would be a fully controllable ornithopter. Although of much lower intellectual and imaginative calibre than Leonardo, Burattini achieved the distinction of building and testing, reputedly with success, the first scale model of a flying machine intended to carry men.

Swedenborg's flying saucer 7

As early as 1714, when he was twenty-six years old, Emanuel Swedenborg (1688–1772) showed an interest in the possibility of building a flying machine. Writing from Rostock on 8 September 1714 to his brother-in-law Erik Benzelius, he ambitiously listed fourteen inventions which he had underway and about which he had prepared manuscript notes. The projected inventions included 'a kind of flying chariot, or the possibility of being sustained in the air and of being carried through it.'[1] Two years later, in 1716, Swedenborg founded Sweden's first scientific journal, *Daedalus hyperboreus,* in the fourth number of which he anonymously published a short article entitled 'Utkast til en *Machine* at flyga i wädret' (Sketch of a Machine for Flying in the Air).[2] Although this article is not illustrated, a rough line drawing (Figure 34) is included in an antecedent manuscript version that also shows a number of important textual variants.[3]

Several short analyses of Swedenborg's article, in both versions, have appeared in print.[4] Inaccuracies, sometimes serious, disfigure all published transcriptions and translations of the manuscript that is in many places rough, hastily written, and difficult to read. My commentary is based on a new study of the original sources, including a photocopy of the manuscript that is in the Stifts- och Landesbibliotek, Linköping, Sweden.

Although it is little more than a suggestion for a practical configuration, Swedenborg's ornithopter is rationally conceived, uses sound design principles, and is based on a properly quantitative approach. He begins with the cockpit or car, a simple rectangular box 4 feet long, 6 feet wide, and 2 feet deep.[5] The car is made wider than it is long so as to give the aeronaut sufficient space to manipulate the wings. Saying that the box should be as light as possible, Swedenborg specifies leather, cork, or, for preference, the birch bark commonly used in Scandinavia for making lightweight containers.

The box is placed in the centre of a single large wing whose area is to be about 600 ft^2. Believing that an oval shape would be best, Swedenborg draws his flying machine accordingly, but other shapes, he says, will serve. The appropriate dimensions are 32 ft × 24 ft for an oval, a diameter of 28 ft for a circle, 25 ft^2 for a square, and 30 ft × 20 ft for a rectangle. The main supports for the wing are four stout spars arranged in two parallel pairs at

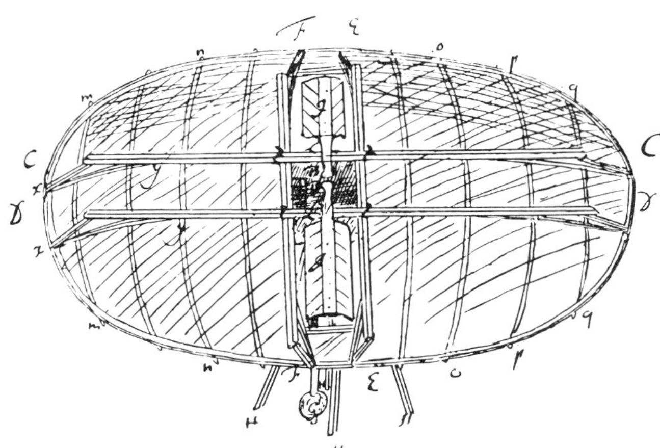

34. Swedenborg's flying machine. Linköping, Stifts- och Landesbibliotek, codex 14a. 1714. Lightly retouched. (Reproduced by permission)

right angles, with the box placed under the rectangle formed by their crossing. A thin wooden rim is bent into shape and fitted around the perimeter of the wing, after which eight wooden ribs are fixed equidistant from each other and running across the wing parallel to the shorter pair of spars.

The whole wing is built so as to be concave as seen from below. Swedenborg says that the four main spars should be bent downward at each end, but in fact in his drawing he arranges the curve by means of wooden brackets (Figures 34, 36, 37). The eight ribs are bent downwards to conform. Underneath this structure is attached a piece of strong sailcloth that must be checked to ensure that it has no rifts or cracks through which the air could pass. On either side of the box a gap is left in the covering so that the flappers can be moved up and down. When completed, the whole looks like a giant inverted flying saucer.

Fixed under the machine, presumably at the corners of the box, are four poles or legs that serve as an undercarriage. Mentioning the idea almost as an afterthought, Swedenborg says that 'it would do no harm if wheels were attached under them.'

At this stage the machine is placed between two horizontal bars so that its centre of gravity can be exactly determined. Under the centre of gravity, projecting vertically downwards, a beam or steelyard is attached, with a weight at the bottom. Serving as a pendulum to ensure lateral stability, the beam is fixed to a narrow spar attached across the bottom of the box. Applying principles of mechanical advantage, Swedenborg calculates that if the beam is made nearly 8 ft long, a weight of 1 *lispund* (= 18.75 lb) will suffice to keep the machine flying level.

The flapping wings are rectangular, measuring about 5 ft × 1.5 ft each. Ribs placed symmetrically across a central spar are bent downward to produce the same curvature as is given to the fixed wing. The frame is covered with sailcloth fixed in such a way as to make flap valves: the cover opens on the upstroke, to allow air to pass through, and closes on the downstroke. Although he could not have known about them, similar flap valves had been proposed more than two hundred years earlier by Leonardo.[6] In the printed version of his paper, Swedenborg suggests an alternative and more complex structure that would cause the wings to fold on the upstroke and open on the downstroke, as Burattini had done with his flying dragon of 1647–48.

Although Swedenborg appears to envisage a simple up-and-down movement of the flapping wings, he suggests that they should 'lie obliquely backwards like the wings of a windmill,' that is, that they should be held at a positive angle of attack so as to produce forward motion. The aeronaut flaps the wings by holding handles fashioned at the inboard ends of the central spars. The spars are hinged, like oars, at the sides of the box, the structure being so designed that the hinges are placed at the centres of gravity. To assist the flapping motion, Swedenborg designs a spring to be attached at the hinge (Figure 35). When the wing is raised, the coiled part of the spring is wound up, storing energy that is released for the downstroke.

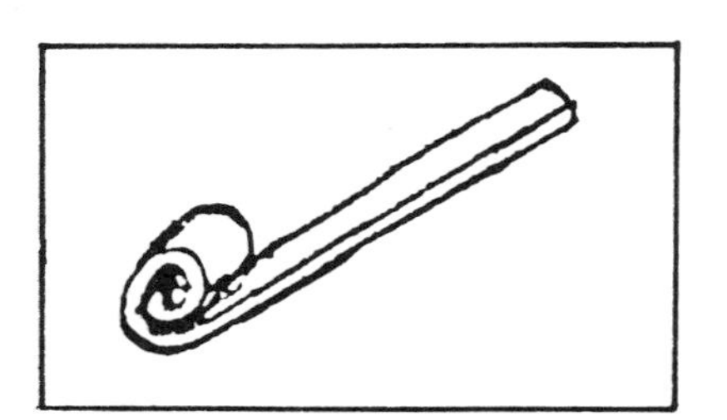

35. Coiled spring used to help power the flappers on Swedenborg's flying machine. Linköping, Stifts- och Landesbibliotek, codex 14a. 1714. (Reproduced by permission)

Apart from the sides of the pilot's box, no vertical surfaces are included in the original design. In a brief note Swedenborg nevertheless says that in due course one should observe whether an additional sail, set perpendicularly to the mainplane, is necessary in order to control the direction of flight. To begin with, however, he is prepared to rely on shifts of the centre of gravity: 'The pilot himself will determine the flight by swaying his body so as to direct the machine downwards, upwards, or to the side.'

The flying machine should be launched from a height and should be flown only when a strong wind is blowing since otherwise it will not be possible to raise it from the ground. By making five appeals to experience and one to authority, Swedenborg attempts to prove that such a substantial structure can be got to fly. He cites: (1) the soaring flight of large birds; (2) the stable flight of heavy paper kites, even in calm weather; (3) the assurances of great scholars such as Athanasius Kircher, despite their having achieved nothing in practice; (4) the astonishing power of the wind, which can blow a door open even when two men are pushing against it; (5) the experience of a student who was buoyed up by his cape when he fell from the Skara church tower in a strong wind and landed on the ground unhurt[7]; (6) the fact that while efforts have to be made to keep a kite flying near the ground, it needs less attention the higher it rises. Despite his many expressions of confidence, Swedenborg is aware of the likelihood that arms and legs may be broken in the initial trials. Accordingly he makes the prudent suggestion that the machine should first be tried by pushing it from a roof after loading it with ballast.

Among the textual variants of the printed version are several significant new ideas. First is a suggestion that the mainsail might be built in such a way as to allow the wing to be reefed like the sails of ships. Swedenborg does not say what purpose might be served by this capacity to vary the tension. More important is a proposal for improving the lifting power by emulating insects. Referring to the hard outer wings of various beetles, he suggests that 'if there were a hood or covering over the wings, as some insects have over theirs, the wing would have greater strength to increase the resistance and make the flapping easier.' As the hard outer wings of beetles do indeed appear to have some lifting function, Swedenborg's idea has a genuine aerodynamic basis. It is in any case of special interest in view of the rarity of an appeal by early designers of flying machines to the structure of insects. One further new

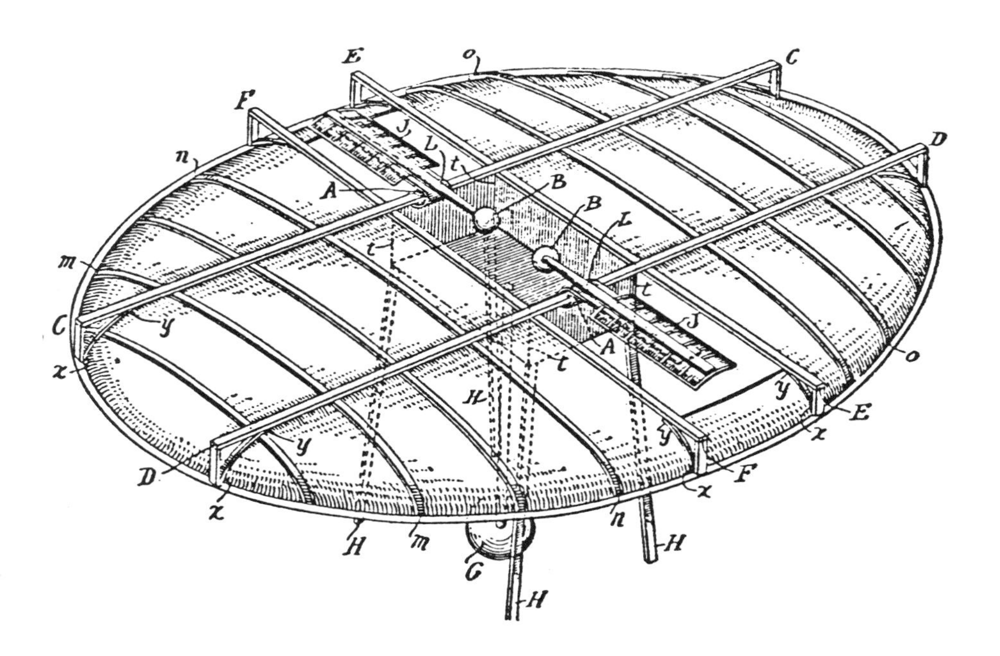

36. Redrawing by Gustav Genzlinger of Swedenborg's sketch for his flying machine, in Alfred Acton, *The Mechanical Inventions of Emanuel Swedenborg* (Philadelphia 1939) 23. Although generally accurate, it misrepresents Swedenborg's ideas in three respects: first, the horizontal dimensions of the box (two ells and three ells) have been reversed, with a consequential change in the positions of the main spars; second, the wheels are not shown on the undercarriage; third, the springs for the flappers are shown inverted. Despite the additional mechanical awkwardness, it is clear that Swedenborg intended the springs to be attached with the coil above the leaf. Only then can they be 'wound up' on the upstroke. (Courtesy of the Swedenborg Scientific Association)

proposal in the printed version is that the machine might be launched by a helper pulling it along with a rope, 'which would achieve as much as strong flapping of the wings.'

Although as a young man Swedenborg read widely in the physical sciences and aspired to be thoroughly up to date, his flying machine owes little to earlier speculators about aviation. The only aeronautical designs offered by Athanasius Kircher are amusing dragon-shaped kites having no bearing on Swedenborg's sketch.[8] In the printed text he quotes his collaborator, the great Swedish inventor Christopher Polhem (1661–1751) who, although he had himself undertaken experiments with the air, has only general cautionary remarks to offer. While the direct up-and-down stroke of the wings may owe something to Borelli's famous discussion of bird flight, published in 1680, the symmetrical shape of Swedenborg's flappers is fundamentally different from Borelli's conception of a wing that flexes about a rigid leading edge. The concave aerofoil section of the mainplane and flappers might derive from intuition, observation, or any of a number of published discussions of bird flight, including Borelli. Pendulous stability had been discussed in 1618 by Hieronymus Fabricius, but Swedenborg's idea of using a long steelyard is probably an independent attempt to apply some of the principles of mechanics that had been so rapidly developing in the seventeenth century. Apart from Leonardo's work, of which Swedenborg knew nothing, the system of control by shifts of the centre of gravity is new in the discussion of man-powered aircraft, though of course the pilot is left far too little room in which to move. Like many others before modern times, Swedenborg assumes that the pilot will be able to extract energy from the wind and therefore that the weakness of his muscles will not prevent him from flying.

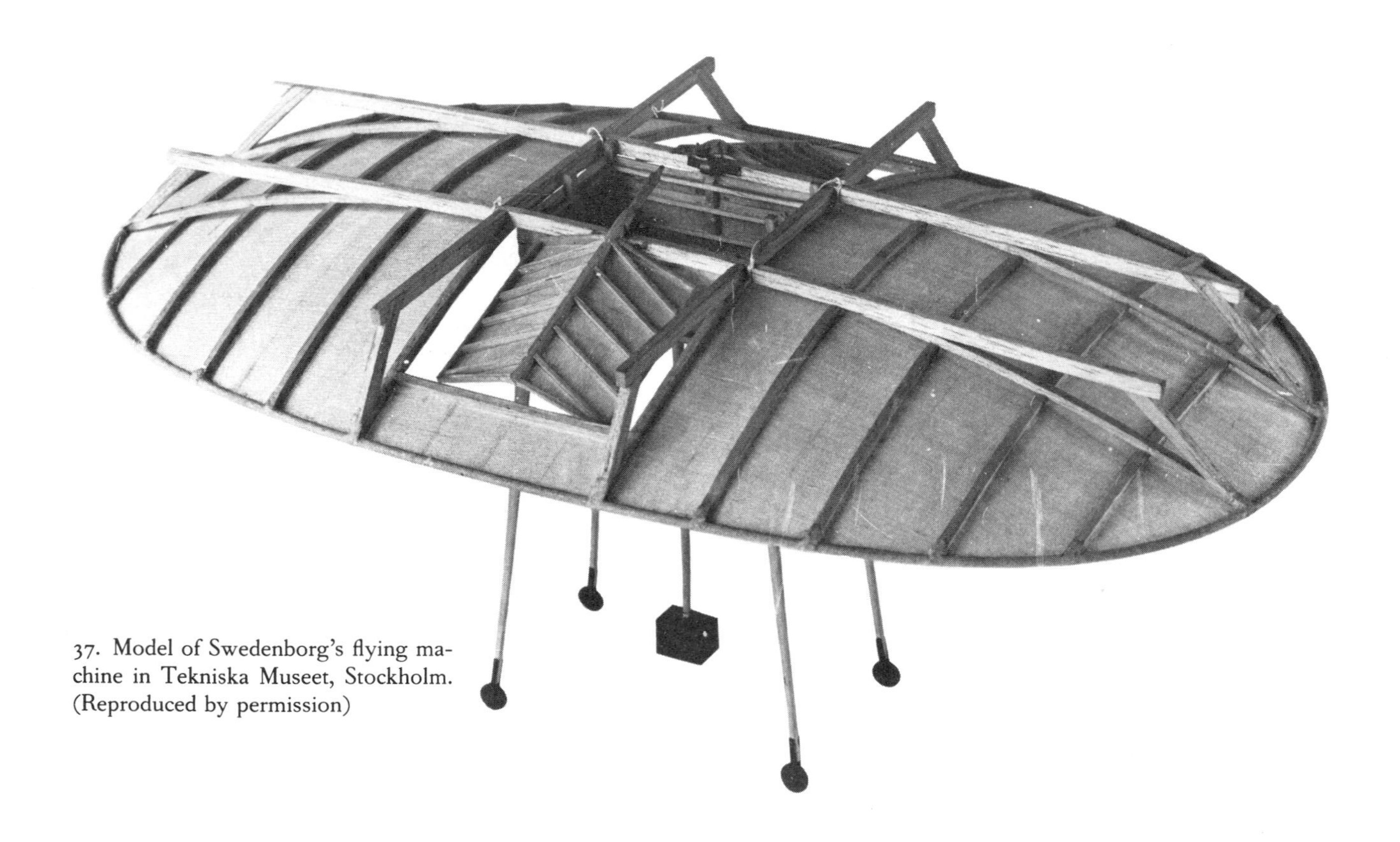

37. Model of Swedenborg's flying machine in Tekniska Museet, Stockholm. (Reproduced by permission)

The product of his youthful secular imagination, Swedenborg's design for a flying machine antedates by several decades the spiritual works for which he is now famous. While he has nothing to say about the possible uses or value of his invention, one may perhaps guess at the general tenor of his attitudes, at least in potential, by reference to his later interpretations of Biblical images and symbols of flight. These attitudes, expressed in several books written in old age, are remarkably consistent. They are, however, by no means simple but reflect the ambivalence towards birds, wings, and flight so often found in the words of mediaeval and Renaissance commentators on the Creation. Throughout his explications Swedenborg identifies birds with things of the intellect and reason. Wings signify spiritual truth, power, and protection, while flight in general is a symbol of perception and circumspection. The following passages are typical:

> Those who are in evil and at the same time in truths may be compared to eagles that soar on high, but drop down when deprived of their wings; for so do those men after death when they have become spirits, who have understood truths, have spoken about them and have taught them, and yet

have had no regard to God in their life. By means of the things of their own understanding they raise themselves on high, and sometimes enter the heavens and feign themselves angels of light; but when they are deprived of their truths and cast out, they fall down to hell. Moreover, eagles signify men of a predatory nature who are endowed with intellectual sight, and wings signify spiritual truths. It was said that such were those men who had no regard to God in their life. By having regard to God in the life is meant considering that this or that evil is a sin against God and therefore not doing it.[9]

Powers are signified by the 'wings,' because by means of them [birds] raise themselves upwards, and wings with birds are in place of arms with a man, and by 'arms' powers are signified. . . . By 'flying' is signified to perceive and instruct, and in the highest sense to look out and provide.[10]

By 'wings' is signified power and protection . . . by an 'eagle' is signified intellectual sight and thought therefrom . . . by 'to fly' is signified to look into and around.[11]

By 'unclean spirit' and 'unclean bird' are signified all the evils that are of will and thence of deed and all the untruths that are of thought and thence of deliberation; and because these are in the hells with them it is therefore signified that they are diabolical; and because they are turned away from the Lord to their own selves the 'bird' is termed 'hateful'. . . . That 'birds' signify such things as are of the understanding and thought and thence of deliberation in both a good and a bad sense is plain from the Word.[12]

Swedenborg also considers the 'birds in the spiritual world,' saying that

Birds of every genus and species also appear there, the most beautiful ones in heaven, birds of paradise, turtle-doves and doves. In hell dragons, screech-owls, horned owls and others like them. All of these are life-like representations of thoughts derived from good affections in heaven, and of thoughts derived from evil affections in hell.[13]

Surrounding his discussions of bird imagery with references to lust, he implicitly associates the evil aspects of flight with uncontrolled sexuality, which he often acknowledged to have been his own major temptation to sin. In his later years he would clearly have had serious reservations about the desirability of 'being sustained in the air and of being carried through it.'

The ornithopters of Grimaldi, Morris, and Desforges 8

Some interest was aroused in the middle of the eighteenth century when an Italian visitor to London exhibited a colourful flying machine that he promised to demonstrate. A full but comparatively little-known report was published by the triweekly *Whitehall Evening-Post* in the issue dated 3–5 October 1751.[1] Soon after its appearance, another Italian plagiarised it, sending a translation to a friend in Italy in the form of a letter, dated 18 October 1751, which purports to be an eyewitness account. This may have been published in Venice in the same year, as a fly sheet.[2] In 1752 the letter and the flying machine were mentioned in a collection of noteworthy events, and in 1753 it was given cursory attention by Clemente Baroni Cavalcabò, in his famous book about the inability of demons to carry men through the air.[3] The celebrated art historian Francesco Milizia included a brief summary in his biography of Paolo Guidotti, the painter and would-be aeronaut, and more recently the plagiarised text has been noticed by Italian historians of aviation.[4] As the English original is not readily available, I give it here in full:

> There is lately arrived in Town from the East-Indies, but last from Lisbon, a Man of the most surprizing unaccountable Genius, that has appeared in the World for these many Ages past; he says he is an Italian, and a Native of Civita Vecchia, named Signore Andero Grimale Volante, aged about Fifty, of a middle Stature, in Holy Orders of the College of Jesus, and went abroad twenty Years since, to travel in the Eastern Nations, by Order of the Father Provincial of the Propaganda Fide. This wonderful Man, after fourteen Years great Labour and Expence, has compleated one of the most astonishing and compleatest Pieces of Mechanism, the World ever yet beheld. It is a Case of a most curious Texture and Workmanship, which, by the Help of Clockwork, is made to mount in the Air, and to proceed with that Rapidity of Force and Swiftness, as to be able to travel at the Rate of seven Leagues an Hour. It is in Shape of a great Bird, the Extent of whose Wings, from Tip to Tip, is 22 Feet; the Body is composed of Pieces of Cork, curiously held together by Joints of Wire, covered with Vellum and Feathers; the Wings of Catgutt and Whalebone Springs, and covered with

the same, and folds up in three Joints each. In the Body of the Machine is contained thirty Wheels, of peculiar Make, with two Rollers, or Barrels of Brass, and small Chains, which alternately wind off from each other a counterpoize Weight, and by the Help of six Brass Tubes, that slide in Grooves, with Partitions in them, and loaded with a certain Quantity of Quicksilver, the Machine is, by Help of the Artist, kept in due Equilibrium and Balance; and by the Friction of a Steel Wheel, properly tempered, and a large surprizing Magnet, the whole is kept in a regular progressive Motion, unless the Temperature of Winds and Weather prevents, for he can no more fly in a Calm than he can in a Storm. This wonderful Machine is guided and directed by a Tail of seven Feet long, which is fastened by Leather Straps to his Knees and Ancles; and by expanding his Legs, either to the Right or Left, he moves the Whole which Way he chuses: The Head is also beautifully formed, and represents that of an Eagle's. The upper and lower Bill is made of a curious Arabian Goat's Horn, transparent, and the Eyes of Glass, as natural as the Life, and turns upon an Axis inward, by the Help of two Wires, fastened to the lower Beak, which keeps all three in perpetual Motion, as long as the Machine flies, (which is but three Hours) and then the Wings gradually close, and he of course lights gently on his Feet, when he winds up the Clockwork, and sets himself again on the Wing. But should any of his Springs or Wheels give Way, he must inevitably fall to the Earth, like a Mill-stone out of the Clouds, for which Reason he never soars above the Height of a common Tree; nor has but once adventured to cross the Seas, which was from Calais to Dover, and the same Morning arrived in London, to which Metropolis he was out of Curiosity drawn, by the Fame of some of our learned and curious Workmen in Mechanics, who at this Day seem to vie, and even out-do, any of the known World for Invention, Beauty, and Elegance; and he has already made his Application to two of the most Eminent in that Science, (who have seen him perform) and have promised to accomplish him by Christmas an entire new Set of Wheels, finished in a more accurate Manner, and not so liable to Accidents; and to be contained in half the Space, with this additional Difference, that it shall move much swifter, and continue for the Space of six Hours, at the Rate of thirty Miles an Hour, without winding up. The delightful Choice of Features that adorn this Bird, as I may call it, surpass Belief or Imagination, and much more the Skill of the most eminent Painter, to imitate the beautiful Diversity of Colours and Shades there represented in the most lively Manner; the Colours consist of Azure, Gold, Scarlets, Greens, Browns, Blues, and White, ranged in such beautiful Form and Order, that the like was never seen. He has made one Tour from Hyde-Park to Windsor-Lodge, and back again, in less than two Hours, and proposes on his Majesty's Birth-Day, to set off from the Top of the Monument, at Nine o'Clock, if Wind and Weather permit, and to make a Tour of the whole City and Suburbs, and settle in Hyde Park, about Eleven. He at present is lodged privately at the House of an eminent Jew in Duke's-Place. When his Machine is newly compleated, he proposes to teach any Gentleman the Art and Use of it in a Month's Time, for fifty Guineas, provided they do not live at above the Distance of a hundred and fifty Miles from London.

It seems clear that the name of the inventor has been corrupted. Although it can be a proper name, 'Volante' in this context is merely the epithet 'flying,' while 'Grimale' should probably be read as either 'Grimaldi' or 'Grimaldo,' the form in which it is given in the Italian letter. Even that may well be false, the inventor having perhaps sought to impress by assuming the name of a famous and noble Genovese family. In an attempt to discover whether any Jesuit called Grimaldi could have been in London in 1751, Venturini consulted the Vatican Archives, the records of the Propaganda Fide, and the archives of the Society of Jesus. In none of these is there any mention of a relevant Grimaldi.[5] The tone of the passage suggests, in any case, an adventurer rather than a priest, and one may well doubt whether he was even Italian.

While Grimaldi himself may not have believed his machine capable of flight, the structure, as described, is of great interest. The use of whalebone, the folding wings, and the system of wheels and pulleys in the body all suggest a possible derivation from Burattini's flying dragon, which had been quite famous in the preceding century (see Chapter 6). Grimaldi could well have heard of it, especially if he was truly Italian. The use of counterpoise weights on brass rollers shows some degree of engineering skill and recalls the application of similar devices in Renaissance handbooks of mechanics. (Leonardo had used rollers in some of his sketches for flying machines, but as the manuscripts were not freely available it is highly unlikely that Grimaldi could have known about them.) While the operation of the quicksilver balance weights is far from clear, their use reveals a proper concern for the adjustment of the centre of gravity. The movable tail, which has been conceived on an appropriate scale, is similar to the one used on Carl Meerwein's ornithopter a generation later (see Chapter 10). Only the friction wheel (perhaps intended to generate static electricity) and the magnet (a favourite with showmen for mystifying audiences) seem to have no real place in the design of a flying machine. The impossible clockwork motor may owe its presence to the combined influence of popular legends and a number of published suggestions for the application of clockwork to flying models.[6]

Grimaldi's advertised demonstration flight from the top of the Monument was to have taken place on George II's sixty-ninth birthday, Wednesday 30 October 1751 (Old Style). By the time the day arrived, Grimaldi seems to have disappeared. No mention of him or his machine is found in *The Whitehall Evening-Post* during the period immediately following 30 October, nor have I come across any other contemporary British reports.

Grimaldi's exploits in London were probably inspired at least in part by a remarkable outburst of imaginative writing about flight that appeared in the same year, 1751. Richard Owen Cambridge published his *Scribleriad,*[7] which contains a passage describing an aerial battle between a Briton and a German. The well-known engraving illustrating their flight (Figure 38) shows the men using artificial flappers, one pair of which is based on the apparatus used by Besnier in 1678.[8] In a more fanciful book, *The Life and Adventures of Peter Wilkins,* Robert Paltock describes and illustrates a strange race of flying men and women, the Glums and Gawries.[9] In some respects the most interesting

38. Imaginary aerial battle, from Richard Owen Cambridge, *The Scribleriad* (London 1751) frontispiece to Book IV. The flappers on the right are based on those of Besnier (see Fig. 27).

of all the flying books published in this year is *A Narrative of the Life and Astonishing Adventures of John Daniel, a Smith at Royston in Hertfortshire,* attributed to 'Mr. Ralph Morris.' A popular book, it was reprinted in 1770 and 1801, excerpted in 1848, and reissued in 1926 as Volume I of the *Library of Impostors.*[10] Cast away on a desert island, Daniel discovers that his only companion, who at first masquerades as a man, is in fact a woman. They marry, have children, pair the children off in marriage and create a whole new civilisation. Only their son Jacob remains unmatched. In lieu of the exertions of paternity, Jacob devotes some energy to the invention of a flying machine that, in an inversion of the Daedalus-Icarus story, he demonstrates to his nervous father. With the help of the machine they inadvertently fly to the moon.

39. Morris's flying machine. *A Narrative of the Life and Astonishing Adventures of John Daniel* (London 1751) opposite p. 178. (Reproduced by permission of the British Library)

Jacob's flying apparatus, illustrated and described in detail in the text, is a flapper of generally rectangular shape operated by the vigorous working of a handle like that of a water pump (Figure 39). Although it is included in the context of a fictional narrative in the familiar eighteenth century mode of the fantastic voyage, with an incidental rewriting of Genesis, the author seems at least half intent on making a serious proposal for a real flying machine, the structure and function of which are fully explained. It is built mostly of iron, 'thin, light, and taper,' and of 'several pieces of wood work too . . . one somewhat like a pump.' To begin the construction, Jacob

> first of all struck four poles into the earth at proper distances, measuring them with four bars, in the ends of the two longest of which, on the flat sides, were four holes, into which the four points of the upright poles were to enter, at about three feet high from the ground. (p. 178)

On this frame is built up a rectangular grating which is covered with wood, so as to make a 'stage or floor.' A wooden trap door about a foot wide is included along one edge of the rectangle to make access possible from underneath. A pipe, about 4 inches in diameter, like the casing of a pump, is fixed into a hole in the centre and held in position by four bracing bars. Through the pipe passes a bar that can be worked up and down by a double-sided pump handle. Around the edges of the platform, at regular intervals, are fixed tapering bars of iron, 3 inches wide at the platform, pointed at the other end, and about three yards long. A 'female screw' (presumably a ring bolt) is fitted on the underside of each of the bars, about 3 feet from the edge of the platform. The whole of the parasollike extension so created is then covered with calico dipped in wax, thus making 'an horizontal superficies of callicoe (including the floor) of about eight yards diameter, but somewhat longer than broad' (p. 155).

Under the centre of the floor is placed an iron ring, 5 feet in diameter, attached to the wood by upright iron members 1 foot long. A set of straight iron bars, equal in number to the ribs, are arranged radially under the ring, parallel to the ribs and hinged to the ring where they pass across it. The bars are of such a length as to end under the ring bolt fixed to the ribs. The inner ends of the bars are attached by chains to the 'sucker iron' that passes down the pipe, while the other ends are attached by chains to the ring bolts. The ribs may thus be pulled down by operating the pump handle so as to raise the 'sucker iron.' The 'upstroke' is automatic as the ribs spring back into shape.

No undercarriage is provided, and since the lever arms attached to the ring need room for movement, it is necessary either to fly the 'eagle' (as Jacob calls it) from the four upright posts on which it is assembled, or from makeshift supports arranged after landing. Although he describes the machine as capable of easy, rapid flight, Morris appears to be aware of the many published caveats about the inadequacy of a man's muscle power and accordingly provides Jacob and his father with a mysterious herb that gives them the necessary strength. Directional control is achieved by variations in the centre of gravity, though Morris's notions of cause and effect seem here a little strange:

> There being a handle on each side the pipe or pump, he could make it go which way he would, by altering his own standing, as he told me, either on the one side or the other of the pump; for the side he stood on being the heaviest, and the other consequently mounting rather the highest; it would always move that way, which end was the highest. (pp. 181–82)

Morris's flying machine was an ornithopter of a rather special kind. As the rectangular or oval wing flaps downwards on all sides simultaneously, the lifting function is presumably conceived of along the lines of the jet propulsion of a jellyfish. The structure is ingenious and original, and the basic idea by no means ridiculous. With a total lifting surface of about 500 ft^2, the wing is conceived on a scale appropriate to the task of raising two men. Although the machine is radically unstable and aerodynamically unsound, Morris's idea that it might be controlled by shifts of the high centre of gravity is entirely rational. It is therefore unfortunate for Morris's reputation that his sober invention should have been misrepresented when, in 1769, it was parodied by the anonymous writer of a satirical attack on Dr Samuel Musgrave (1732–80), who had made serious allegations about the conduct of diplomatic relations

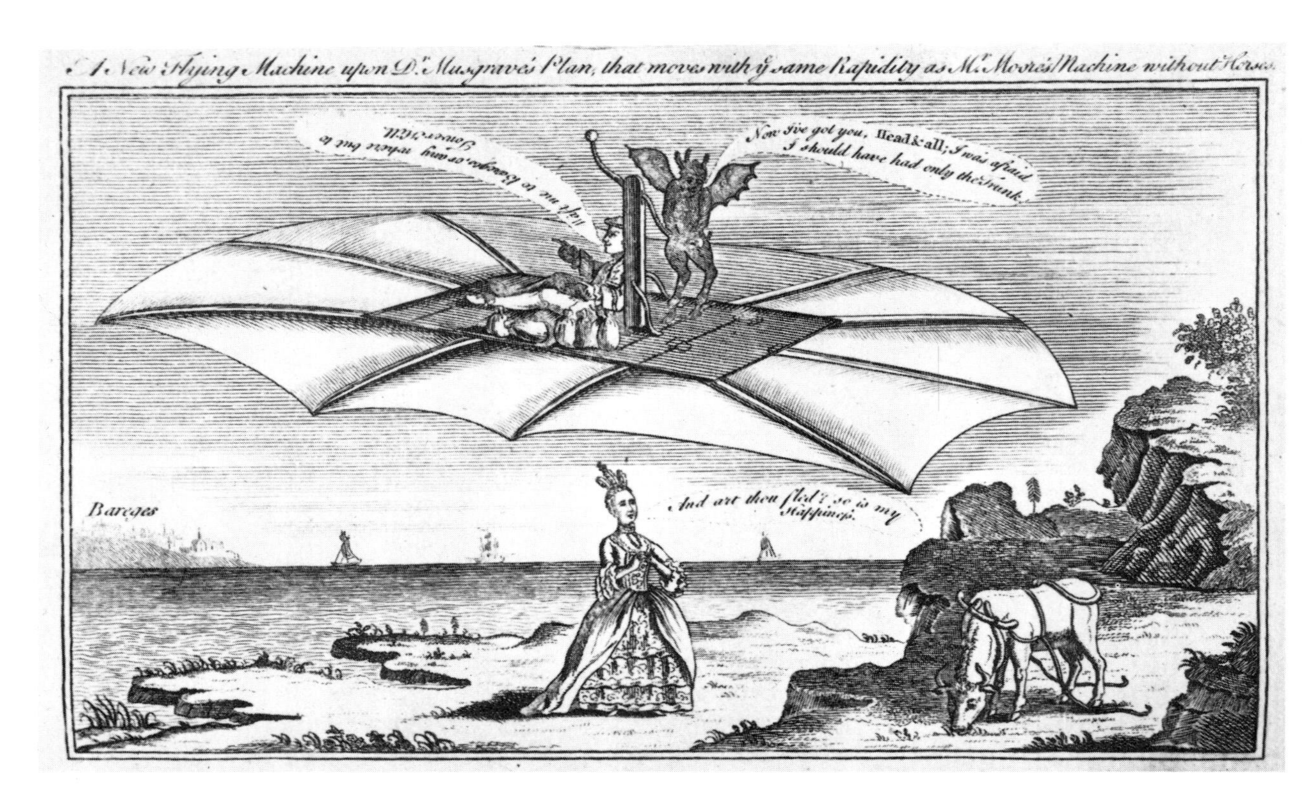

40. Satirical version of Morris's flying machine. *The Oxford Magazine* 3.3 (September 1769) opposite p. 108.

between France and England. The lampooning version of Morris's 'eagle' (Figure 40) shows the devil in charge as sacks of gold are exported across the Channel. The misrepresentations are not in the engraving, which is carefully copied, but in the text. Morris's practical waxed calico covering is replaced by a symbolic mixture of 'the feathers of a Phoenix, intermixed with those of the Bird of Paradise; those of the latter are used in a sextuple ratio of those of the other to make them float upon the air the better, because that bird having no feet, must always be suspended in the air.'[11] More serious, however, is the mistaken account of the action of the pump handle. Evidently having failed to read Morris's text, the satirist takes the machine to be kept flying by the action of a hand-driven jet engine, air being vigorously pumped from above to below:

> The pump in the center operates upon the principle of hydraulics, and by discharging the air above the superficies of the machine, prevents its descent, in the same manner as other solids remain in equilibrio, or suspended in fluids, whose specific gravity is the same.[12]

This fundamental error about the function of Morris's machine has been frequently reiterated by later commentators who have seen the Musgrave lampoon but who have not had access to the imaginative and carefully considered design proposals in the novel.

As far as I know, neither Grimaldi nor Morris ever entrusted himself to the air. Twenty-one years later, however, the wits and letter writers of Paris were amused by the practical, if also absurd, flying experiments of a genuine clergyman, the Abbé Pierre Desforges, canon of the church of Sainte-Croix at Etampes, south of Paris.[13] Born about 1723, Desforges had already drawn unwelcome attention to himself by writing a two-volume treatise that attempted to prove that Catholic priests and bishops had a duty to marry: *Avantages du mariage, et combien il est nécessaire & salutaire aux prêtres & aux évêques de ce tems-ci d'épouser une fille chrétienne.* Published in Brussels in 1758, the treatise was condemned to be burned on 3 October of the same year after 2,000 copies had been confiscated from Desforges's lodgings. The book nevertheless excited sufficient interest to be republished in 1760 and 1768, and an Italian translation followed in 1770. Desforges's heresy earned him a spell in the Bastille (26 September 1758–9 May 1759),[14] where, according to the notorious and malicious gossip Baron von Grimm, he profited from his incarceration by studying the mating habits of swallows, about which he wrote a poem. Although this new literary effort was not found heretical, he was forbidden to publish it because of its many follies and lubricious details, and he was threatened, should he persist in his aberrant behaviour, with further and permanent imprisonment.

Following these skirmishes with matters of morality, Desforges turned to the less controversial field of mechanics. About 1770, still pursuing his interest in flight, he made a pair of wings that he was nevertheless prudent enough not to try himself. Instead, he fixed them to the arms of a peasant, whom he clad from head to toe in feathers. Having led the peasant to the top of a belfry, Desforges ordered him to throw himself boldly into the air. When the peasant firmly refused, Desforges abandoned the attempt and set about planning a different experiment. In a number of provincial newspapers he announced that he would make a flying machine that would enable one to 'rise into the air from a deep valley and fly as one wishes, to right or left or straight ahead, without the least danger, and so easily that one can cover more than a hundred leagues at a time without fatigue.'[15] It would be so simple to manipulate that even young ladies could fly at their ease. Like many an obsessed inventor before and since, Desforges sought financial support from the public. Anyone interested in having a flying machine was invited to deposit 100,000 livres with a notary, for payment to Desforges after the invention had been tested and proven in the presence of the backer. As soon as the money was deposited, Desforges would undertake to build the machine, at his own expense, in a maximum of six weeks.

Desforges's announcement immediately aroused interest. Three weeks after his proposal had been published, an anonymous letter appeared in the *Annonces* offering slightly ironic moral support:

> It would be of real benefit to the State and to the Public if it were possible to establish an aerial postal service and have deliveries made in an hour that normally take a week. The packages in the charge of the flying courier for delivery to the towns on his route will not stop for an instant. He would

> only need to place himself perpendicularly over the town to which the package is to be delivered, and let it fall into the main square, prepared by the inhabitants to receive it without accident. The State would gain immeasurably from the suppression of post horses. In addition, by means of your machine . . . the Sovereign would immediately know what was happening in the interior and at the ends of his Realm. (pp. 161–62)

The writer goes on to make interesting suggestions about navigation:

> It remains to me to ask you whether the compass and map will be of use to a man travelling by your form of transport. I believe it possible. Each town can attach one or more flags to the highest belfry, and this will form the daytime signal. But at night one or more lights, according to the tariff, will indicate the spot over which one passes like a flash, or where one may descend like a bomb, and from which one may rise like a rocket. (p. 162)

Finally, he adopts a satirical tone:

> One other difficulty which I ask you to resolve, and I fear it will not be the last, is whether a hunchback, who, because of his shape, is incapable of making certain efforts, could handle the flying chariot with as much assurance as any other man. I confess to you that I am tired of falling, and, having deformed myself by stealing birds' nests, I should not want to risk being crushed by flying through their empire. (p. 162)

It is not clear exactly when Desforges began to formulate the proposals that stimulated this and other, more pointed sarcasm. Although the call for subsidy first appeared, as far as I know, on 4 September 1772, in the *Annonces, affiches, nouvelles et avis divers de l'Orléanois,* the editorial offices of which were only about 60 kilometres from Desforges's home in Etampes, there seem to have been significant delays between writing and publication.[16] On 2 October 1772 (a week after the anonymous letter) the *Annonces* published another article by Desforges which showed that he had begun to think his appeal for money unlikely to succeed. The article is in the form of an open letter to 'a provincial lady who had asked him for information about his flying carriage.' Since no one seemed willing to put up the sum that he had stipulated, Desforges announced that he had decided instead to give a demonstration:

> I shall leave from Etampes for Paris, but without landing there, for fear of being held back by the crowds. But after having flown five or six times around the Tuileries, with the same uninterrupted flight, I shall return to Etampes. As soon as I have arrived there, I shall burn the machine, and shall not make another until I have been recompensed for my trouble. (p. 165)

By the time this article appeared, however, at least two and a half weeks had passed since Desforges had completed his machine and attempted to fly in public, with the ludicrous results described later.

The belated article of 2 October describes how, with the assistance of a basketweaver, Desforges had been hard at work for a week constructing an example of the machine that would, he said, be ready after another twelve days. The gondola, made of osier (costing 40 sous), and sallow wood (costing four livres), was 6 ft long, 3 ft 8 in wide, and 6.5 ft in depth from the 'feet' to the top of a canopy that covered the whole and afforded protection from rain. The canopy was 8 ft long and 6 ft wide. No nails were used anywhere in the construction. If painted with nut oil, the gondola ought, Desforges believed, to last for about eighty years.

Two wings, to be manipulated like oars, were attached by hinges to the sides of the gondola. These could be removed, for ease of storage, when the machine was not in use. The total wingspan was 19.5 ft. Both the wings and the canopy were first to be covered with English waxed taffeta ('the most expensive material'), after which, because Desforges apparently thought that smooth taffeta might slide through the air too easily, feathers should be added to the wings to give some friction. Without that addition, 'one would fly too fast.' In Desforges's estimation, the hinges, the least durable part of the structure, would need to be replaced whenever they were half worn out, which would be every three months, assuming that one flew 300 leagues a day.

Designed to be inherently stable and, like Burattini's dragon, to serve if necessary as a boat, this flying machine weighed 48 lb in all and was large enough to carry a pilot weighing 150 lb. As long journeys were envisaged, Desforges made provision in his calculations for lifting a luggage bag weighing up to 15 lb. The construction was so light that 'if someone fired two cannon balls at it to remove the two wings when it was as much as 200 ft up, it would not fall, but would sink ten times more lightly than a shuttlecock.'[17]

Although he at first envisaged only a single seater, Desforges had plans to make a version of the flying machine capable of carrying a passenger on a seat fixed, to ensure stability, immediately beneath the pilot. This arrangement would be 'rather like an eagle carrying off a lamb in its talons.' If anyone feared that the air might be too thin at high altitudes, he should consider that the forward motion would increase the supply to the lungs. When fast flight was necessary, the pilot would be protected against the air by a large sheet of pasteboard extended across his stomach and by a pointed pasteboard bonnet or flying helmet resembling a bird's head and equipped with glass goggles.

In trying to estimate the performance of his flying machine, Desforges made comparisons with the observed speed and endurance of swallows. These had been timed by Réaumur (1683–1757), who had found them capable of covering 64 leagues in an hour (= approx. 150 mph). Making more modest claims for his gondola, Desforges believed that when the wind was favourable it would be capable of covering 30 leagues an hour, this figure dropping to 24 leagues in time of calm, and to as little as 10 leagues if the wind were adverse. Comparisons that Desforges makes with the efforts of a boatman rowing

41. The Tour Guinette in Etampes, France, from which Canon Pierre Desforges tried to fly in 1772.

upstream show that, for his time, he had a rare capacity to understand the notion of relative velocities. With favourable winds, one might thus hope to cover 36,000 leagues in four months, working for 10 hours each day to travel 300 leagues. Gliding was also possible. Desforges points out that although birds can glide no farther than about 60 paces, his machine has the advantage of a canopy as well as wings, allowing it to cover an estimated one-eighth league (= approximately one-third mile) without flapping. Desforges concludes his open letter, which was, as far as I know, the last thing he wrote about aeronautics, with the proud words: 'And thus it is I who shall have the pleasure of being the first to travel through the regions of the air.'

The machine appears to have been finished during the first half of September 1772. At some time before 15 September, when the Baron von Grimm wrote a letter about the attempt, Desforges found four peasants to help him carry the gondola to the top of the Tour Guinette, not far from his church. The choice of the Tour was entirely in keeping with Desforges's naive flamboyance. Originally built in the twelfth century, it was a remarkable fortress whose dominating position above the town is shown in one of the illuminations of the *Très riches heures* of the Duke of Berry. Under the reign of Henri IV (1553–1610) it was partially demolished at the request of the townsfolk, who rightly blamed it for being the indirect cause of an endless succession of battles. Still a prominent feature in 1772, as it is today, the stump of the Tower, whose name means 'lookout,' rises about 100 ft above the mound on which it was built. The walls are vertical on all sides (Figure 41).

After having boarded the machine and taken control of the wing mechanism, Desforges gave the peasants the signal to drop him over the edge in full view of a large crowd. To everyone's surprise, he escaped from the resulting fall with no more than a bruised elbow. As Grimm wrily observed: 'They will never burn the Canon of Etampes as a sorcerer.'[18] Rather, Grimm said, the idea of the gondola would be likely to lead him straight to the madhouse.

Desforges became famous overnight. In the 1770s his flying attempts were widely publicised, stimulating a great deal of salon conversation and journalistic irony. Responding to a letter about him from Madame d'Epinay, Ferdinand Galiani commented on 24 October 1772: 'Your Canon of Etampes took up too much space in your letter and too little in the air.'[19] Despite his participation in the universal raillery, Galiani nevertheless took the opportunity to suggest that he thought human flight might one day be achieved:

> I believe we might be able to fly in the air if we could discover a spring having an almost infinite force. I believe that a man's wings should be eighty feet wide. A machine weighing as much as a man with, in addition, a man on board, would require one hundred and sixty feet. It is difficult to make a wing stiff and light if it is only half as wide as that; and thus it will be a long time before we fly. I haven't the time this evening to tell you more on this subject.[20]

A comparison with modern man-powered aircraft shows that Galiani's estimates are not far from the truth.

Desforges's attempts were taken at least semiseriously by the *Dictionnaire des origines* (1777), in which he is discussed in the article 'Voler.' As with so much else in that dictionary, the account is rather noncommital, ending with the lame words: 'It is easy to understand how difficult it would be to carry out such a project.'[21] We should nevertheless be careful, I believe, not to treat Desforges too lightly. He was clearly a man of vigorous and original mind who really wanted to fly. He applied a carefully quantitative approach to his work and used figures published by one of the most eminent scientists of the recent past. Had constructional details of his flying machine survived, we might have seen ideas as worthy of respect as those of Burattini, who was treated with similar irony by those ready to scoff at unusual enterprise.

Melchior Bauer's cherub wagon

9

No eighteenth century designer of flying machines surpassed Melchior Bauer in imagination and aerodynamic insight. Born on 19 October 1733 at Lehnitzsch, a village near Altenburg, he was the son of a gardener, Hans Bauer.[1] Little is known of his life, but in 1763 he went to London, where he sought patronage from the recently crowned King George III for the construction of a man-powered aircraft. According to his own account, he got no further than the official scribe who refused to copy out the submission, saying that even if he were offered £500 he would never agree to send such folly to the Court. 'If you could do all that you claim,' said the scribe, 'you would certainly be the foremost and greatest man on earth' (f. 1r).[2]

Shortly afterwards, in July 1763, he tried his fortunes in his native Germany, approaching Frederick the Great (1712–86) in Potsdam. Although he managed this time to send a written account of his invention, he again failed to find a scribe willing to undertake the task and had to draw up the letter himself. It was not a propitious time to ask favours of Frederick, who was busy with the aftermath of the Seven Years' War and with the serious economic crisis of July 1763. Although Frederick's officials normally sent the plans of any promising new invention to the Academy of Science for investigation and assessment,[3] Bauer was not offered this mark of respect and fared even worse in Potsdam than he had done in London. War Counsellor Kiper, who delivered the short text to the King, was instructed to dismiss him rudely. This the Counsellor did with relish, haranguing him at length and itemising his follies:

> 'The fiery fever has turned your head. If you could do that, the King would have you driven around all the days of your life in a golden coach and you would never again need to go on foot unless you wished. For you can imagine, fool, that it would be worth more than a kingdom; for by this means the King could make the whole world subject to him. There have been many cleverer than you in this matter, who have studied and learned more than you, you foolish man; and nevertheless they have failed to bring it to fruition.

'My dear man, are you not now fearful for your sanity? But I pity you from the heart for having fixed such a mad scheme into your head, for to all appearances you are a sensible fellow. If you had not given me your text I would not have believed you to be so great a fool.' And he brought me into ill repute with the King, representing me as an advisor who rakes together all kinds of muck and drags it before his king, and who should be dismissed because of it. And so my hopes were at an end. [f. 1^{v}][4]

Having failed to impress two kings, Bauer tried to find favour nearer to home. For his third and last attempt he appealed to the ruler of his local corner of Thuringia, Count Heinrich XI of Reuss (1722–1800). In a fresh submission, almost certainly written in 1764 (see later), he recounted his earlier disappointments, copied out his unsuccessful letter to Frederick the Great, and now for the first time drew diagrams of his invention. The style of the manuscript is unsophisticated and a little repetitive, but the sense is always clear.

An ardent evangelical Protestant with an obsessive hatred of Catholicism, Bauer had written to Frederick in an exalted tone, appealing to Scriptural authority and recalling especially the winged and wheeled cherubim of Ezekiel, chapters 1 and 10. It is not altogether surprising that Counsellor Kiper should have treated him as a madman:

Most Serene Almighty King—
Most Gracious King and Lord—[f. 1^{v}]

I wish most respectfully to reveal to Your Royal Majesty a hitherto undiscovered invention, as follows:

That which by God Himself is called the Mercyseat, or Cherub Carriage [Exodus 25:20; Hebrews 9:5], the instrument which Moses, the servant of the Lord, and also after him King Solomon, made an omen of the latter day, on which the race of men can soar in the air, like flying animals, which serve us as teachers and models:

This carriage I can now make with my own hands as it should be made, just in the shape in which the holy prophet Ezekiel saw it with his godly eyes and of which he writes in his first and tenth chapters.

The various animals,[5] however, give an indication of the many races that will be associated with this instrument, for just as all mankind now travels over the water on the instrument of Noah, so will it be with this carriage. It is indeed on the Mercyseat that the whole of mankind will alike be presented: on the right the godly-minded, like judges strong as lions, on the left the dull and brutish, with minds like cattle and yet sent to fly like eagles to execute their Lord's command.

The things from which I can make this carriage are fir wood, woven silk, and brass wire; and I can finish it in some three or four months' time. [f. 2^{r}] But I need a house, perhaps built of planks, in which I can make this carriage. The house should stand on a high hill, and should be 16 ells long, 8 ells wide, and 6 ells high; but there must be no pillars in it. When the work is finished it will in all things most clearly resemble what the holy prophet Ezekiel saw and described. [cf. plate XII]

Bauer then proceeds to give a general description of the configuration and structural principles of his flying machine, a parasol-winged monoplane of low aspect ratio. Standing in a short, four-wheeled fuselage fixed below, the pilot provides power by flapping an additional pair of compound wings whose span is about half that of the mainplanes. Unlike most other ornithopter wings, Bauer's flappers are firmly fixed from tip to tip on rigid spars, so that one wing rises while the other descends, the action being much like that with which one handles a double-ended canoe paddle.

As Bauer says, the flying machine is intended to be, as far as possible, a practical realisation of Ezekiel's vision. Not only does he call the fixed wing a 'canopy,' using the German word *Himmel,* but many details of the configuration depend on the Biblical source:

> There is a canopy stretched out over the man's head. This canopy is like two outspread wings of a bird, each 7 ells long and 5 ells wide. But in place of feathers is woven silk or paper sewn to the wood and wire. For the canopy is tightly stretched with wire. And thus the whole carriage can soar into the air with a man in it, for it is driven along through the air with two other movable wings, by the man's hands, who steers it wherever he wishes.
>
> For if one rocks the wings with one's hands, the carriage, which stands on four large but light wheels, will go straight forward. For the carriage frame must have no flexible joints in it, as Ezekiel tells us, for otherwise the carriage would not go straight ahead. [f. 2^v]

That the canopy should stretch out over the man's head probably reflects Ezekiel 1:8: 'And they had the hands of a man under their wings on their four sides.' The four wheels certainly derive from the vision: 'behold the four wheels by the cherubims' (Ezekiel 10:9). The emphasis on straight and level flight echoes Ezekiel's repeated insistence that the cherubim always headed directly towards their goal, without deviating: 'they turned not when they went; they went every one straight forward . . . And they went every one straight forward: whither the spirit was to go, they went; and they turned not when they went' (1:9, 12). The requirement that the cherub wagon have no flexible joints is probably an interpretation of Ezekiel's stress on the tight unity of the cherubim in his vision.

Bauer continues his description with an implicit recall of the noise made by the cherubs' wings: 'And when they went, I heard the noise of their wings, like the noise of great waters, as the voice of the Almighty, the voice of speech, as the noise of an host: when they stood, they let down their wings' (1:24):

> This carriage also makes a great rushing and thundering when the wings are moved. This is because everything is light and drawn tight with the wire, and is made so taut that the wires ring.
>
> And when the man who stands in the carriage flaps the wings strongly, the carriage moves on its [f. 2^v] wheels so quickly that with its outstretched

canopy it cuts its way up into the air and soars as a bird does when it holds its wings steadily spread. For the canopy over the man's head is very smooth, like frozen ice, and rigid like two outstretched bird's wings and soars away lightly into the air. For these two vaned wings are made like grasshopper wings and when they are moved they draw the whole carriage forwards through the air. For because the carriage stands on four large wheels it departs from a smooth hill or from a broad, smooth expanse of ice and flies lightly into the air like an ascending swan, at a sharp angle, carrying all the rest of the structure with it, as the prophet says: 'And their whole body, and their backs, and their hands, and their wings, and the wheels' [Ezekiel 10:12], for when the Cherubim beat their wings and raised themselves from the earth, the wheels did not depart from them; just like the feet of an ascending swan the wheels went with them, higher and higher into the air ['and when the living creatures were lifted up from the earth, the wheels were lifted up,' 1:19].

For it is the counsel of the righteous God that we, mankind, should go in three ways: namely on the earth, on water, and in the air. His word is sufficient witness for us, as also the creatures and animals on the earth. For should silly flies, gnats, and grasshoppers have an eternal advantage over reasonable men, the children of God? For are men not worth as much as ravens, geese, swans, and storks? With God's help, should such things not be as easy for men as to travel over [f. 3^r] the water? For just as God has given us the means of travelling over water, He can also give us the means of travelling in the air, for He is mighty and wise enough for that.

But because I lack not only the money but also the opportunity, I have not yet been able to bring this work to its full completion.

Let it therefore stand to the judgement of Your Royal Majesty's Grace, whether you will be pleased to help me with it.

Melchior Bauer, born in Lehnitzsch,
A village near Altenburg.

Having copied his unsuccessful submission to Frederick the Great, Bauer goes on to stress the practicality and value of his invention before offering the Count the detailed drawings and constructional notes:

Because Your Grace has understood my thought from these words and wishes to see this miracle carriage, I can certainly and accurately complete it with God's help, for I can wager my life upon it that the carriage will withstand any trials which may be prescribed.

Instead of silk, it can be made of paper, so that in itself it would hardly cost 10 talers. The house would therefore be the most expensive item.

Because, according to my most certain knowledge as the Master, it is nothing other than a victory or triumphal carriage over the antichristian Pope who is still ruling with his accomplices, and over the unbelieving, idolatrous heathen, I should therefore prefer, if it were possible, to make your carriage in secret.

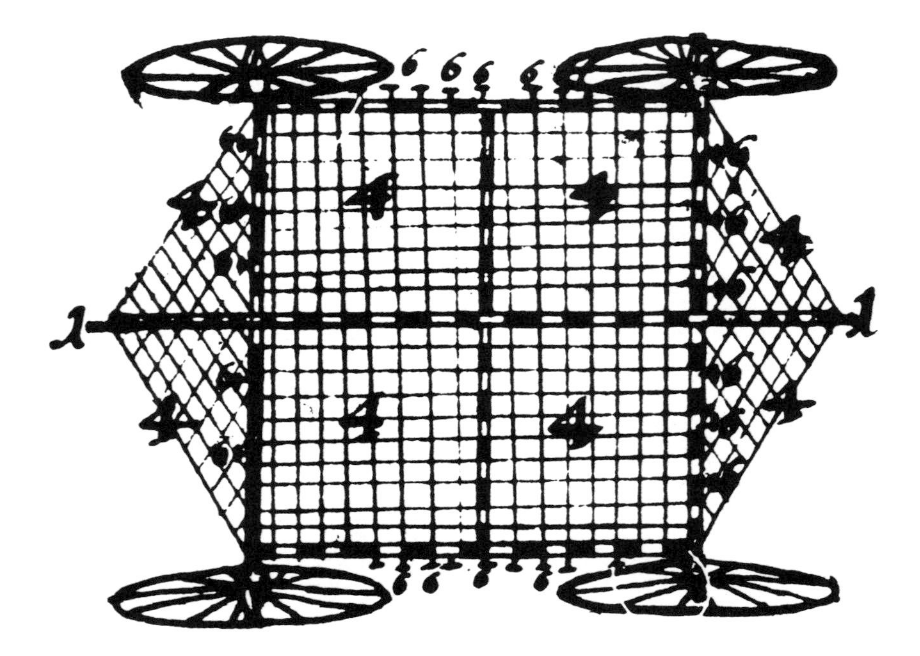

42. Fuselage of Melchior Bauer's cherub wagon, plan view. Wooden members are indicated by the double lines. Wire mesh, drawn with single lines, is made by twisting brass wire around the pegs numbered 6. DDR–Staatsarchiv Weimar, MS a. Rep. A Greiz Rep 41 Nr. 12a, f. 4^{v}. 1764. (Figs. 42–49 reproduced by permission of Staatsarchiv, Weimar, German Democratic Republic)

For we know that the popes and their accomplices promote the devil's teaching: that is, they hinder and [f. 3^{v}] forbid people from reading God's word, and they forbid priests to marry. As a result the heathen are seduced into a loathing of Christ and his teaching, and are led to think that Christ himself must have been just such a teacher as the marriage-banners are now.

For by means of this carriage the word of God and the pure gospel will be kindled and spread throughout the whole world, as God himself promised, among the Jews, the Turks, and the heathen. For as all commentators explain, it is a carriage of victory, as Isaiah (30:17) and Moses (Deuteronomy 32:20) say. One of them will chase away a thousand, and two will make ten thousand fly, and before five all the enemies of the Lord will flee.

Therefore this may with justice be called the greatest of all the arts, since by its means God will destroy and overthrow the kingdom of Antichrist and will help mankind to reach complete salvation and righteousness.

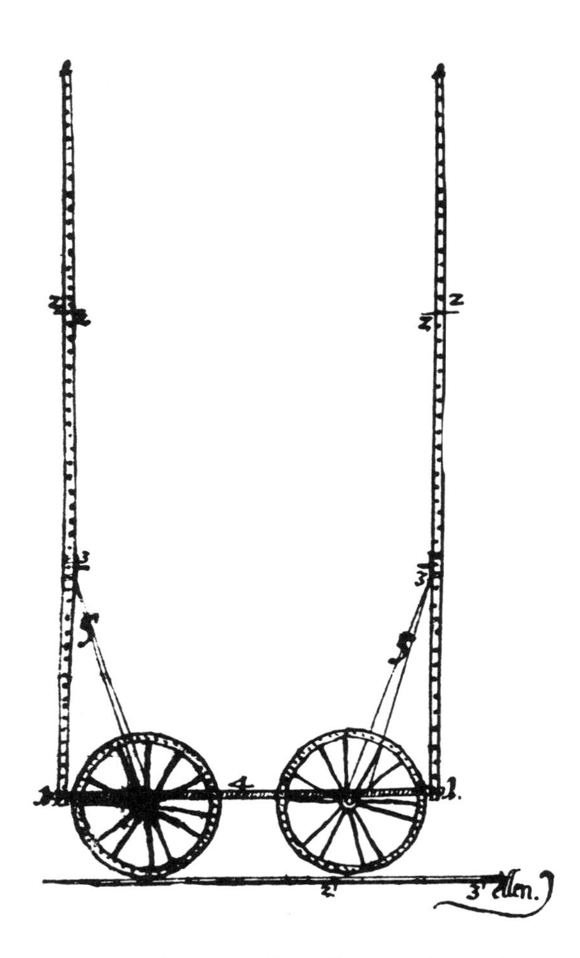

43. Side elevation of the cherub wagon, showing 'the canopy pillars' and the stay wires, numbered 5. F. 4ᵛ.

And so since Your Grace wishes to be a counsellor and a patron in this great matter, there can be no doubt about the truth of what I have written.

Turning now to his diagrams, Bauer keys a set of notes to the arabic numbers with which he indicates important structural points. The machine consists, he says, of three main parts: the carriage proper, the movable wings, and the fixed wings or canopy. First he deals with the four-wheeled carriage (Figures 42, 43) that serves as cockpit or fuselage. At each end are two tall, upright posts, the 'canopy pillars.' These are tied on at point number 1 by cord of hemp or flax that is then coated with size. Bauer explains that everything is bound in this way and coated with size both to add strength and to prevent the cord from stretching. Number 2 shows where the fixed wing is attached. Lower down the pillars, at 3, are the iron hinges to which the flappable wings are fitted. Number 4, on the plan view, shows the base of the wagon, a grid made of wire. The oblique wires shown at 5 on the elevation prevent the base

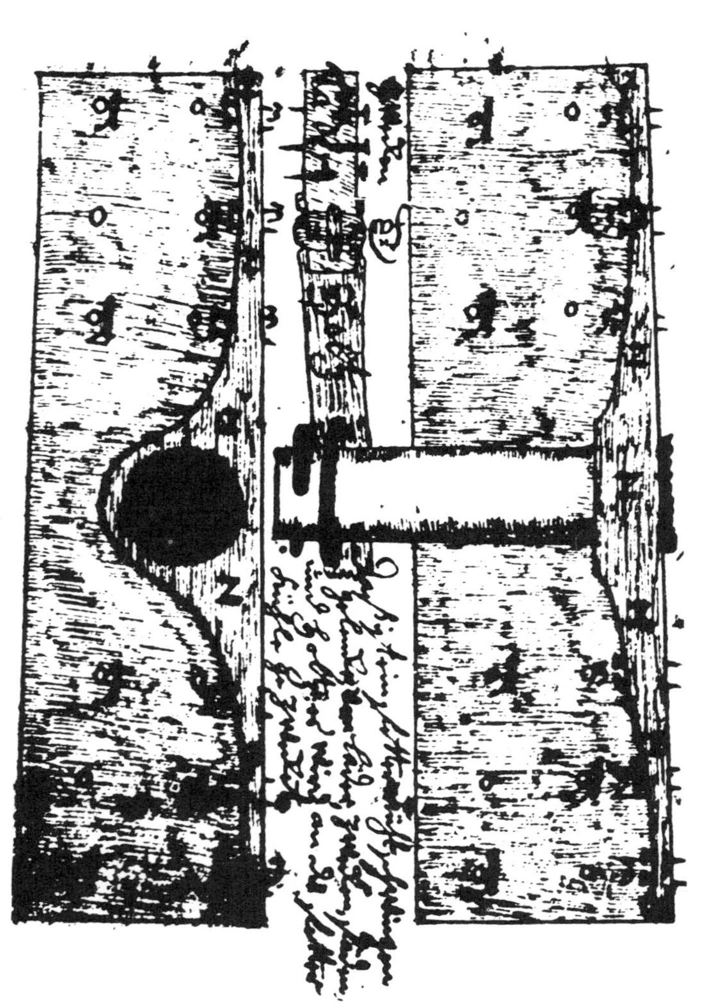

44. Structure of the hinges used to mount the flapping wing. Between the halves is a diagram showing how the stiffeners for the wing covers are made. F. 4^{v}.

from twisting about the pillars, while at 6 on the plan are pegs about which the base wires are wound and drawn tight.

Figure 44 shows a pair of wing hinges, drawn full size. It is not easy to interpret their structure, but each half appears to consist of a piece of 90° angle iron, the two being joined by a spindle. Through the small holes marked 1 in each half, nails are driven into the pillar and into the flappable wing. Hemp threads are tied over the nails and then sized to ensure that the hinges cannot move. The supplementary diagram between the halves of the hinge, with notes running from top to bottom, shows a 'vane swinger': a strengthener or spreader for the wings (see later).

Coming to the second main part, the flapper, Bauer draws a most interesting compound structure with eight pairs of wings arranged in tandem (Figure 45). Once again, the source of Bauer's inspiration is Ezekiel. The rigid construction, making it necessary for one wing to be raised while the other is

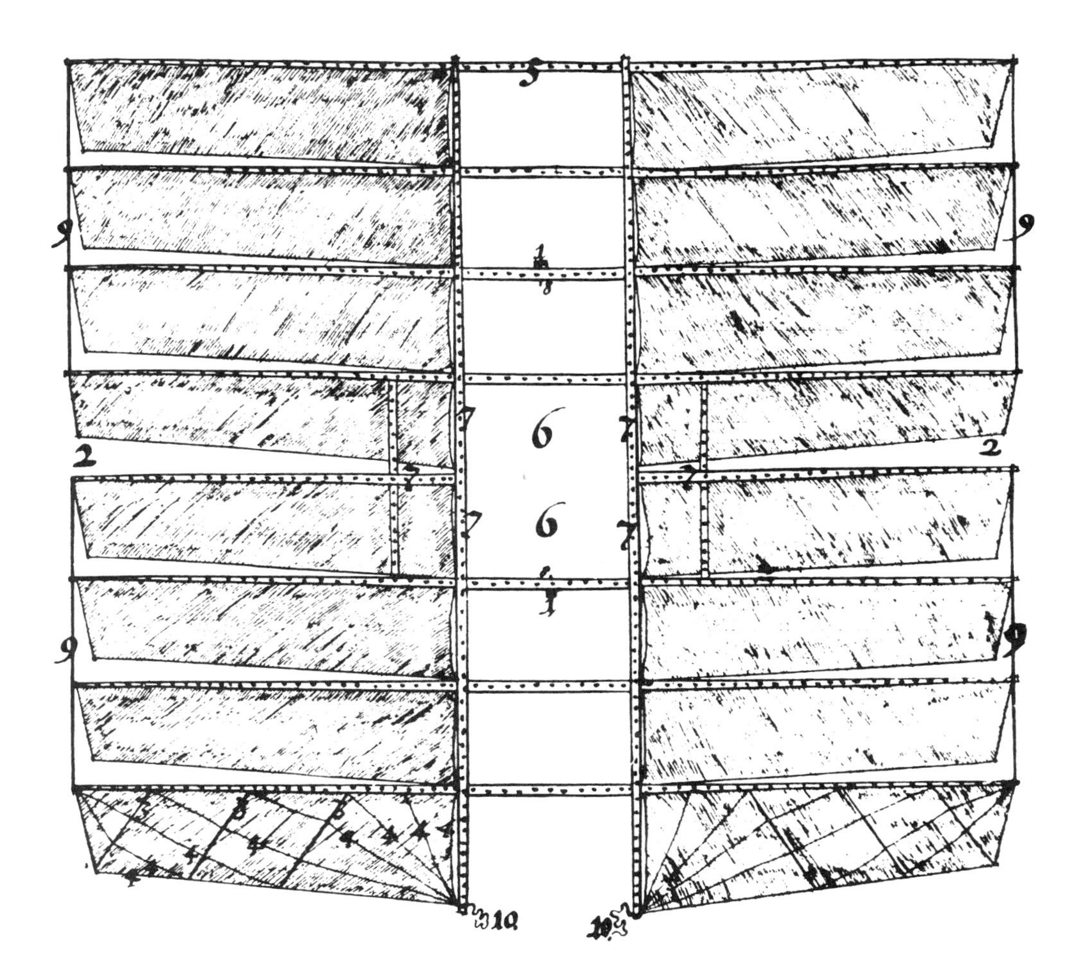

45. Plan view of the flapping wing. F. 5^{r}.

lowered, appears to depend on Ezekiel 1:6 and 1:11. Each of the cherubim had four wings, making sixteen in all, and 'their wings were joined one to another,' suggesting both the compound structure and the rigid horizontal spars. The gaps marked 2 prevent the wings from fouling the canopy stay wires. Numbers 3 and 4, shown on the rearmost vanes only but meant to apply to the whole wing, refer to structural details, revealing how carefully Bauer had thought about his machine. At 3 is to be attached a series of articulated 'vane swingers' or 'wing swingers,' each made up of several lengths of wood joined end to end by pieces of leather and attached to the wings by pegs. Running approximately at right angles to them are wire stiffeners marked 4. These members keep the vanes spread while allowing them sufficient flexibility. The flier stands at the space marked 6, rocking the wings which he holds at 7. Narrow tie beams, for holding the vanes apart, are shown at 9. At 10 are the ends of the cords used to fasten the inboard rear ends of the vanes to the central longerons.

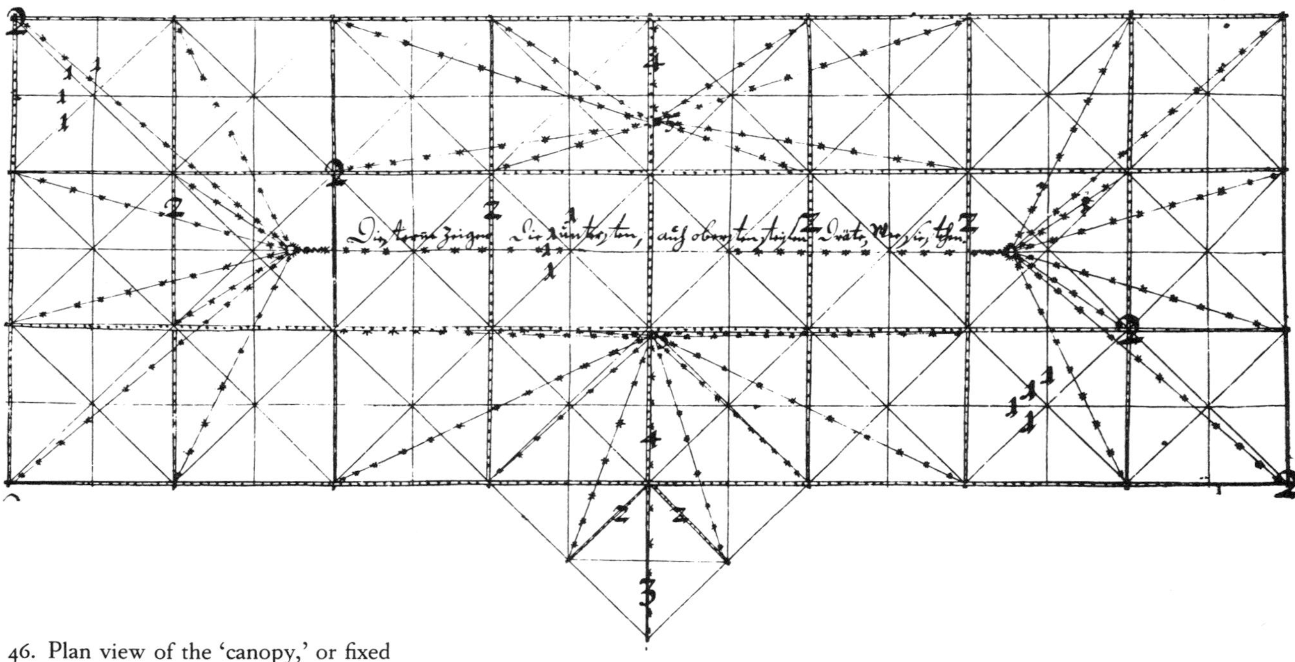

46. Plan view of the 'canopy,' or fixed wing. F. 5ᵛ.

The third main part of the machine is the canopy, or fixed wing (Figure 46). The solid single lines marked 1 show narrow wire bracing to give rigidity to the main wooden frame, 2. At 3 is a gesture towards a horizontal tail. The wooden members are notched where they cross, the joints being bound with hemp threads painted over with size. Before the wing is attached to the carriage, the paper cover is sewn on underneath. A system of external bracing wires, above and below the wing, is shown marked with asterisks. These are attached to the canopy pillars (numbered 5) and also to outboard queen posts passing through the canopy. Number 4, along the central rib, indicates what Bauer calls 'the back wheel, because the tail and the long and short ribs of the canopy are attached there.' The 'tail' is shown in the elevation drawing (Figure 47). It is a ventral fin stiffened, like the vanes of the flapping wing, with wood and wire. Bauer's curious use of the term 'back wheel' probably owes something to Ezekiel's mysterious account of the cherubim: 'And their appearance and their work was as it were a wheel in the middle of a wheel' (1:16). The spokelike radiation of the surfaces from the central rib suggests to the innocent eye the image of an awesome wheel with which Ezekiel attempts to convey the impression left by his vision.

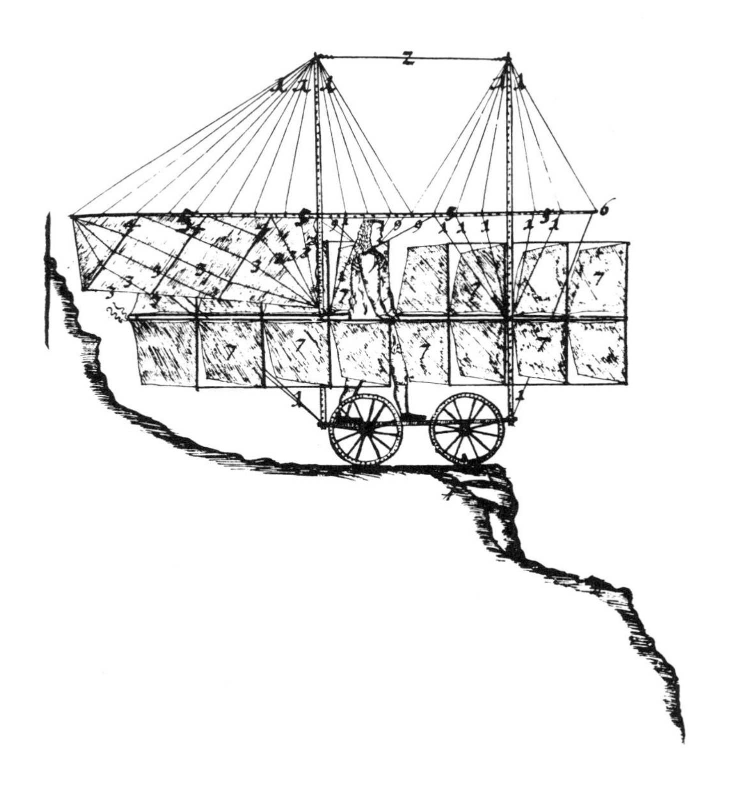

47. Side elevation of the completed machine, standing on a 'smooth hill,' ready for takeoff. Not contemplating an initial dive to gain airspeed, Bauer places the pilot to the rear of the wagon, ready for the climbout. F. 6r.

Two further interesting details are shown on the side elevation. The aeronaut, who stands on the wire grid of the carriage holding the flapping wing with his hands, is provided with a 'chest girdle,' a harness attached to the canopy by cords marked 9. These not only ensure the man's safety but also serve as 'steering cords,' controlling turns to right and left as well as ascents and descents. Bauer now describes the handling of the completed machine:

> If the wheels leave a high hill, constant moving of the wings will cause the carriage continually to ascend into the air. And if the man remains standing at the rear pillar and moves the wings strongly, the canopy will cut continuously upwards high into the air. If he steps to the forward pillar, it will descend, because of the weight. And if he steps to the right side of the carriage, it will automatically turn to the right in the air. Similarly if he goes to the left side it will turn to the left. For it steers itself through the air with the tail; for the weight which the man makes when he steps pulls at the canopy. Thus it will gently sink or turn, just as the man wishes, without further effort or movement. [f. 6r]

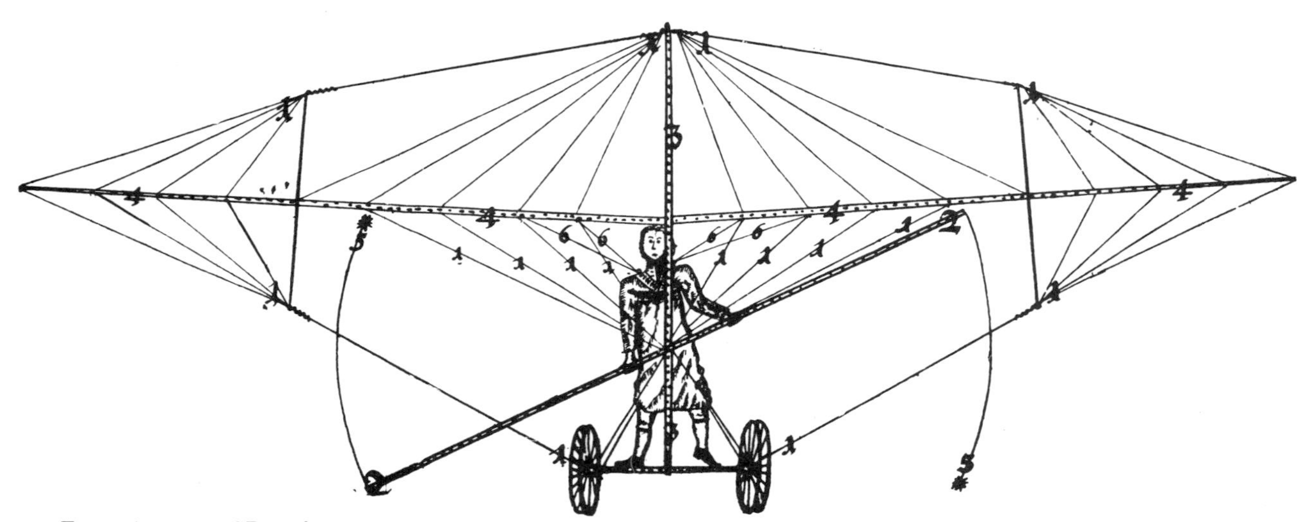

48. Front elevation of Bauer's completed machine. F. 6^{v}.

The last drawing of the machine shows a front elevation (Figure 48). Bauer points out that the bracing wires must be attached to eyelets of hemp, which are fixed to the canopy when the ribs are being bound and sized. The chest harness is clearly shown, and Bauer indicates (numbers 2 and 5) the path of the wing's movement. The slight dihedral angle on the canopy is now mentioned for the first time:

> The carriage cannot turn over in the air because it is light at the top. The ends of the canopy are also bent upwards a little, and everything heavy hangs below it, which may be seen as a model and example in all flying things: that is, the wings are above and the body hangs below. For in this case the canopy serves instead of two wings held stiffly out. [f. 6^{v}]

Bauer calculates that the finished machine will weigh some 45 or 50 lb and that it will lift a man plus another 100 lb. The all-up weight is thus about 300 lb. Assuming Bauer's 'ell' to be approximately the 'Flemish ell' of 27 inches, and ignoring the surface of the flappable wing, the loading on the canopy is very light, at about 4 kg/m^2. The construction is exceedingly, indeed impossibly, fragile. The wing ribs and spars are made of spruce or pine about as thick as a finger. The main fore-and-aft supports for the flapper are as thick as a man's wrist. At the wing tips and on all outboard parts of the structure the wood is fined down to a point or sharp edge, 'as the feathers of bird wings teach us and demonstrate' (f. 7^{v}). Realising that this all seems rather frail, Bauer points out that most of the strain is taken by the bracing wire.

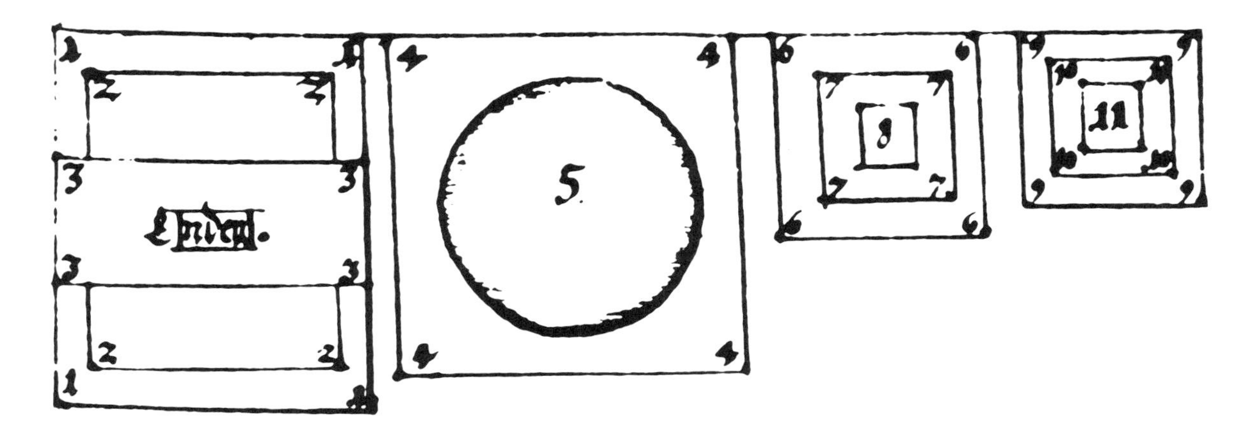

49. Sections of the wooden members: (1) longerons for the flapping wings, section at the centre; (2) longerons, section at the end; (3) wing spars, section at the junction with the longerons, showing the size of the hole through the longerons, with (insert) the end section; (4) axles, and canopy pillars at their base; (5) axle holes and axle ends; (6), (7), and (8) maximum, middle, and end sections of the smaller canopy spars. F. 8^{r}, reproduced same size as the original.

Having by now altogether forgotten about silk, the more expensive alternative for the covering, Bauer says that the sheets of paper for the canopy must first be glued together and that the glue and size used on all parts of the machine should be further coated with oil varnish or white lead to protect it from rain. Trestles or scaffolds are to be used for the construction, the canopy and wings being built on two sets of supporting planks placed at appropriate heights above the floor, namely, 2 ells and 3 ells. After the basic structure of the canopy has been completed, further planks should be laid on top of it to allow the workman to walk over it while making the eyelets and fixing the bracing wires. This second set of planks will also prevent the canopy's being pulled out of shape when the bracing wires are tightened. In a brief appendix included as an afterthought to fill up his last page, Bauer draws cross-sections of the various wooden members, showing their actual size at three points: at their widest, halfway along their taper, and at their ends (Figure 49).

Bauer did not want to fly merely for the joy of it. He expressly and emphatically believed that he was inventing a new and virtually invincible war machine that would enable any Protestant prince to eradicate the curse of the ungodly Catholics. He supports his argument with references to Scripture:

> If I am not mistaken we shall in future times, in my opinion, be able to hurl fire, brimstone, and stones the weight of a talent on to the antichristian and idolatrous peoples, places, and states which choose to rebel against the true Christian Kingdom, as Saint John saw as a prophecy in his holy vision (Rev. 16:17–21, 20:9; Ps. 11:6, 38:22).

And throughout the whole world there will no longer be any safety for the antichristian party and armies. They will all have to turn to the true Christ, who will offer safety to such people as root out all the vexations of the earth (Matth. 13:40–43). [f. 7r]

Recommending himself and his invention in his closing lines to the Count, Bauer hopes that he has shown the construction of his flying machine to be a relatively simple matter: 'It certainly demands less effort and art than the familiar windmill, which had wings as long ago as 1364' (f. 8r).[6]

Little need be said about the technical shortcomings of the design. It is hopelessly flimsy, unstable in pitch, and underpowered by its inefficient oscillating propeller. For its day it is nevertheless an invention of true genius. Although in 1618 Hieronymus Fabricius had pointed out that pendulous stability is an important factor in bird flight, [7] Bauer was the first to describe how hang-gliders might be controlled by shifts of the centre of gravity. Good sense and observation, rather than mathematical analysis, appear to have led him to place his aeronaut in approximately the right position, relative to the canopy's centre of pressure, from which to make the necessary movements. The combined safety and control harness is a brilliant idea, though in this case Bauer does not seem to have analysed the system with sufficient rigour. In the form in which he has drawn it, the tensions exerted by movements of the pilot's body would counteract rather than assist the effects of shifting the centre of gravity. Although clumsy and inefficient, the oscillating propeller applies sound principles. Perhaps based in part on Borelli's diagram showing how a bird's wing flexes about the leading edge during the up- and downstrokes (Figure 11), Bauer's flapper would to some degree function like a pair of heavily emarginated, slotted wingtips, the sixteen vanes bending up and down at each stroke, no doubt to the accompaniment of much rushing and thundering noise, as he describes. Although there is no positive yaw control, the inclusion of vertical surfaces of any kind is rare in designs of flying machines before those of Cayley. Bauer's incorporation of a large ventral fin shows an intuitive sense of the need for inherent stability in yaw as well as in roll. Only in pitch stability are his design principles grossly inadequate.

We do not know whether Bauer received sufficient support to enable him to build his cherub wagon in secret. If it was completed, we may hope that he never tested it, as he had intended, from the top of a steep hill. No documents survive to tell us of his subsequent fate, even the date of his death being unknown. It is perhaps unfortunate that the manuscript remained in obscurity until comparatively modern times. Had it been more widely read, its many insights might well have accelerated the design of heavier-than-air machines in the late eighteenth and nineteenth centuries.

Carl Meerwein's ornithopter 10

In 1781–84 Carl Friedrich Meerwein (1737–1810), a south German builder and architect to his namesake, Karl Friedrich, prince of Baden, worked on ideas for a practical ornithopter. Living in comparative obscurity in Emmendingen, a small town near Freiburg, he began in August 1781 by building a simple pair of ellipsoidal wings. He might have taken the matter no further had his interest not been caught by reports of Jean Pierre Blanchard's abortive flying attempts in Paris, published in a Basel journal in July 1782. During the following months, when everyone in Europe was talking about the French balloonists, Meerwein decided to set out his ideas for a heavier-than-air machine. He accordingly wrote a little book entitled *Der Mensch: sollte der nicht auch zum Flügen gebohren seyn?,* which was issued in serial form in January 1783.[1] An expanded and improved version, which was widely read, was published as a book, with a new title, in the summer of 1784.[2] A modified French translation was published in the same year and reissued in 1785, and in 1812 a Portuguese translation appeared.[3]

Unlike Melchior Bauer, Meerwein set down very few particulars of the dimensions or structural details of his machine, offering only a general description of the configuration and the principles to be applied. A large wing, as drawn in the first of the book's two illustrations (Figure 50), is hinged about its centre line so that it can be flapped downwards by a prone pilot held underneath in some form of sling. In his text Meerwein specifies a low aspect ratio of 4:1 (2:1 for each half of the wing), although the illustrations show an overall aspect ratio of 3:1 relative to the maximum chord. The framework is made of a suitably tough wood, such as lime or fir. Before assembly, the wooden members are strengthened by being bound with cords of flax, special attention being given to the points where nails are to be used. The framework is then covered with strong fabric, such as linen, cotton, or oilcloth. Writing with the experience of a professional engineer, Meerwein points out that provided the covering is made and attached with care, the strain will be evenly distributed and the cloth will not tear. Using the small south German foot measure (*Werkschuh*) as his unit of length, he calculates that the area of a wing such as the one illustrated will be at least 240 ft^2. (The diagram shows a span

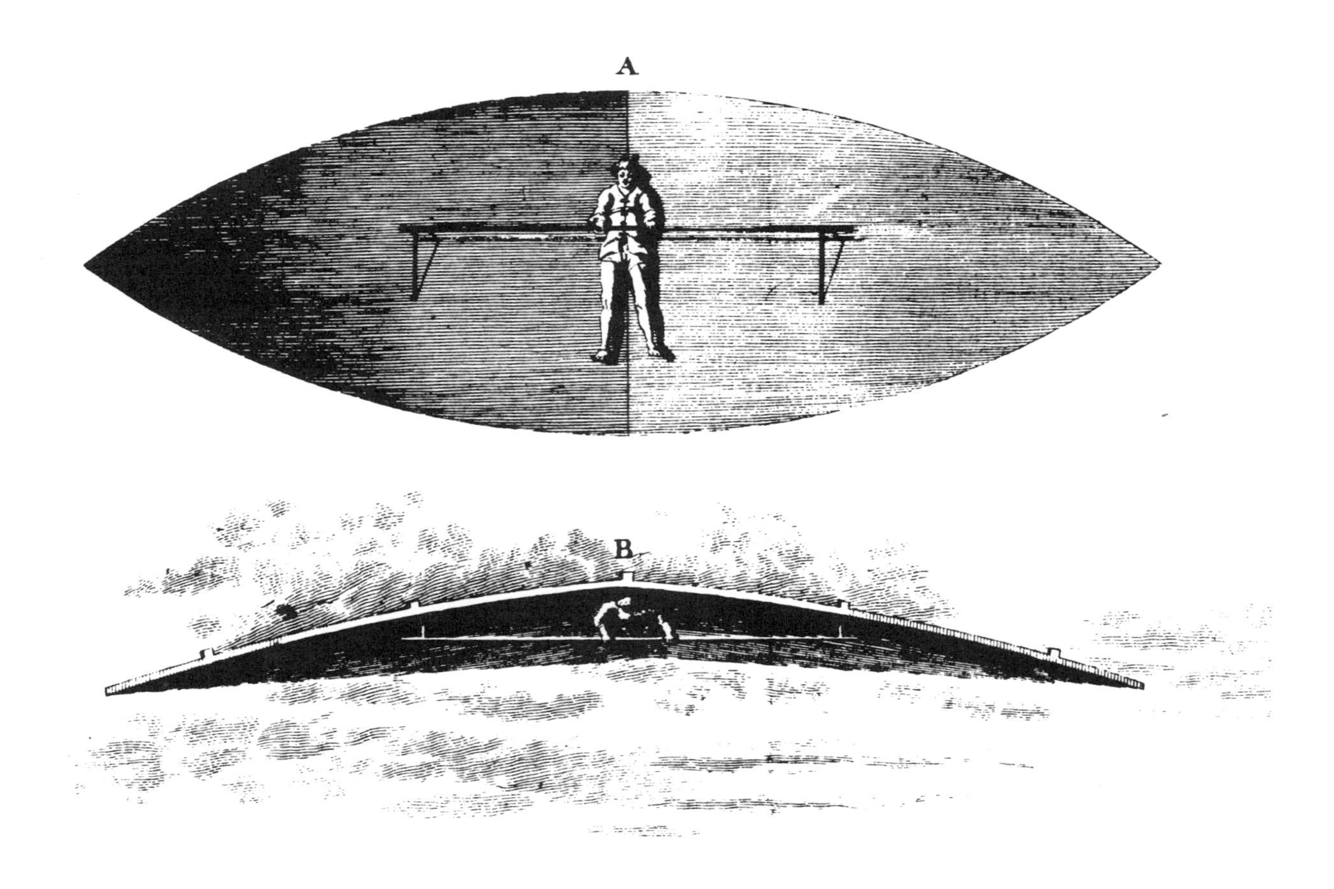

50. A general impression of Meerwein's ornithopter in flight. The absence of the tail from the engraving shows how little importance Meerwein attached to it. *Die Kunst zu fliegen nach Art der Vögel* (Frankfurt und Basel 1784) end plates. (Reproduced by permission of the British Library)

of 30 ft and a maximum chord of 10 ft.) The all-up weight will be about 200 lb, allowing 150 lb for the pilot and about 50 lb for the machine. The loading is thus about 4.5 kg/m^2.

Meerwein's design is for a simple up-and-down flapper. The wing is so rigged that when fully raised it lies at an anhedral angle of about 9°. The pilot creates the downstroke by pushing with his hands on a 'balance rod' 10 ft in length. The ends of the rod are hinged approximately at right angles to two shorter rigid members which in turn are hinged to the framework of the wing. When the pilot pushes the rod forward a distance of $1\frac{3}{4}$ ft, the wing is fully depressed through a further 12°. Although Meerwein does not say so, the upstroke is presumably automatic. The front elevation drawing (Figure 51) appears to indicate that a cambered shape is maintained by air pressure.

A fan-shaped horizontal tail, not shown in the illustrations, is to serve as a steering device. Showing little interest in problems of flight control, Meerwein gives only scant attention to its construction and use. The pilot is to wear a special pair of trousers reaching to below his feet. Between the legs of the trousers is to be sewn a piece of cloth sufficiently wide to allow the legs to be spread. In case the pilot should find it tiring to hold his legs apart for long periods, Meerwein suggests that they be supported by a horizontal rod or rods so hinged that when the pilot is lying prone the centre of the tail can fold

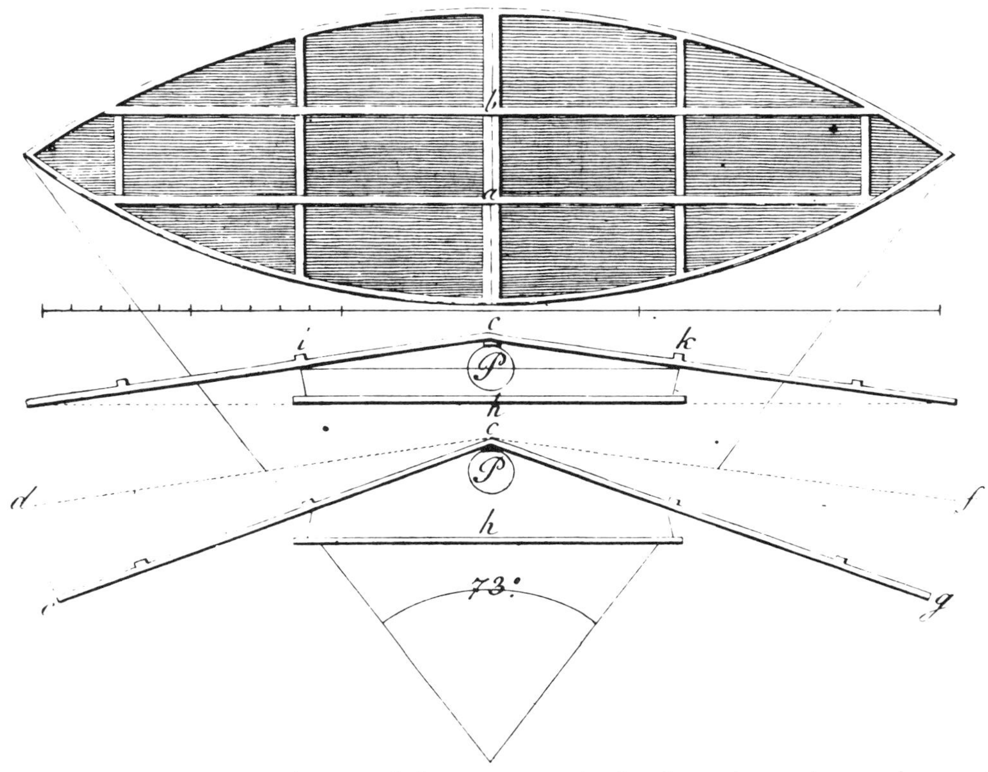

51. Plan and elevation drawings showing the structure of Meerwein's wings. The lines drawn at 73° show the arc of the circle represented by the leading and trailing edges. *Die Kunst zu fliegen*, end plates. (Reproduced by permission of the British Library)

downwards but not upwards. An alternative to these rods, mentioned in the French text but not in the German, is an iron triangle to which the legs are attached. Apparently envisaging that the tail would extend well below the pilot's feet, Meerwein says that it will add a further 20–40 ft^2 to the total area of the machine. The only other provision for the pilot's safety and comfort is a face mask to enable him to breathe more easily while speeding through the air.

Meerwein based the dimensions of his ornithopter on a study of the weight and wing area of nine species of bird.[4] At first, in 1781, he chose the wild duck, an excellent flier, which weighed $2\frac{5}{8}$ lb and had a wing area of about 165 in^2. Scaling this up for a man-carrying aircraft with an all-up weight of 200 lb, the necessary wing area was found to be 126 ft^2. Although convinced that these figures provided enough working information, during the next three years Meerwein examined other birds, from whose dimensions he derived necessary wing areas ranging from 99 ft^2 (based on the bustard) to 634 ft^2 (based on a species of owl). The ornithopter of August 1781 was a good deal smaller than the one he illustrates, and he speaks of it as more like a model than a real man-carrying aircraft. In comparison with the bustard, it should nevertheless have been big enough, having an area of 111 ft^2, excluding the tail. Meerwein describes it as having been made from coarse materials allowing for a good deal of refinement. Covered with cotton ticking and sized, it weighed some 56 lb, plus an estimated 4 lb for the harness and minor fittings.

Considering the problems facing the designer of a heavier-than-air machine, Meerwein says that the potential difficulties are of four kinds, arising from

1. The shape and structure of a man's body
2. A man's excessive weight
3. The lack of a force sufficient to enable the machine to be handled
4. The lack of suitable constructional materials

He hopes that his notes on the structure and handling of the wings, and his discussion of the principles of flight, will show that none of these difficulties is insuperable.

A part of what Meerwein has to say about the theory of flight depends on a curious idea that he owes to the German scientist, economist, and physiocrat Johann August Schlettwein (1731–1802).[5] Maintaining the outmoded Aristotelian idea that the air is inherently light rather than heavy, Meerwein says that it is denser near the earth's surface because of the mutual attraction between the particles of water and fire that are mixed with it and with the earth. As water and fire are more heavily concentrated in the earth than in the air, the air closest to the earth is pulled down strongly and condensed. The next layer up is attracted to the relatively heavy concentration immediately below it, and so on up in ever thinner layers. Not only is it untrue, states Meerwein, that the air is pressing down on itself from top to bottom of the atmosphere, but the upper layers are volatile and are prevented from flying away into space only by the force of this attraction downwards to the denser layers. Idiosyncratic though this seems, explanations of gravity appealing to broadly comparable ideas had sometimes been proposed by earlier writers. In a famous letter of 28 February 16$\frac{78}{79}$ to Robert Boyle, Newton speculated on the possibility that gravity was caused by gradations in the kinds and qualities of *aether* contained in the air, in the earth, and in other bodies:

> I shal set down one conjecture more which came into my mind now as I was writing this letter. It is about ye cause of gravity. For this end I will suppose æther to consist of parts differing from one another in subtilty by indefinite degrees: That in ye pores of bodies there is less of ye grosser æther in proportion to ye finer then in open spaces, & consequently that in ye great body of ye earth there is much less of ye grosser æther in proportion to ye finer then in ye regions of ye air: & that yet ye grosser æther in ye Air affects ye upper regions of ye earth & ye finer æther in ye earth ye lower regions of ye air, in such a manner yt from ye top of ye air to ye surface of ye earth & again from ye surface of ye earth to ye center

> thereof the æther is insensibly finer & finer. Imagin now any body suspended in ye air or lying on ye earth: & ye æther being by the Hypothesis grosser in ye pores wch are in ye upper parts of ye body then in those wch are in its lower parts, & that grosser æther being less apt to be lodged in those pores then ye finer æther below, it will endeavour to get out & give way to ye finer æther below, wch cannot be wthout ye bodies descending to make room above for it to go out into.[6]

First published in 1744 in Thomas Birch's five-volume edition of Boyle's works and issued separately, with a commentary, in 1745, this letter was much discussed by scientists in the middle of the eighteenth century.

Despite important differences of detail, Schlettwein's theory of gravity, on which Meerwein bases his aerodynamics, has the same general character as Newton's and is clearly influenced by the discussion. According to the theory, all weight is to be attributed to the layered structure of the atmosphere. Using this principle to explain flight, Meerwein says that it is the air under a bird, relatively heavier than the layer above it, that pulls the bird down. If the air underneath can be pushed downwards at sufficient speed, the bird's weight will be abolished because the attraction, or adhesion, to the lower layer will then be no greater than that to the upper, and the bird will be in equilibrium. Meerwein says that the effect is proportional to the product of the quantity of air and its downward velocity produced by flapping. Although far from understanding even the elementary aerodynamics of the situation, he shows by his comment some intuitive awareness of the relationship between lift and the rate of downward transport of air. When he comes to consider how a bird rises, Meerwein does not attempt to extend his theory but falls back on elementary mechanics, saying that the bird uses the underlying air as a fulcrum against which to lever itself up.

Having observed birds in flight, Meerwein is aware that the larger species flap their wings more slowly than the smaller, the biggest of them being able to remain aloft for long periods without having to move their wings at all. When he tries to account for these observations, Meerwein ceases to be altogether logical. If a large bird holds its wings out and curves them somewhat down to make a concave undersurface, its weight will automatically cause enough air to escape downwards so that its tendency to fall will be counteracted. Turning to the smaller species, he is nearer the mark. Noticing that they often close their wings between bouts of flapping, he says that they first begin to fall more quickly but then, when they again spread their wings, they are able to convert their fast descent into fast forward motion. Meerwein stresses that the supportive effect is entirely due to the action of the wing in moving the air downwards and is at pains to deny the popular mediaeval idea that a bird can decrease its weight by spreading its wings.

Analysing how a bird moves forward, Meerwein considers the shape of the wing and the junction of wing and body. The wingtip, moving up and down at a faster linear rate than the inboard parts, produces most of the lift. The breadth of the wing, coupled with its concave shape near the body, simulated by his machine, causes the depressed air to move backwards as well as down, thus creating a reaction which results in forward motion. When considering how the backward movement of air might be ensured in his ornithopter, Meerwein suggests that a shallow triangle at the front (lettered *c-i-k* in Figure 51) should be covered with cloth, thus inhibiting the forward flow of air. An additional and mysterious surface is 'a kind of sail' that is to be attached to the pilot's back. This sail, whose shape and purpose can only be surmised, does not appear to serve any essential aerodynamic function. Mentioned briefly in a footnote (p. 39n), it nowhere figures in Meerwein's main discussion of his machine.

Although Meerwein built a small version of his ornithopter, he appears never to have tried it. He confesses that he was dissuaded by fear of the bovine ridicule of ignorant bystanders. By their jeering they would create an entirely wrong psychological atmosphere that would be far from conducive to the learning of a new and delicate art. Meerwein fully understood that success would not be achieved at the first attempt and was resigned to the likelihood of breaking a few bones. Practice over water would help, since we well know how to save ourselves from drowning. The ideal would be a secluded low cliff overhanging a deep river, for example at the Rheinsprung in Basel. Although it is not so good as the one in his illustration, a simple machine such as he made in August 1781 would serve, with a few minor modifications, for the initial trials.

If Meerwein did ever attempt to fly, he would have needed all his resolution and patience to arm himself against inevitable disappointment. Apart from the inadequacy of his arm muscles, the ornithopter would have been virtually uncontrollable, even in a glide. The large anhedral angle would have created marked lateral instability, and although a large tail would have helped stability in pitch, the drawings appear to show the pilot's centre of gravity a good deal too far to the rear. If he had ever launched himself, as he had intended, from a cliff, his machine would almost certainly have pitched nose up, stalled, and fallen in a side slip.

Meerwein writes somewhat impatiently of those who believe that we shall never learn to fly. Confident that both man and materials will be strong enough, he foresees that success will soon be achieved and that mankind will be the happier for the perfection of his invention. Arguing for its potential social value, he makes an analogy with gunpowder: although many evils have resulted from its introduction, the good effects outnumber the bad. Are we not better off, he asks, when even a small child can now protect itself by shooting a robber? The use of firearms has not caused men to grow physically flabby, as some believed it would. We are still as strong, but with the refinement of the art of warfare we have learned to prize the power of the spirit and the understanding above brute strength. The invention of the flying machine will bring similar benefits, both spiritual and physical. Taking an example from recent experience, he points out how the availability of wings would have helped alleviate suffering during the recent flooding in many parts of Europe. Considered in relation to any such practical application, his wings have, in

Meerwein's estimation, several decided advantages over balloons. They are less costly, more durable, less affected by weather, and better able to resist damp. Moreover, they enable the aeronaut to direct his flight wherever he wishes and might even be used to pull and guide balloons.

While the original German edition of Meerwein's book ends with the section celebrating the invention of flying as a great boon to mankind, the French edition contains, as an appendix, a letter by a 'M. de G***' who attacks the idea on social grounds. Reworking an old theme, he asks how moral values might be preserved in a world of flying lovers. Imagining the feelings of a rich and avaricious father with a daughter to dispose of in marriage, M. de G*** writes: 'If the girl's lover manages to fly other than by the wings of his desire, how warmly do you think her father will thank you for your invention? Already his limbs are shaking; and those of his daughter tremble with delight.' To this is added a short reply 'sous le nom de M. Merrwein [*sic*]' but clearly not by him. It is no more than a somewhat ponderous *jeu d'esprit* taking up the theme of the hypothetical daughter's theft by the lover and punning throughout on *voler* (to fly) and *voler* (to steal). The frivolous end to the French version brings into sharp relief the care and seriousness of Meerwein's original. The anonymous French translator was inclined to treat the whole proposal as a joke; Meerwein, by contrast, shows quiet faith in the practicality and value of his ideas. Having written his book, he was apparently happy to wait patiently for the day when others would perfect his invention.

Conclusion: flying ships 11

Most accounts of attempted flight, and most discussions of its possibility, are couched in personal, individual, even idiosyncratic terms. At times, indeed, there is a strong element of secretiveness, as when Leonardo shut himself away in a large attic room so as to be able to experiment unobserved. Usually flight has, it seems, something to do with the individual will. There is nevertheless an important class of aeronautical legend and speculation in which the enterprise is communal: these all include, in various forms, the notion of the flying ship. Equally significant, the flying ship is distinguished from almost all other flight imagery by a fundamental shift of point of view, of perspective. Winged men, parachutists, mysterious floaters in dreams are usually seen and reported from the point of view of the flier. *We* are flying in imagination; the earth is below us, the heavens are within reach. In the case of the flying ship, we remain behind, on the ground: the aeronaut is not ourself, he is the other fellow, seen and often envied from below. Most dreams of flight take us, potentially, away from earthly bounds, set us free from earthly bonds. The flying ship, on the other hand, comes to us; and, in the typical story, it is briefly ensnared by our mundane entanglements. A tombstone (death) or a church door (established moral law) momentarily restrains the freedom of the fliers. The bond is quickly broken, but while it lasts we have a hurried opportunity to establish contact. Above and below are linked, as they were in the dream of Jacob's ladder, allowing human access through the medium of air to the heavenly regions.

It is not surprising that, despite the tantalizing glimpse of higher things, the image of the flying ship often has sinister connotations. In the reports of the old chroniclers they are early manifestations of UFOs: bearers of remote humanoid races having some special relationship with the heavens. The intriguing but rather frightening remoteness of the aeronauts is emphasised by their strange appearance in the work of a number of early pictorial symbolists, of whom Hieronymus Bosch is by far the best known. It is, of course, through the air that the gods make contact with man, and commonly in association with spectacular weather phenomena, or 'meteors': Zeus descends in his shower of gold, Thor wields his thunderbolt, the Valkyries ride through the storm, the shaking of their horses' manes showering the earth with hail and dew. Not only the gods but also flying men or giants have long been associated with disturbances in the air, sometimes mediating the gods' power, sometimes acting as an independent species making up another rung on the scale of nature. In his beautifully rhythmic description of the shows performed for Elizabeth I, Henry Goldwel reminds us of these aerial apparitions, linking them with the unhappy flight of Icarus:

> The Gyaunts long agoe did scale the cloudes men say, in hope to winne the fort of IUPITER. The wanton youth, whose waxed wings did frie with soaring up aloft, had scapt unscorcht if he had kept a meaner gale below.[1]

In his *Roman History,* Dio Cassius implies that flying giants were responsible for the eruption of Vesuvius and the associated darkening of the air in A.D. 79:

> Numbers of huge men quite surpassing any human stature—such creatures, in fact, as the Giants are pictured to have been—appeared, now on the mountain, now in the surrounding country, and again in the cities, wandering over the earth day and night and also flitting through the air.[2]

The tradition continues into the Middle Ages: writing of the violent hailstorms of southern France, Thomas of Cantimpré (thirteenth century) attributes them to the work of demons in the shape of beasts running in and out of the clouds to meet each other. Thomas tells a very circumstantial story about a storm near Limoges that devastated most of the crops. While it was raging, voices were heard in the air as the demons warned each other not to damage the vines of one Pierre Richard. Richard, Thomas points out, was known in the region to be an evil man who was clearly in league with the spirits of the air.[3]

The flying stormmakers, or *tempestarii,* were so important and so permanent a feature of the popular imagination that laws were sometimes passed against them, as, for instance, by Charlemagne.[4] Taking an unusual line, the seventeenth century *fantaisiste* l'Abbé Montfaucon de Villars thought such laws unjust. The sylphs of the air, in whom he took great delight, were usually innocent and maligned:

> Seeing the people, the pedants, and even crowned heads armed against them, they resolved, in order to dispel this bad opinion held of their innocent flotilla, to take some men from all regions, to show them their beautiful women, their republic, and their government, and then to replace them on earth in various parts of the world. They did as they had planned. The people who saw these men descend from the sky ran from all parts; and believing that they were sorcerers separating from their companions so as to poison the fruit and the fountains, they were carried away by the fury inspired by such imaginings and dragged these innocent men off to punishment.[5]

52. (a, b) Magical flying ships from Girolamo Graziani, *Il conquisto di Granata* (1650), in *Parnasso italiano* 39 (Venezia 1789) I, 175.

Often the celestial spirits were to be seen sailing, as Villars puts it, 'in aerial ships of marvellous construction, the flying flotilla gracefully following the breeze.' Villars's ready acceptance of the sylphs was uncommon, since their appearance usually portended trouble. According to Stefano Breventano, celestial ships were seen in Rome in 215 B.C., just before Hannibal entered Italy to defeat the Romans at Lake Trasimeno, near Perugia.[6] About A.D. 800 Bishop Agobard of Lyons reported a common belief, which he scorned, that the aerial sailors acted not only as sinister harbingers of evil but as pillaging mercenaries in the hire of evil gods:

> We have seen and heard many, so struck with madness, so crazed with stupidity, that they believe and say there is a place called Magonia, from which ships sail into the clouds and in these the crops knocked down by hail and ruined by storms are carried to that region, the aeronauts paying the gods of the tempests, and receiving in exchange the corn or other crops. We have seen several, so blinded by profound stupidity that they believe such things to be possible, exhibiting in an assembly four people, three men and a woman, who had been bound because they were thought to have fallen from such ships. These, having been kept prisoner for some days, were at last presented to the whole assembly of men, in our presence, as I have said, as people fit to be stoned.[7]

Flying ships are often associated with demons and bad weather in poetic and pictorial fantasies of the Renaissance.[8]

Not all of the early UFOs were thought to be destructive. Some chroniclers report the appearance of flying ships whose crews seemed no worse than strange, if also a little pathetic. The aerial travellers in these stories wish to do no harm. Their contact with the human race is more or less accidental and involuntary, a neutral meeting of two independent and sometimes incompatible worlds of experience. An archetypal account of one such flying ship is given by Gervasius of Tilbury, in his *Otia imperialia,* composed around 1211. One feast day in Britain, the people were leaving church after mass. The sky was heavily overcast and the light was dim. As the worshippers passed through the churchyard, they saw an anchor caught around a tombstone, the anchor rope rising into the air. As some of the astonished people stood discussing what to do, they saw the rope being vigorously shaken as if someone were trying to free the anchor. When these efforts proved unsuccessful, a voice was heard above in the thickened air, sounding like the shouts of airborne sailors arguing about how to pull up the anchor. Soon afterwards one of the sailors, sent to free it, climbed hand over hand down the rope. As this sailor was disengaging the anchor, he was seized by the churchgoers, who manhandled him. He then suffocated as if drowned in the sea, unable, as Gervasius puts it, to withstand the humidity of the heavy air we breathe here below. The sailors above, deprived of their companion, cut the rope after an hour and sailed off leaving their anchor behind. In memory of this event, iron ornaments for the church door were fashioned from pieces of the anchor.[9]

A variant version by Geoffroi de Vigeois sets the story in London some time during the years 1122–24. In this version, the sailor drowned after having been thrown roughly into the river Thames.[10]

Less violent versions are associated with Ireland. In one of them the ship is said to have appeared over Clonmacnois, where the anchor was caught in the church door. The sailor who came to free it dived through the air, moving his hands and feet as if swimming. As he was about to be seized, the Bishop ordered him to be left in peace, saying that otherwise he would drown. In a further variant set in Teltown, County Meath, the sailor was freed by order of the king. The ecclesiastical or royal intervention is an implicit gloss on the function of the flying ships as a link between heaven and earth, a link that only a man of authority was in a position to understand.[11]

Prompted, perhaps, by these common folktales, some mediaeval and Renaissance scientists from at least as early as the fourteenth century discussed the theory of flying ships, conceiving of them as lighter-than-air machines, or (from a twentieth century point of view) as forerunners of the balloon. Given that in Aristotelian science the element of the air has an upper surface, contiguous with the lower surface of the region of fire, it should be possible to float a lightweight ship there, using principles of displacement entirely analogous with those that keep a ship afloat on the water. Albert of Saxony (c. 1316–90) makes an explicit suggestion:

> Fire is much subtler and more tenuous and lighter than is the air, for it is related to air as air is to water. Now air is much more tenuous and much subtler and lighter than is water; therefore the same is true of fire with respect to air. From this follows what may be readily demonstrated from the science of relative weights: that the upper air, where it is contiguous with fire, is navigable, just as the water is where it is contiguous with the air. Hence if a ship is placed on the upper surface of the air, filled, however, not with air but with fire, it will not sink through the air; but as soon as it is filled with air it will sink. Just as, if a ship is filled with air rather than with water, it will float on the water, and not sink; but when it is filled with water, it sinks.[12]

This thought experiment appears to have become a commonplace. It turns up again a little later in the fourteenth century in Nicole Oresme's *Le livre du ciel et du monde:* 'A vessel of heavy material loaded with heavy objects such as a man or several men, standing upon the nearly spherical convex surface of the element of the air, and with no perturbation, could remain up there as naturally as a ship rests on the Seine.'[13] Oresme (c. 1320–82) illustrates his idea with a charming little line drawing of a somewhat surprised-looking mariner sailing a skiff shaped like the crescent moon (Figure 53).

Immeasurably more imaginative and sensitive than Albert of Saxony, Oresme ruminates on the relationship between his flying ship and the natural

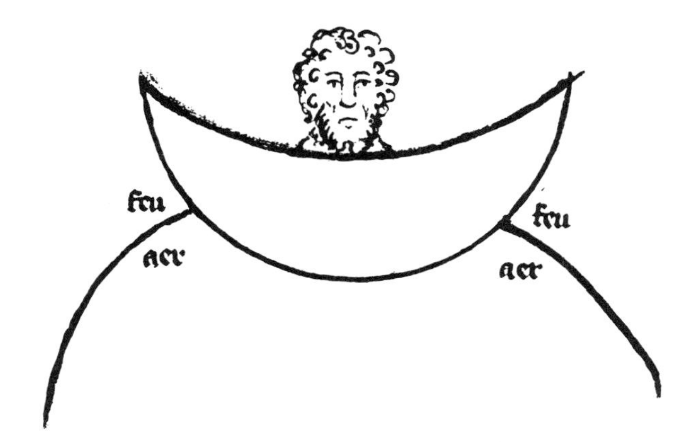

53. Ship floating on the surface of the air. Paris, Bibliothèque nationale, MS français 1082, f. 103^{r-b}. 1377. Nicole Oresme, *Le livre du ciel et du monde.* (Reproduced by permission)

order. Exploring physical questions analogous to moral doubts about the acceptability of human flight, he asks whether the idea of the flying ship is a violation of universal harmony. Could the ship be placed there, could it remain there, only by the exercise of violent forces? If placed on the surface of the air, would its unnatural position not result in its being burned up by the region of fire? Taking care not to be dogmatic, Oresme tentatively concludes that although in his day it was not possible in practice to raise a ship above the air, there was no theoretical objection to the achievement of such a feat at some future time. Unlike the fire here below, the elemental fire was probably *innoxius,* and a ship floating on the calm upper air would be able to remain there for ever:

> A vessel would remain up there as naturally as a boat floating at rest in the water and probably without requiring any violence. It seems that, if the receptacle were perpetual, it would remain there eternally.[14]

Oresme's flying ship is neither a threat nor a boon. Ignoring social matters, he is as little troubled by ancient fears of flight as he is excited by the possibility of improving man's lot. Although he speaks of human cargo, he does not imagine a vehicle for manned flight in the ordinary sense. His astronaut is not perceived as one of us, rising above the earth for practical, aesthetic, or metaphysical reasons and ready to return in due course with news good or evil. The man staring out of the crescent moon is rather another sort of creature altogether, permanently, eternally stationed above us in lonely orbit, suspended between heaven and earth. He is not swimming, he makes no effort, he is neither angel nor demon. Natural forces function alone; science prevails.

Both Albert of Saxony and Nicole Oresme wrote of flying ships in general cosmological contexts.[15] A later, more literary version of the same idea relates it once again to the more specifically human concern of meteorology. The principles of a hypothetical and grandiose flying ship were described in 1755 by Father Joseph Galien (1699–?1762), who was born at Saint-Paulin near Le Puy. By 1726 Galien was already a professor, at Bordeaux, and in the 1740s he moved to take up a similar post at the University of Avignon. A respected father of the Dominican order, he published in 1745 a volume of *Lettres théologiques*. Like many other learned men of his time, he was also concerned to find an explanation for the hailstorms that so often devastated the crops in southern France. In a book that he first published anonymously[16] but acknowledged in the second edition, of 1757,[17] Galien revives and develops the old Aristotelian idea that there are several distinct regions of the atmosphere. Using the evidence of sharply defined cloud bases and the sudden cessation of the rise of smoke, he proves to his own satisfaction that the air in each region is different in kind, and especially in density, from that in the others. The various layers, he suggests, are sharply divided from each other on the analogy of the boundary between oil and water.

Having described the mechanism of hail formation based on this theory, Galien concludes with an addendum in which, as a *jeu d'esprit,* he proposes a flying ship using principles similar to those of Albert and Oresme. Since by 1755 the region of fire had been, if not quite put out, at least reduced to a doubtful glimmer, Galien relies on the difference of density between the layers of air. Making what he hopes is a reasonable assumption that the second and third layers differ by a ratio of two to one, he calculates the necessary size of a flying ship large enough to carry a substantial cargo. Leaving the shape to be decided by skilled mechanics, he settles on a ship whose volume would be the equivalent of a cube with a side of 1,000 fathoms (6,000 ft). Conscious of the awesome scale of what he is proposing, Galien says that this will be no matter, since 'it will cost no more, so long as we are only making it in our imaginations.'[18] Such a ship, he points out a little sheepishly, would be longer and wider than the city of Avignon, and as high as a mountain of moderate size (p. 78). Obviously enjoying the opportunity to play with impressive figures, an amusement he shares with many other writers of the eighteenth century, Galien shows that it would be ten times as heavy as Noah's Ark (p. 79), that it would carry a cargo of 58,000,000 hundredweight (as much as two thousand merchant ships could hold), and that 4,000,000 people could be accommodated, together with their baggage (pp. 84–85). Giving no attention at all to problems of engineering, he specifies, for the construction, skins, cables, cords, and strong double cloth that should be waxed or treated with pitch (p. 77). As on Swift's flying island, a whole civilisation is to be made airborne. An order of magnitude greater than the Ark, Galien's flying ship could provide refuge from another flood, or perhaps from a world engulfed in pestilential air.

As the air in which the ship sailed would probably be too cold and too tenuous to support human life, Galien explores ways of accommodating the crew in comfort. The hull would provide protection from the wind, and in any case the hold would be made deep enough to reach down into temperate regions of the atmosphere. In the case of a really high-flying ship, he proposes that the crew might manage it from a number of little skiffs suspended from the hull on ropes and pulleys. Two or three small dinghies would be attached in turn to each of the skiffs so that the aeronauts could descend to and rise from the earth at will, like glaziers using cages to work on the upper windows of churches (p. 60).

Despite its gigantic size, such a ship would, if accidentally 'submerged,' crash to the earth only very slowly. Weighing, when 'sunk,' only one-third as much again as its volume of air, the ship would descend much less rapidly than even the lightest feather. Having perhaps grown bored with sailing between earth and sky, the aeronauts would probably welcome the chance to tell us of what they had seen while flying (pp. 86–87).

Galien makes almost no attempt to discuss the practical details of his flying ship. Although he makes a passing reference to the difficulty of launching such a structure, he says nothing about propulsion or means of control and is typical of his century in focussing on cost as the greatest single problem. He is almost equally silent about the advantages of flying, not seeming to realise, in particular, how his ship might be used to help prevent the occurrence of the hailstorms with which he began. While the familiar moral questions so often raised in connexion with the possibilities of human flight are not expressly discussed, Galien's similes and allusions several times evoke them by implication. The vast human cargo might, he suggests, be 'a numerous army, with its weapons and provisions.' Such an army would be engaged on an imperial enterprise, sailing to 'the middle of Africa, or other countries no less unknown' (p. 76). Travelling for such practical purposes, the passengers would find no delight in flying for its own sake. Of all the ships, indeed, Galien's is the one that 'flies' least. As the appeal to the church glaziers suggests, it is more like a floating structure than a truly aerial carriage. There is in his book a striking lack of spiritual awareness, of real imaginative energy, of zest for life. Conceiving of his addendum as a 'physical and geometrical amusement,' Galien treats his flying ship with a lack of seriousness bordering on cynicism. The contrast with Oresme's cosmological sweep of mind is marked.

Although after the end of the Middle Ages flying ships were no longer held to be implicated in the creation of storms, they continued to be connected with the gods and the weather in the imaginations of some creative writers. In 1770 the theatregoers of Paris were amused by a farcical play called *Le cabriolet volant,* by Jean François Cailhava (1731–1813).[19] A work of no literary importance, it consists of a series of episodes in which Harlequin outwits others by the use of a flying chariot which he is given by an inventor called Musco. When he presents it to Harlequin, Musco explains how to fly it: 'By lowering this spring, you will rise; by raising it, you will descend; and you can go to right or left by turning it to the direction in which you want to go.'[20] Musco's cursory instructions read like a parody of the technical details offered by real eighteenth century inventors. For the maiden flight, Harlequin arranges a scene that provides an ironic commentary on the traditional fears of flying robbers. He gathers together all of his creditors, leading them to suppose that he is about to pay his accounts. He and his companion Pierrot mount the chariot, 'Pierrot, or a false Pierrot, sitting behind, with a parasol in his hand.' Instead of paying what he owes, Harlequin depresses the spring and flies away, jeering gleefully. Elsewhere in the play, and in the sequel performed in the same year,[21] Harlequin's mastery of the air leads frightened earthbound people to think him a god. While relying for some of its effects on *The Thousand and One Nights,* on which it is loosely based, *Le cabriolet volant* also appeals to the audience's awareness of contemporary talk about flying machines.

The ghostly galleon of the moon forms the basis of the famous and more significant chariot in Act IV of *Prometheus Unbound* (1820). Going even further than Oresme in conceiving of flight as harmonious with the natural order, Shelley makes the wheels of the chariot from clouds and invokes the *tempestarii* with no implication of moral disapproval. A transformation of Ezekiel's cherub wagon, the vision is described in a tone of entranced celebration:

> *I see a chariot—like that thinnest boat*
> *In which the mother of the months is borne*
> *By ebbing light into her western cave—*
> *When she upsprings from interlunar dreams*
> *O'er which is curved an orblike canopy*
> *Of gentle darkness, and the hills and woods,*
> *Distinctly seen through that dusk airy veil,*
> *Regard like shapes in an enchanter's glass;*
> *Its wheels are solid clouds, azure and gold,*
> *Such as the genii of the thunderstorm*
> *Pile on the floor of the illumined sea*
> *When the sun rushes under it. . . .*[22]

While poetry and the arts, folklore, and science contain many expressions of the urge to fly, I believe that the image of the flying ship is to be numbered among them in only an ambiguous sense. Although it reveals an intense longing for the ultimate achievement of flight—ascent to another and higher world—it also powerfully expresses the fear of entrusting oneself to the air. Common usage reminds us that to fly is to be free, and visions of flight seen from the point of view of the aeronaut traditionally stress the sense of release. The flying ship, seen from below, reveals man in a state of inevitable bondage from which he can merely glimpse a dream of freedom. Even on Galien's grandiose Ark, ropes and pulleys ensure, as did the anchor ropes on the legendary craft of the chronicles, some positive physical contact with the ground. His vision of a vast army of slightly reluctant airborne soldiers, human successors to the *tempestarii,* gives us good cause to fear, as Pluche had done, the invention of the art of flying. Among my examples the only significant exception is the ecstatic passage from Shelley, in which faith and hope are victorious. In Act IV of *Prometheus Unbound* all bonds of the imagination are broken, higher worlds are available to spiritual experience, the eye of poet and reader are everywhere. The flying ship merges with both the natural and the supernatural worlds, fear of the unknown is overcome, and man is for a time released from his deep-seated obsession with the air.

Appendix I: a directory of heavier-than-air flying machines in Western Europe, 850 B.C.–A.D. 1783

This directory attempts to list all heavier-than-air flying machines, whether models or of man-carrying size, that are said to have been built and tested in Western Europe prior to the Montgolfiers. While I do not include machines that are known to have been totally imaginary, I have been fairly liberal in my sifting of the evidence. Thus the list includes items ranging from those about which there is no historical doubt whatever (e.g., the ornithopters of Pierre Blanchard) to others that may never in fact have existed (e.g., the zany structure conceived by d'Alcripe's drunken Norman labourer).

In selecting material from the many passing allusions to flight in the work of early historians and chroniclers I have, in general, accepted items only when there is some indication of both time and place. Thus, while I ignore phrases such as 'recently attempted by two men' (Cardanus), I include Campanella's 'a certain Calabrian a few years ago.' Although my aim has been to list all examples of winged flight, I must inevitably have missed some.

A few well-known examples from outside Western Europe are added for comparison, but here I have made no attempt to be comprehensive.

Rotorcraft and parachutes are excluded, as also are wingless rockets and all forms of nonaerodynamic lift. Francesco Lana and others refer to an inventor named Mögling who, in the seventeenth century, is said to have raised a man and lowered him again safely by means of hidden bellows. As Erasmus Francisci says: 'Even if this is true, to have been lifted and let down again is not to have flown.'[1]

In 1709 Joannes Ludovicus Hannemann expressed his scepticism of the whole enterprise:

> Hitherto we have had no examples or experimental evidence to show that any man has ever flown successfully: on the contrary, we have examples of men who have broken their legs in the attempt.[2]

While there were certainly no successful flights in the full sense of the words before 1783, some primitive aviators may have managed semicontrolled glides followed by crash landings. In the notes I comment on one or two such possibilities.

In most cases I give brief details of at least one primary source, but I do not attempt to list all sources. A suggested reference number is included for each flight, based on the date of the event insofar as it is known.

After the directory I have added a checklist of unadopted items, which I have so far been unable to confirm, and a further checklist of spurious flights, with brief comments on my reasons for rejection.

The successful flights of the first hot air balloons in 1783 had an enormously stimulating effect on aeronautical experiments, especially in France. Not only balloons but also fixed-wing aircraft, flappers, parachutes, and a variety of eccentric, improbable machines were announced, designed, and sometimes tested in public. So much was going on that it is not practicable to continue the list so as to close the gap between 1783 and the earliest entries in Charles Gibbs-Smith's *A Directory and Nomenclature of the First Aeroplanes 1809 to 1909.*[3] I hope to give a full account of that period in a separate monograph: *Aviation: The First Twenty-Five Years.*

A Directory of Heavier-Than-Air Flying Machines

Date	*Place*	*Identity*	*Flying Machine*	*Duration/Distance*	*Sources and Notes*	*Ref. No.*
c. 850 B.C.	Troja Nova (London), England	King Bladud	Wings attached to the arms.	Fell on to the Temple of Apollo and was killed.	Fabyan, *The Chronicle* (1516) f. viii^{r}. See also H. C. Levis, *The British King Who Tried to Fly* (London 1919). Although Bladud is legendary, the story of his flight may have some distant factual basis.	−850.1
4th century B.C.	Greece	Archytas of Tarentum (fl. c. 400–350 B.C.).	A wooden dove, worked by 'a current of air hidden and enclosed within it.' Model.	?	Aulus Gellius, *Noctium atticarum libri xx,* X.12.8–10. Sometimes interpreted as a kite. 'Dove' should perhaps be read 'small flying object.' Cf. modern 'big birds' = airliners.	−375.1
c. A.D. 60	Rome, Italy	Actor at a feast given by Nero	Feathered arms?	0/0 (fatal).	Suetonius, VI.xii.2. Such spectacles appear to have been comparatively frequent.	+60.1

A Directory of Heavier-Than-Air Flying Machines *(continued)*

Date	*Place*	*Identity*	*Flying Machine*	*Duration/Distance*	*Sources and Notes*	*Ref. No.*
c. 875	Andalusia, Spain	Abu'l-Kāsim 'Abbas b. Firnās	Feather-covered wings; body covered in feathers; no tail.	'A considerable distance,' alighting at his starting point.	al-Makkari, *The History of the Mohammedan Dynasties in Spain* I, trans. de Gayangos (London 1840) 148.	+875.1
1002–3 or 1009–10	Nīsābūr, Arabia	al-Djawharī	Wings made of wood.	0/0. Threw himself from the top of a mosque and fell to the ground, where he was killed.	A. Zéki Pacha, 'L'aviation chez les Arabes,' *Bulletin de l'Institut égyptien* 5th s., V (1911) 92–101. See also 'al-Djawharī,' by L. Kopf, in *The Encyclopaedia of Islam, new edition* (Leiden and London 1960 etc.).	+1002.1
c. 1010	Malmesbury, England	Eilmer, a monk (c. 980–1066)	Wings attached to the hands and feet; no tail.	More than a *stadium* (= 606.75 ft) from the top of a tower; broke his legs.	William of Malmesbury, *De gestis regum anglorum* I, ed. Stubbs (London 1887) 276–77.	+1010.1
1162	Constantinople	A Turk	Saillike wings made of 'a long and large white garment, gathered into many pleats and foldings.'	He had planned to fly a furlong from a high tower but fell immediately to the base.	Nicetas Choniates, *Historia* (Basileae 1557) 60. Frequently quoted in the Renaissance and later.	+1162.1
c. 1232	Bologna, Italy	Buoncompagno, a Florentine	Wings (to be flapped by the arms?).	Flight abandoned at the last moment.	Salimbene de Adam, *Cronica* (13th century) I, ed. Scalia (Bari 1966) 109–10.	+1232.1
c. 1250	?Oxford, England	A friend of Roger Bacon	Flying boat or carriage, with wings flapped by turning a crank handle.	Presumably 0/0.	Roger Bacon, *De mirabili potestate artis et naturae* (c. 1260) (Lutetia Parisiorum 1542) f. 42^{r-v}.	+1250.1

A Directory of Heavier-Than-Air Flying Machines *(continued)*

Date	*Place*	*Identity*	*Flying Machine*	*Duration/Distance*	*Sources and Notes*	*Ref. No.*
c. 1420	Venice	Giovanni da Fontana (c. 1395–c. 1455)	Model dove powered by a rocket.	Apparently about 100 ft for each flight.	Fontana, *Metrologum de pisce cane et volucre* (c. 1420), Bologna, *Biblioteca universitaria,* MS 2705, ff. 95ᵛ–104ᵛ.	+1420.1
c. 1474	Nuremberg, Germany	Regiomontanus (1436–76)	Mechanical fly made of iron.	A circuit around the dinner table.	Petrus Ramus, *Scholarum mathematicarum libri unus et triginta* (Basileae 1569) II.65. Although the account as given must be inaccurate, Regiomontanus may have experimented with some kind of flying model. (His other flying invention, an 'eagle,' was a kite.)	+1474.1
c. February $14\frac{98}{99}$	Perugia, Italy	Giovanni Battista Danti (c. 1477–1517)	Feathered wings on a structure of iron bars.	Trial flights over Lake Trasimeno, followed by a flight from a tower across the city square, crashing on to the roof of Saint Mary's Church.	Cesare Alessi, *Elogia civium perusinorum* II (Romae 1652) 204–07.	+1498.1
?1505	Monte Ceceri, Italy	Leonardo da Vinci (1452–1519)	Complex ornithopter.	?	Leonardo, *Sul volo degli uccelli* (1505) f. 18ᵛ. Whether Leonardo ever undertook a flight with one of his ornithopters is uncertain. He could perhaps have succeeded in making a short glide, but in 1550 Cardanus wrote that Leonardo had tried 'in vain.'	+1505.1

A Directory of Heavier-Than-Air Flying Machines *(continued)*

Date	*Place*	*Identity*	*Flying Machine*	*Duration/Distance*	*Sources and Notes*	*Ref. No.*
September 1507	Stirling Castle, Scotland	John Damian, an Italian adventurer	Wings made of hen feathers.	A very short distance from the walls of the Castle.	John Lesley, *The Historie of Scotland* (1568–70) (Edinburgh 1830) 76.	+1507.1
1536	Troyes, France	Denis Bolori (d. 1536)	Wings flapped by a spring mechanism.	Two or three kilometres from the tower of the cathedral, ending in a fatal crash after a spring broke.	Pierre Jean Grosley, *Oeuvres inédites* I (Paris 1812) 84–88.	+1536.1
20 June 1540, at 5:00 P.M.	Viseu, Portugal	João Torto	Two pairs of wings, of which the upper were larger than the lower, covered with calico, and joined by iron hoops lined with cloth. The flier was provided with a helmet representing an eagle's head with open beak.	From the tower of the cathedral, intending to fly to the nearby Saint Matthew's fields. Crashed on to a roof when the helmet slipped over his eyes and died a few days later.	Donna Maria da Glória, *Probenda* (17th century). Quoted in a number of later Portuguese books on aviation history, e.g., Albino Lapa, *Aviação portuguesa* (Lisboa 1928) 12. According to Donna Maria, Torto had his attempt announced by the town crier on 1 June 1540.	+1540.1
c. 1550	Tour de Nesles, Paris, France	An Italian	Wings, possibly of cloth.	According to the poet Augié Gailliard, he dropped 'like a pig' close to the base of the tower and broke his neck.	Pierre de Saint Romuald, *Trésor cronologique et historique* III (Paris 1669) 583.	+1550.1
16th century	Saint Mark's, Venice	?	Wings.	?	J. Sturm, *Linguae latinae resolvendae ratio* (Argentoraci 1581) 40. Probably imaginary.	+1550.2

A Directory of Heavier-Than-Air Flying Machines *(continued)*

Date	*Place*	*Identity*	*Flying Machine*	*Duration/Distance*	*Sources and Notes*	*Ref. No.*
16th century	Nuremberg, Germany	An old church cantor	Wings flapped by a mechanism including wheels.	'Flew here and there,' but broke his arms and legs when the mechanism failed.	J. E. Burggravius, *Achilles* (Amsterodami 1612) 52.	+1550.3
16th century	Normandy, France	A French labourer	Wings made of the two halves of a winnowing basket, with a coal shovel for a tail.	0/0. Fell from the top of a pear tree into a drain and broke his shoulder.	Philippe d'Alcripe, *La nouvelle fabrique des excellens traits de vérité* (16 century), ed. Gratet-Duplessis (Paris 1853) 178–79. Probably fictional.	+1550.4
1557–58	San Yuste, Spain	Giovanni Torriano (d. 1580 or 1581)	Wooden sparrows (models).	A circuit around the dining room of Charles V's retreat near the monastery.	F. Strada, *De bello belgico* (Romae 1632) 8.	+1557.1
c. 1589	Conway, Wales	John Williams	Long coat used as a sail or wings.	Fell almost immediately on to a stone which emasculated him.	John Hacket, *Scrinia reserata* (London 1693) 8. Williams was about seven years old at the time.	+1589.1
c. 1600	Lucca, Italy	Paolo Guidotti (c. 1560–1629)	Wings made of whalebone and covered with feathers; springs were used to give them curvature.	About $\frac{1}{4}$ mile, starting from 'a height.' Fell through a roof after his arms grew tired. Broke a thigh and was left in 'a sorry plight.'	F. Baldinucci, *Notizie de' professori del disegno* IV (Firenze 1700) 248–50.	+1600.1
c. 1600	Venice	Giovanni Francesco Sagredo (1571–1620)	Wings based on those of a falcon.	Threw himself from a height and arrived 'many yards from his starting point.'	Paris, *Bibliothèque nationale,* MS Latin 11195, f. 57^{r-v}.	+1600.2

A Directory of Heavier-Than-Air Flying Machines *(continued)*

Date	*Place*	*Identity*	*Flying Machine*	*Duration/Distance*	*Sources and Notes*	*Ref. No.*
c. 1610	Calabria, Italy	?	?	Broke his legs on landing.	Tommaso Campanella, *De sensu rerum* (Francofurti 1620) 280: 'a certain Calabrian, a few years ago.'	+1610.1
c. 1620	Schussenried, Germany	Kaspar Mohr, a monk (1575–1625)	Wings made from goose-feathers held together by whipcord.	Practice flights.	S. B. Wilhelm, 'Schweikart und Mohr, zwei schwäbische Flieger aus alter Zeit,' *Illustrierte aeronautische Mitteilungen* 13 (1909) 441–45.	+1620.1
c. 1640	England	Gascoyn	Winged arms?	?	Robert Hooke, 'An Account of the Sieur Bernier's [*sic*] Way of Flying,' *Philosophical Collections* I.1 (1679) 15.	+1640.1
c. 1640	? Near Vauxhall, England	An English boy	Winged chariot made from farming machinery.	Said to have flown the length of a barn. The inventor, the Marquis of Worcester, said that he knew 'how to make a man fly; which I have tried with a little Boy of ten years old in a Barn, from one end to the other, on an Hay-mow.'	Edward Somerset, Second Marquis of Worcester, *A Century of . . . Inventions* (London 1663) 54–55 (invention 77). Although perhaps entirely apocryphal (Worcester was a gross exaggerator), some kind of experiment may have been made.	+1640.2
1647–8	Krakow, Poland	Tito Livio Burattini (1617–c. 1680)	Flying dragon: a complex ornithopter of which at least three working models were made.	Successful indoor flights reported.	Paris, *Bibliothèque nationale,* MS Latin 11195, ff. 50^r–61^r. The projected man-carrying machine appears never to have been built.	+1648.1 +1648.2 +1648.3

A Directory of Heavier-Than-Air Flying Machines *(continued)*

Date	*Place*	*Identity*	*Flying Machine*	*Duration/Distance*	*Sources and Notes*	*Ref. No.*
c. 1650	Augsburg, Germany	Salomon Idler, of Cannstatt, (c. 1610–c. 1670), a cobbler	Wings made of iron and feathers.	Dissuaded from flying from a tower, he flew instead from a low roof on to a bridge covered with mattresses. He broke the bridge, killing some hens nesting under it. Later he took his wings to Oberhausen and chopped them to pieces.	J. J. Becher, *Närrische Weiszheit und weise Narrheit* (Franckfurt 1682) 164–68; C. J. Wagenseil, *Versuch einer Geschichte der Stadt Augsburg* IV.2 (Augsburg 1822) 485–87.	+1650.1
c. 1650	Scutari (Üsküdar), Turkey	Hezârfen Ahmed Çelebi	Wings, like those of an eagle, attached to the arms.	Began with training flights, 'turning round and round in the air'; then, starting from Galata Tower, flew several kilometres, landing in Dŏgancılar Square, the marketplace of Scutari.	Evliyâ Çelebi (1611–83), *Seyahatnâme* I (Istanbul A.H. 1314 [= 1896]) 670. See also *Türk Ansiklopedesi* XIX (Ankara 1971) 207.	+1650.2
17th century	London, England	A Frenchman	Bat-shaped wings of leather, with wooden ribs and iron hinges; no tail.	Managed a safe descent from the roof of St Paul's, London. Broke his neck at a second attempt when one of the hinges failed.	Georg Heinrich Büchner, *Merkwürdige Beyträge zu dem Weltlauf der Gelehrten* III (Langensalza 1766) 542–66. Account may be fictional.	+1650.3 +1650.4

A Directory of Heavier-Than-Air Flying Machines *(continued)*

Date	*Place*	*Identity*	*Flying Machine*	*Duration/Distance*	*Sources and Notes*	*Ref. No.*
17th century	The Netherlands	Adriaen Baartjens	Improved wings like those of the Frenchman above, but with the addition of eagle feathers and a tail like that of an eagle.	A successful trial flight from the highest tower in Rotterdam was followed, about a year later, by a trial flight from a tower in The Hague, with a safety line attached. A further free flight ended in a crash that broke his arm.		+1655.1 +1656.1 +1656.2
1658–59	Oxford, England	Robert Hooke (1635–1703)	Model bird, powered by 'springs and wings.'	Hooke says it 'rais'd and sustain'd it self in the Air.'	Richard Waller, 'The Life of Dr. Robert Hooke,' prefaced to *The Posthumous Works of Robert Hooke* (London 1705) iii–iv.	+1658.1
c. 1660	Nuremberg, Germany	Johann Hautsch (1595–1670), a skilled mechanic and coach builder	Flying carriage?	?	Becher, *Närrische Weiszheit* 164–68. Little is known of the work, which appears to have been carried out late in Hautsch's life.	+1660.1
15 January $16\frac{72}{73}$ at 7:00 P.M.	Regensburg, Germany	Charles Bernouin, a surgeon of Grenoble	Wings, described as a 'well-tensioned sail.' The flight was assisted by rockets.	Said to have flown from a high tower. A variant account in *Le journal des sçavans* (12 December 1678) (ed. of Amsterdam 1679, 455) says that he broke his neck flying at Frankfurt.	*Le mercure hollandois* (Amsterdam 1678) 98–99. Bernouin (spelled 'Bernovin' in the *Mercure*) had spent eight years in Germany. He was reputed to be good at flying.	+1673.1

A Directory of Heavier-Than-Air Flying Machines *(continued)*

Date	*Place*	*Identity*	*Flying Machine*	*Duration/Distance*	*Sources and Notes*	*Ref. No.*
1678	Sablé, France	Besnier, a locksmith	Hinged wings made of taffeta stretched over frames; flapped alternately using both arms and legs.	Did not claim to be able to rise from the earth but, starting from a height, to be able to sustain himself sufficiently to cross a wide river.	*Le journal des sçavans* (12 December 1678) (ed. of Amsterdam 1679, 452–55).	+1678.1
11 February 16$\frac{78}{79}$	Venice	?	?	A flight from a high tower in Venice, on the occasion of the annual banquet of the Duke and Counsel.	Erasmus Francisci, *Der Wunder=reiche Uberzug unserer Nider-Welt* (Nuremberg 1680) 370.	+1679.1
3 April 1680 and again soon afterwards	? Moscow, Russia	A Polish peasant	Wings made of mica.	0/0. Both attempts having failed, the peasant was obliged to refund subsidies he had received; he was also severely beaten.	See Savorgnan di Brazzà, *La navigazione aerea* (Milano 1910) 15, who cites the Russian chronicler Zheliabuzhskii (b. 1638).	+1680.1 +1680.2
c. 1710	Halle, Germany	Johann Gabriel Illing, a locksmith	Artificial eagle, with a wing span of 5 ells (= approx. 12 ft). The pilot was to have sat inside while the feather-covered wings were operated by a perpetual motion machine.	Probably never finished.	Johann Gottfried Zeidler, *Der fliegende Wandersmann oder philosophische Untersuchungen der Fliegekunst* (Halle 1710). Possibly fiction elaborated from some factual basis.	+1710.1

A Directory of Heavier-Than-Air Flying Machines *(continued)*

Date	*Place*	*Identity*	*Flying Machine*	*Duration/Distance*	*Sources and Notes*	*Ref. No.*
c. 1712	Saint Germain, France	Charles Allard (c. 1650–after 1711), an actor and acrobat at the French court	Wings attached to the arms.	An attempt to fly from the Terrasse de Saint Germain to the Bois du Vésinet.	Traditionally said to have been gravely injured in the attempt and to have died from his injuries.	+1712.1
c. 1730	Turin, Italy	Abbé don Falco	? Possibly a lighter-than-air machine, like Lana's.	?	Johann Georg Keyssler, *Neüeste Reise* I (Hannover 1740) 252.	+1730.1
1742	Paris, France	The soi-disant Marquis de Bacqueville (c. 1680–1760)	Wings attached to the arms and legs.	A short distance across the Seine, crashing into a boat.	Pierre-Mathias Charbonnet, *Eloge prononcé par La Folie devant les habitans des petites-maisons* (Avignon 1761).	+1742.1
1750	Wildberg (Württemberg), Germany	Schweikart, a miller	Two large wings of taffeta.	0/0. Stood on a high mountain before many observers and tried to fly over the town in the valley below; fell, smashed the wings, and hurt himself.	*Franz Lana und Philipp Lohmeier von der Luftschiffkunst,* trans. anon. (Tübingen 1784) sig. $\chi\ 5^v$–[$\chi 6^r$].	+1750.1
October 1751	London, England	Andrea Grimaldi	Bird-shaped flying carriage of complex construction; wingspan 22 ft.	0/0. Probably never tested.	*The Whitehall Evening-Post* (3–5 October 1751) 1. Grimaldi, posing as a widely travelled priest, appears to have been a charlatan.	+1751.1
c. 1770	Etampes, France	Canon Pierre Desforges (b. c. 1723)	Feathered wings.	0/0. The wings were fixed to a peasant, who refused to make the attempt.	*Annonces, affiches, nouvelles et avis divers de l'Orléanois* 36, 39, 40, (4, 25 September and 2 October 1772) 147–48, 161–62, 165–66.	+1770.1

A Directory of Heavier-Than-Air Flying Machines *(continued)*

Date	*Place*	*Identity*	*Flying Machine*	*Duration/Distance*	*Sources and Notes*	*Ref. No.*
1772	Etampes, France	Desforges	Wickerwork gondola with flapping wings of 19.5 ft span, and with an overhead canopy of 8 ft × 6 ft.	0/0. Immediate fall from the top of the Tour Guinette (c. 100 ft).	See *Annonces* and Laurent Gaspar Gérard, *Essai sur l'art du vol aérien* (Paris 1784) 40–45.	+1772.1
August 1781	Emmendingen, Germany	Carl Friedrich Meerwein (1737–1810)	Ornithopter of wood and fabric; wing area 111 ft^2; triangular tail; weight 56 lb.	Probably never tested.	Carl Friedrich Meerwein, *Die Kunst zu fliegen nach Art der Vögel* (Frankfurt und Basel 1784).	+1781.1
Autumn 1781	Saint Germain, France	Jean Pierre Blanchard (1750–1809)	A nacelle 4 ft × 2 ft, with four wings, each 10 ft long.	0/0.	*Journal de Paris* for the period. See also Jules Duhem, *Histoire des idées aéronautiques avant Montgolfier* (Paris 1943) 174–76.	+1781.2
End of autumn 1781	Saint Germain, France	Blanchard	*Vaisseau volant*: a larger machine, similar to the first. Four oval wings, hinged along their central spars, like those of Besnier (see earlier).	Never tried in public.	See *Journal de Paris* and Duhem, *Histoire des idées aéronautiques.* There are also four contemporary illustrations that have often been reproduced.	+1781.3

A Directory of Heavier-Than-Air Flying Machines *(continued)*

Date	*Place*	*Identity*	*Flying Machine*	*Duration/Distance*	*Sources and Notes*	*Ref. No.*
End of 1782–1783	Saint Germain, France	Blanchard	A machine for producing vertical lift on the jellyfish jet propulsion principle (cf. Morris's fictional flying machine of 1751).	Some lift demonstrated in trials.	Duhem, *Histoire des idées aéronautiques* 176. Blanchard continued to experiment with his heavier-than-air machines until the success of the hot-air balloon converted him to lighter-than-air craft. See Léon Coutil, *Jean-Pierre Blanchard: physicien-aéronaute* (Evreux 1911).	+1782.1
1783	? Paris, France	?	Wings based on measurements of a large number of birds.	Said to have glided down from a height of c. 500 yards before falling on to the top of an open well. The wing structure saved him from falling in.	*Journal politique de Bruxelles* (= part II of *Mercure de France*) (18 October 1783) 127.	+1783.1

A Checklist of Unadopted Items

(I have provisionally excluded the following because I have thus far been unable to confirm the details in primary sources.)

Date	*Place*	*Identity*	*Notes*	*Source*	*Provisional Ref. No.*
c. 1700	Rozoy Abbey, France	Canon Oger	Among the several seventeenth century abbeys and priories called Rozoy, the one most likely to have been intended is either Rozoy-le-Grand, at Oulchy-le-Château, or Rozoy-le-Bellevalle, near Château-Thierry, both in Aisne. A search of the documents relevant to the abbeys has failed to confirm the story.	Charles H. Gibbs-Smith, *Aviation* (London 1970) 12.	+1700.1

A Checklist of Unadopted Items

Date	*Place*	*Identity*	*Notes*	*Source*	*Provisional Ref. No.*
c. 1700	Péronne, France	A priest	Owing to the destruction during the world wars of many mss relevant to the region, it may never be possible to document the story.	Gibbs-Smith, *Aviation* 12.	+1700.2
c. 1765	Romania	A peasant named Kostic	Said to have built a rustic machine covered with bark and based on the idea of the kite. Kostic is credited with some successful hops, including one of about 100 m.	Jules Duhem, *Histoire des idées aéronautiques avant Montgolfier* (Paris 1943) 229.	+1765.1

A Checklist of Rejected Items

Date	*Identity*	*Notes*	*Source*
c. 1265	Griffolino d'Arezzo	Griffolino, an alchemist of Arezzo, is placed in Hell by Dante. He says that he jestingly claimed to be able to fly and that he offered to teach the art to Albero of Siena. When he failed to do so, Albero had him burned at the stake. It is doubtful whether there is any substance behind the story.	Dante, *Inferno* XXIX.109–20.
1607	William Bush	Bush invented a ship that could travel through air, on land, and in water. Wheels were used on land, while travelling through the air meant no more than being hauled up ropes attached to a tower.	[Anthony Nixon], *A True Relation of the Travels of M. Bush, a Gentleman* (London 1608).
c. 1650	Saint Joseph of Copertino (1603–63)	Said to have experienced levitation many times, Saint Joseph is the patron saint of fliers.	See *Bibliotheca sanctorum* VI, 1300–03.
17th century	Daniel Mögling	See introduction to this appendix.	
2 February $17\frac{39}{40}$	Robert Cadman	A funambulist showman, Cadman attempted to slide down a cord from near the top of the spire of Saint Mary's church, Shrewsbury, to the other side of the Severn. The inscription on the west wall of the church appears to say that he was trying to fly. He fell and died, aged twenty-eight, when the cord broke as a result of having been overtightened.	H. Owen and J. B. Blakeway, *A History of Shrewsbury* II (London 1825) 409–11n.

Appendix II: parachutes, sponges, and tenuous air

Two related and well-executed drawings in British Library Add. MS 34113 (Figures 54, 55), one of which depicts what is probably the first European parachute, have been perceptively and wittily analysed by Lynn White.[1] Among the most impressive of White's insights is his understanding of the artist's treatment of the mouth in the first drawing: 'At first glance the expression of the mouth is odd; the reason is that he is gripping a sponge between his teeth to protect his jaws from the shock of landing. All that is braking his fall is a pair of cloth streamers. He looks scared. He should be.' Commenting on the second drawing, he points out that 'the sponge is now secured in his teeth by a strap that runs around his head, so that if he cries out in terror he will not drop it.'[2]

Perceptive though this analysis is, it omits an additional and equally important reason for the use of the sponge. The tenuousness of the air in the upper regions was among the most familiar features of the classical world model, and commentators frequently mentioned the difficulty of breathing in such conditions. To render the air denser and more breathable, men walking on high mountains appear sometimes to have held damp sponges before their mouths.[3] It seems likely that the parachutists are trying to create a similar effect.

What the artist intended to be the function of the cloth streamers in the first drawing remains a mystery.

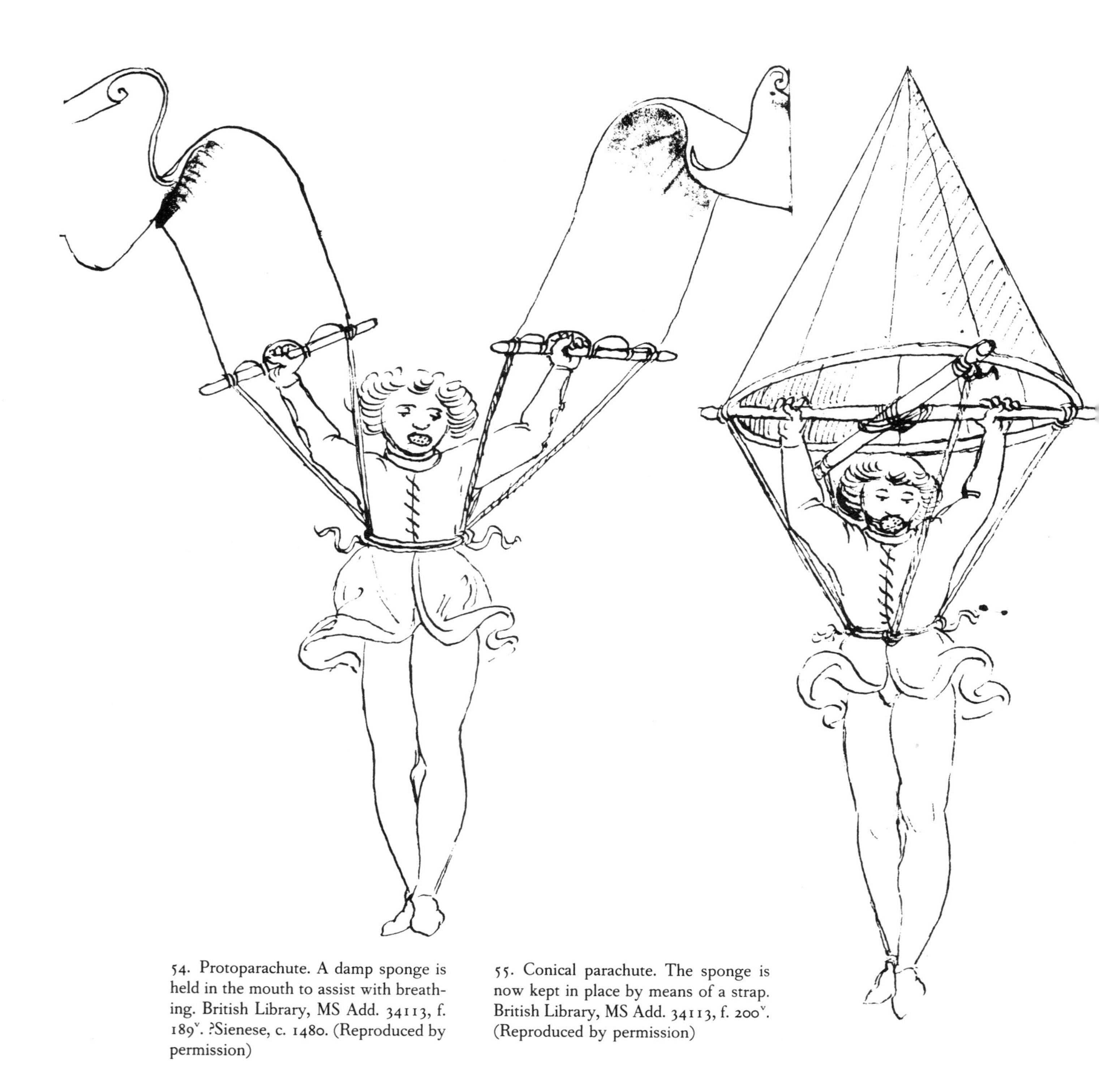

54. Protoparachute. A damp sponge is held in the mouth to assist with breathing. British Library, MS Add. 34113, f. 189ᵛ. ?Sienese, c. 1480. (Reproduced by permission)

55. Conical parachute. The sponge is now kept in place by means of a strap. British Library, MS Add. 34113, f. 200ᵛ. (Reproduced by permission)

I. Creation of the birds and fishes. The wavy lines used to depict the air suggest its kinship with water. London, British Library, MS Cotton Claud. B. IV, f. 3^{v}, eleventh century. (Reproduced by permission of the British Library)

II. Adam naming the animals. MS Cotton Claud. B. IV, f. 6^{r}, eleventh century. (Reproduced by permission of the British Library)

Lignum q: faciens fructum & hab
unumquodq: sementem scdm spe
suam. Et vidit deus qd esset bonu
factumq: e vespe et mane dies ter
Dixit autem deus. fiant lumin
a in firmamento celi ut dividan
em et noctem. & sint in signa & tem
ra & dies & annos ut luceant i firm
mento celi et illuminent terram
factum est ita. fecitq: deus duo m
gna luminaria. Luminare ma
ut pesset diei & luminare min
ut pesset nocti. Et stellas. Et po

III. Creation of the birds and fishes. The symmetry of the creatures' movement towards their proper regions is well indicated. London, British Library, MS Add. 15425, f. 3ᵛ, late fourteenth century. (Reproduced by permission of the British Library)

IV. Creation of the birds and fishes. God's lively, balletic movement is depicted in only the best illustrations of the *Bible historiale.* Paris, Bibliothèque de l'Arsenal, MS 5059, f. 5ʳ, 1370. (Reproduced by permission)

V. Creation of the birds and fishes. The birds, following the line of God's torso, contribute to the expressive formalism of this picture as they fly away along the diagonal network of the diapered background. Paris, Bibliothèque nationale, MS franç. 8, f. 5^{v}, fourteenth century. (Reproduced by permission)

VI. Creation of the birds and fishes. God blesses the bird tightly held in His left hand before releasing it to join in the gay dance of its fellows, painted in white on a gold leaf ground. Amiens, Bibliothèque municipale, MS 21, f. 7^{r}, thirteenth century. (Reproduced by permission)

VII. Creation of the birds and fishes. This rather decadent version of the scene, with its tamed parkland and farmyard animals, is decidedly unaerial. Paris, Bibliothèque nationale, MS franç. 20087, f. 3^v^, fifteenth century. (Reproduced by permission)

VIII. Creation of the birds and fishes. The farmyard domestication is here less significant than the joyous aerobatics of the eight birds around God's head. Although the birds share in the mysterious and stiff formality of their Creator, they express delight in the freedom of the air. London, British Library, MS Royal 19. D.III, f. 5^v^, 1411. (Reproduced by permission of the British Library)

IX. Composite Creation scene. Set simultaneously in night and day, this charming celebration of the created world is an interesting combination of realism and symbolism. The vastly out-of-scale phoenix is an emblem of the resurrection to come. The birds are the only unenclosed creatures. Circular depictions of created space as a whole are a familiar feature of German books in the Middle Ages and the Renaissance. Heidelberg, Universitätsbibliothek, MS pal. germ. 471, f. 56^{v}, 1425. (Reproduced by permission)

X. Bat-winged demons falling right through the surface of the earth to their damnation. Paris, Bibliothèque nationale, MS franç. 6260, f. 156r, fifteenth century. (Reproduced by permission)

XI. Gerfalcon and tethered butterfly. Gluing a live butterfly to the end of a thread was a common Renaissance pastime. Pierre Belon, *L'histoire de la nature des oyseaux* (Paris 1555) 95 (from a coloured copy). (Reproduced by permission of the British Library)

XII. Christ ascending, carried by the cherub wagon of Ezekiel's vision. Cf. Melchior Bauer's eighteenth century realisation of the cherub wagon in Chapter 9. Florence, Biblioteca Mediceo-Laurenziana, MS Plut. 1.56, f. 13^{v}, A.D. 586. (Reproduced by permission)

Notes

Notes to chapter 1

1. Plato, *Cratylus* 404C.
2. Cicero, *De natura deorum* II.xxvi.
3. Macrobius, *In somnium Scipionis* I.xvii.15.
4. Geoffrey Stephen Kirk and John Earle Raven, *The Presocratic Philosophers: A Critical History with a Selection of Texts* (Cambridge 1957) 21–22.
5. Charles H. Kahn, *Anaximander and the Origins of Greek Cosmology* (New York and London 1960) 119–65.
6. Hesiod, *Opera et dies* 547–56.
7. Kahn, *Anaximander* 146–48.
8. Kathleen Freeman, *Ancilla to the Pre-Socratic Philosophers* (Oxford 1948) 19.
9. Arthur Fairbanks, *The First Philosophers of Greece* (London 1898) 19–20. For a Renaissance view of how all things may be generated from air, see Fig. 4.
10. Kirk and Raven, *The Presocratic Philosophers* 435n.
11. Kathleen Freeman, *The Pre-Socratic Philosophers* (Oxford 1946) 282. The argument presented here is a summary based on the extant fragments of Diogenes' writings.
12. Geoffrey Stephen Kirk, *Heraclitus: The Cosmic Fragments* (Cambridge 1954) 339–44.
13. Fairbanks, *The First Philosophers of Greece* 223–24.
14. Diodorus Siculus, *Bibliothecae historicae libri xv* I.12.7.
15. Cicero, *De natura deorum* I.x.26.
16. Seneca, *Naturales quaestiones* II.11.1. Cf. pseudo-Aristotle, *De mundo* 392 b 9–10: 'Air . . . admits of influence and undergoes every kind of change.'
17. Pseudo-Aristotle, *Problemata* XXV, 939 b 27.
18. Hieronymus Cardanus, *Commentarii, in Hippocratis de aere, aquis et locis opus* (Basileae 1570) 30b.
19. Antoine Baumé, *Chymie expérimentale et raisonnée* I (Paris 1773) 63.
20. Pseudo-Aristotle, *Meteorologica* IV, 379 a 15–16.
21. Seneca, *Naturales quaestiones* IV.10.1.
22. Thomas of Cantimpré, *Liber de natura rerum* 19.v, ed. H. Boese (Berlin and New York 1973) I.411. In *Polychronicon* II.v [1327], Ranulph Higden says that the waters of the Flood rose to 15 cubits above the highest mountains because it was to that height that men had polluted the air with fire worship.
23. Seneca, *Naturales quaestiones* II.4.1.
24. Philo Judaeus, *De somniis* I.xxii.134.
25. Philo, *De somniis* I.xxii.144.
26. Philo, *De somniis* I.xxiii.146.

27. Martianus Capella, *De nuptiis philologiae et mercurii libri VIIII* VII.732, ed. Adolfus Dick, rev. Jean Préaux (Stutgardiae 1969) 368.

28. Richard C. Dales, 'A Twelfth-Century Concept of the Natural Order,' *Viator* 9 (1978) 179–92. For the Aristotelian version of an animate cosmos, see, for example, *De caelo* 292 a 20.

29. Philippus Aureolus Paracelsus, 'De elemento aeris' [part I of a fragmentary *Buch meteororum* of uncertain date], in *Achter Theil der Bücher und Schrifften des . . . Paracelsi*, ed. Johannes Huserus (Basel 1590) 289.

30. Erasmus Francisci, *Der Wunder=reiche Uberzug unserer Nider=Welt, oder Erd=umgebende Lufft=Kreys* (Nürnberg 1680) 417.

31. *De generatione et corruptione* 331 a 14ff; Plutarch, *De communibus notitiis adversos Stoicos* 1085D. See also Friedrich Solmsen, *Aristotle's System of the Physical World: A Comparison with His Predecessors* (Ithaca 1960) 353–67.

32. Seneca, *Naturales quaestiones* II.6.5–6.

33. Seneca, *Naturales quaestiones* III.14.2.

34. Seneca, *Naturales quaestiones* III.15.1.

35. Saint Augustine, *De genesi ad litteram libri duodecim* III.x.14.

36. Thierry of Chartres, *Tractatus de sex dierum operibus* XVII.4ff [1130–40], ed. Nicholaus M. Häring (Toronto 1971) 562.

37. Pico della Mirandola, *De elementis*, in *Opera omnia* II (Basileae 1573) 170.

38. Pico, *Heptaplus*, in *Opera omnia* I (Basileae 1557) 18; see also Franz Cumont, *After Life in Roman Paganism* (New Haven 1922) 107, 187.

39. Robert Fludd, *Utriusque cosmi maioris scilicet et minoris metaphysica, physica atque technica historia* III (Francofurti 1619) 197.

40. Petrus Clausenius, *Disputationum physicarum de aere prima* (Havniae 1712) 7–9.

41. Seneca, *Naturales quaestiones* II.10.1. Cf. also pseudo-Aristotle, *De mundo* 392 b 6: 'Air, which is in its nature murky and cold as ice. . . .'

42. Philo, *De opificio mundi* VII, IX. Similar arguments appear in later hexaemera. Cf. Thierry of Chartres, *De sex dierum operibus* XXIII, 565.

43. Philo, *De somniis* I.xxii.145.

44. Plutarch, *De facie quae in orbe lunae apparet* V, 921F–922A.

45. Theophilus of Antioch, *Ad Autolycum* II.13, trans. Robert M. Grant (Oxford 1970).

46. Saint Gregory of Nyssa, *In hexaemeron, PG* 44.85–88. (*PG*=J. P. Migne, *Patrologiae cursus completus, series graeca* [Paris 1857 etc.].)

47. Johannes Philoponus, *De mundi creatione* II.viff.

48. Cardanus, *Commentarii* 18a.

49. Aristotle, *Meteorologica* 342 a 34–342 b 24.

50. Pseudo-Aristotle, *De coloribus* 791 a 1–3, 794 a 7–15.

51. Anonymous, *De elementis*, ed. Richard C. Dales, *Isis* 56.2 (Summer 1965) 186, lines 159ff.

52. Thierry of Chartres, *Tractatus de sex dierum operibus* XI.43–44, 560. Not everyone in the twelfth century thought the air green. Marius speaks of it as naturally colourless though sometimes, as in Aristotle, acquiring accidental colour. See *Marius: On the Elements*, ed. and trans. Richard C. Dales (Berkeley Los Angeles and London 1976) 58–61.

53. Philippus Aureolus Paracelsus, *Philosophia de generationibus et fructibus quatuor elementorum* I.x [c. 1525], in *Achter Theil der Bücher und Schrifften des . . . Paracelsi*, ed. Johannes Huserus (Basel 1590) 62.

54. Paracelsus, 'De elemento aeris' 295.

55. A further source may be found in the comparatively rare meteorological phenomenon known as 'green flash.' See D. J. K. O'Connell, S.J., *The Green Flash and Other Low Sun Phenomena* (Amsterdam and New York 1958).

56. Aristotle, *Meteorologica* 339 b 3ff.

57. *Meteorologica* 340 b 20ff.

58. Aristotle, *Physica* 213 a 1-3. See also *De caelo* 310 a 33-310 b 16.

59. Aristotle, *Physica* 255 b 18-21. See also *De caelo* 310 b 10-16, 311 a 1-6, and cf. *De generatione et corruptione* 318 b 28-29: 'Wind and Air are in truth more real—more . . . a "form"—than Earth.' For some useful introductory essays on Aristotle's ideas about form and primary matter, see Ernan McMullin, ed., *The Concept of Matter in Greek and Medieval Philosophy* (Notre Dame 1963).

60. *Physica* 255 a 24-256 a 3. See also *De caelo* 311 a 1-312 a 22. For a thorough discussion of these and related matters, see Sarah Waterlow, *Nature, Change, and Agency in Aristotle's 'Physics'* (Oxford 1982).

61. Aristotle, *Meteorologica* 340 b 35.

62. Ferdinand Columbus, *The Life of the Admiral Christopher Columbus by His Son Ferdinand*, trans. and ann. Benjamin Keen (London 1960) 147. See also 243.

63. Aristotle, *Meteorologica* 340 b 4ff.

64. Alanus de Insulis, *Anticlaudianus* IV.6, trans. James J. Sheridan (Toronto 1973).

65. Marin Mersenne, *Correspondance* IV, ed. Cornelis de Waard (Paris 1955) 118.

66. Erasmus Darwin, *The Botanic Garden* I (London 1791) canto I, additional note 1.

67. Aristotle, *Meteorologica* 361 a 23ff.

68. William Fulke, *A Goodly Gallerye* . . . (Londini 1563) 18^{r}.

69. Federigo Buonaventura, *De causa ventorum motus* (Urbini 1592) 24-25. Buonaventura summarises the views of Gaietanus de Thienis.

70. Leonardo da Vinci, Windsor drawings 12671^{r}. MacCurdy 415.

71. Leonardo, MS K 113(33)^{v} (1504-09). MacCurdy 411.

72. Leonardo, MS G 10^{r} (1510-15). MacCurdy 409.

73. 'Hevyn and erth and all that here me plain,' line 22.

74. See Chapter 5.

75. Jacques Peletier, *La Savoye* (Anecy 1572) 16.

76. Marie Boas Hall [Marie Boas], *Robert Boyle and Seventeenth-Century Chemistry* (Cambridge 1958) 185.

77. Stephen Hales, *Vegetable Staticks* (London 1727) 316-17.

78. Robert Boyle, *The General History of the Air* (London 1692) 1, 2-4.

79. Antoine Lavoisier, *Essays Physical and Chemical*, trans. Thomas Henry (London 1776) 210. The quotation is from Baumé's *Chymie expérimentale et raisonnée* III (Paris 1773) 694.

80. See Edward Grant, *Much Ado About Nothing: Theories of Space and Vacuum from The Middle Ages to the Scientific Revolution* (Cambridge 1981).

81. Hero of Alexandria, *Pneumatica et automata*, trans. Joseph G. Greenwood, ed. Bennett Woodcroft (London 1851) 3.

82. Lucretius, *De rerum natura* I.329ff.

83. See G. N. Cantor and M. J. S. Hodge, eds, *Conceptions of Ether: Studies in the History of Ether Theories 1740-1900* (Cambridge 1981) 1-60.

84. Robert Hooke, *Micrographia* (London 1665) 217-40.

85. Robert Boyle, *The General History of the Air* (London 1692) 4-5. See also Giovanni Alfonso Borelli, *De motionibus naturalibus a gravitate pendentibus* (Regio Julio 1670) 254-63.

Notes to chapter 2

1. Plato, *Phaedrus* 246D–E.

2. Origen, *Homiliae in Genesim* I.8.

3. Saint John Chrysostom, *Homiliae XXI de statuis* XV.3 [A.D. 387 or 388], trans. W. R. W. Stephens and others (New York 1889) 441.

4. Chrysostom, *Homiliae XXI de statuis* XI.4, p. 416.

5. Johannes Scotus Eriugena, *De divisione naturae* III.36–40.
6. Palladius, Bishop of Helenopolis, *Historia lausiaca* XXVII.
7. Plato, *Timaeus* 91D–E, trans. Francis MacDonald Cornford (London 1937) 358.
8. Plato, *Timaeus* 39E–40A 118.
9. Aristotle, *De generatione animalium* 761 b 13f.
10. Aristotle, *Historia animalium* 490 a 11.
11. Aristotle, *De incessu animalium* 710 a 16ff.
12. Aristotle, *De incessu animalium* 709 b 24.
13. Aristotle, *Historia animalium* 487 b 22.
14. Pseudo-Aristotle, *Meteorologica* IV, 382 a 2ff.
15. Aristotle, *De respiratione* 477 a 27ff. For a discussion of alternative theories about the constitution of animals, see D. O'Brien, *Empedocles' Cosmic Cycle: A Reconstruction from the Fragments and Secondary Sources* (Cambridge 1969) 189–95, 301–13.
16. Pseudo-Aristotle, *De mundo* 398 b 22ff.
17. Aristotle, *De generatione animalium* 761 b 17.
18. Aristotle, *De generatione animalium* 737 a 1.
19. Pseudo-Aristotle, *Meteorologica* IV, 382 a 2ff.
20. Aristotle, *Historia animalium* 552 b 10ff.
21. *Libellus de natura animalium: A Fifteenth Century Bestiary,* facsimile with an introduction by J. I. Davis (London 1958) C2^{v}.
22. Aelian, *De natura animalium* II.31.
23. *Physiologus latinus versio Y, University of California Publications in Classical Philology* 12.7, ed. Francis J. Carmody (Berkeley and London 1941) 132.
24. Aristotle, *De generatione animalium* 761 b 16ff.
25. Plato, *Symposium* 202E–203A.
26. Lucius Apuleius, *De deo Socratis* VIII.138–40, in *Apulei platonici madaurensis opera quae supersunt* III, ed. Paulus Thomas (Stutgardiae 1970) 16–17; trans. in *The Works* (London 1899) 358.
27. Possibly that of Posidonius; see Werner Jaeger, *Aristotle: Fundamentals of the History of His Development,* 2nd ed., trans. Richard Robinson (Oxford 1948) 146n.
28. Apuleius, *De deo Socratis* IX.140, in *Opera* III.17 and *The Works* 359.
29. Apuleius, *De deo Socratis* X.141–43, in *Opera* III.17–18 and *The Works* 359.
30. Philo Judaeus, *De gigantibus* II.7–11. A parallel passage is found in his *De plantatione* III.12.
31. Philo, *De somniis* I.xxii.
32. Philippus Aureolus Paracelsus, *Liber de nymphis, sylphis pygmæis et salamandris, et de cæteris spiritibus* [1529–32], in *Neundter Theil der Bücher und Schrifften des . . . Paracelsi,* ed. Johannes Huserus (Basel 1590) 55.
33. Sir Thomas Browne, *Religio Medici* I.30 [1643], in *The Works of Sir Thomas Browne* I, ed. Geoffrey Keynes, new ed. (London 1964) 40.
34. Bernardus Silvestris, *De mundi universitate libri duo* II.vii, trans. Winthrop Wetherbee, (London and New York 1973) 107–08.
35. Origen, *De principiis* I.viii.1.
36. *New Catholic Encyclopedia* IV, 'Demon (Theology of).'
37. Saint Augustine, *De genesi ad litteram libri duodecim* III.x.15.
38. Johannes Scotus Eriugena, *De divina praedestinatione* XIX.
39. Alexander Neckam, *De naturis rerum* I.iii, ed. Thomas Wright (London 1863) 21.
40. Anders Sunesøn, *Hexaëmeron libri duodecim* I.391–440.
41. Alanus de Insulis, *Anticlaudianus* IV.5, trans. James J. Sheridan (Toronto 1973) 127–29.
42. See, for example, Dio Cassius, *Roman History* LXVI.22–23 (epit.).

43. Thomas of Cantimpré, *Miraculorum, & exemplorum memorabilium sui temporis, libri duo* II [mid-thirteenth century] (Duaci 1597) 466-68.

44. Thomas Heywood, *The Hierarchie of the Blessed Angells* (London 1635) 505.

45. Saint Augustine, *De genesi ad litteram libri duodecim* III.ix–x.

46. Saint Augustine, *De civitate Dei* XXI.10, trans. Rev. Marcus Dods (Edinburgh 1872) 434-35.

47. For a compendium of statements about the matter, from Genesis to Saint Thomas, see P. Glorieux, comp., *Autour de la spiritualité des anges* (Tournai 1959).

48. Saint Fulgentius, Bishop of Ruspe, *De trinitate liber unus* IX. For a somewhat awestruck account of the superior powers and sharper senses of bodies made from air, see Rabanus Maurus, *De magicis artibus* [ninth century], *PL* 110.1102. (*PL* = J. P. Migne, *Patrologiae cursus completus, series latina* [Paris 1844 etc.].)

49. Henry Lawrence, *Of our Communion and Warre with Angels* ([Amsterdam] 1646) 15. Perhaps the best known literary work dependent on this idea is Donne's 'Air and Angels.'

50. Calcidius, *Commentarius* 133. Translation adapted from J. Den Boeft *Calcidius on Demons* (Leiden 1977) 32.

51. Pierre de Ronsard, *Les amours* I.xxxi (1553), in *Oeuvres complètes* I, ed. Hugues Vaganay (Paris 1923) 36-37. The following English version is offered to the dedicatee of this book:

> Delicate demons who keep the middle way
> Between the starry sphere and heavy sod,
> Messengers divine, divine messengers of God
> Whose secrets you so rapidly convey,
>
> You who skim our fields so nimbly, say,
> Couriers, tell me (lest some warlock's rod
> Encircle you in fire), so lightly shod,
> Have you not found my amazon, my Kay?
>
> If perchance she see you here below,
> Her sweetness will abuse your forces so
> You'll be no longer free for airy flight.
>
> Like slaves in chains she'll have you bending low,
> Or change your lissom limbs to stony show
> With one Medusa glance from her quick sight.
>
> —C. H.

For a fuller expression of Ronsard's attitude to demons, see his hymn 'Les daimons,' in *Oeuvres complètes* VI, 59-68.

52. For a general introduction to the relationship among demons, fairies, and other middle-order beings, see C. S. Lewis, *The Discarded Image* (Cambridge 1964) passim.

53. Saint Augustine, *De divinatione dæmonum liber unus* III.7.

54. Aurelian Townshend and Inigo Jones, *Albions Triumph* (London $16\frac{31}{32}$) B1r.

55. Alanus de Insulis, *Anticlaudianus* IV.5, p. 128.

56. See, for example, Giovanni Lorenzo Anania, *De natura daemonum libri iiii* (Venetiis 1581) 148-51 and passim.

57. Georges Louis Leclerc, comte de Buffon, *Histoire naturelle des oiseaux* III (Paris 1775) 152.

58. See, for example, John Ray, *The Wisdom of God Manifested in the Works of the Creation* (London 1691) 111-12.

59. Buffon, *Histoire naturelle des oiseaux* I (Paris 1770) 46.

60. Buffon, *Histoire naturelle des oiseaux* III (Paris 1775) 151-54.

61. Pseudo-Aristotle, *Problemata* X.54, 897 b 11.

62. Aristotle, *De incessu animalium* 710 b 31ff.

63. Plato, *Sophist* 220B. A similar point is made in passing by Aristotle, *De partibus animalium* 642 b 10–20, 644 a 15–23.

64. Saint Basil, *Homilia VIII in hexaemeron* II, trans. Sister Agnes Clare Way (Washington, D.C. 1963) 120–21.

65. Saint Ambrose, *Hexaemeron libri sex* V.xiv.44–45, trans. John J. Savage (New York 1961) 197–98.

66. Saint Augustine, *De genesi ad litteram, imperfectus liber* XIV.44.

67. Saint Augustine, *De genesi ad litteram libri duodecim* III.iv.5. For a delightful mediaeval symbol of the air as honey surrounding the earth imagined as a piece of honeycomb, see Saint Hildegard [1098–1179], *Liber divinorum operum simplicis hominis* I.iv.lxxx. For a discussion of Hildegard's highly idiosyncratic cosmology, see Hans Liebeschütz, *Das allegorische Weltbild der heiligen Hildegard von Bingen* (Leipzig/Berlin 1930).

68. Saint Augustine, *De genesi ad litteram libri duodecim* III.vi.8.

69. Johannes Scotus Eriugena, *De divisione naturae* III.40.

70. William of Conches, *De philosophia mundi* I.xxii.

71. Saint Thomas Aquinas, *Summa theologiae* I.lxviii.3.3, I.lxxi.1.3.

72. Ludovicus Buccaferrea, *In quartum meteororum Aristotelis librum* (Venetiis 1563) 146.

73. See, for example, Saint Bonaventure [1221–74], quoted along with other commentaries on the matter in Glorieux, *Autour de la spiritualité des anges* 66.

74. See, for example [Pseudo-Sydrach], *Het Boek van Sidrac in de Nederlanden* [early fourteenth century] I, ed. J. F. J. van Tol (Amsterdam 1936) 98; *Il libro di Sidrach* [fourteenth century], ed. Adolfo Bartoli (Bologna 1868) 167; *Das Buch Sidrach. Nach der Kopenhagener mittelniederdeutschen Handschrift v. J. 1479*, ed. H. Jellinghaus (Tübingen 1904) 111. Not all versions make the same points. One late fifteenth century text that I consulted says that birds fly because it is God's will, because they are naturally light, and because they have more air in them than any other creatures: see *Sydrak efter Haandskriftet ny kgl. saml. 236 4*^{to}, ed. Gunnar Knudsen (København 1921–32) 111.

75. *The History of Kyng Boccus, & Sydracke how he confoundyd his lerned men . . .* (London ?1510) Ti^{v}–Tii.

76. Andrew Marvell, 'Upon Appleton House,' *The Poems and Letters of Andrew Marvell* I, 3rd ed., ed. H. M. Margoliouth (Oxford 1971), lines 671–72.

77. Marvell, 'Upon Appleton House,' lines 673–76.

78. Saint Gregory of Nyssa, *In hexaemeron, PG* 44.81–86.

79. Lambertus Danaeus, *Physices christianae pars altera* V.xxviii, in *Opuscula omnia theologica* (Genevae 1583).

80. Saint Thomas Aquinas, *Summa theologiae* I.lxxi.1.5, trans. Fathers of the English Dominican Province (London 1911–22).

81. Saint Thomas Aquinas, *Summa theologiae* II (1).cii.6.resp.

82. Saint Thomas Aquinas, *In octo libros Physicorum Aristotelis expositio* VIII.xxi.3, ed. P. M. Maggiòlo (Taurini 1965) 611. My translation.

83. Erasmus Francisci, *Der Wunder=reiche Uberzug unserer Nider=Welt, oder Erd=umgebende Lufft=Kreys* (Nürnberg 1680) 399.

84. Buffon, *Histoire naturelle des oiseaux* I. 46–47.

85. Aristotle, *Historia animalium* 535 b 27ff.

86. Hieronymus Cardanus, *De subtilitate libri xxi* (Lugduni 1554) 400.

87. Julius Caesar Scaliger, *Exotericarum exercitationum liber quintus decimus* (Lutetiae 1557) 294^{r}.

88. John Nashe, *Lenten Stuffe* 192, in *The Works* III, ed. Ronald B. McKerrow, rev. F. P. Wilson (London 1958).

89. See, for example, Ambroise Paré, *Des monstres et prodiges* (1573), ed. Jean Céard (Genève 1971) 114.

90. John Swan, *Speculum mundi* (London 1644) 375.
91. Thomas of Cantimpré, *Liber de natura rerum* 7.xli.4–7, ed. H. Boese (Berlin and New York 1973) I.263.
92. Terence Hanbury White, *The Book of Beasts* (London 1954) 199.
93. Aristotle, *De incessu animalium* 714 b 13.
94. E.g., Albertus Magnus, *De animalibus libri xxvi* [thirteenth century], ed. Hermann Stadler, II (Münster i. W. 1920) 1512.
95. See, for example, Pierre Gassendi, *Opera omnia* II (Lugduni 1658) 537a.
96. E.g., William Blake: 'The bat that flies at close of eve / Has left the brain that won't believe' ('Auguries of Innocence,' lines 25–26). Bram Stoker's *Dracula* probably supplies the best known example.
97. *Metamorphoses* IV.391–415.
98. Saint Ambrose, *Hexaemeron libri sex* V.xxiv.87, trans. John J. Savage (New York 1961) 223.
99. *Biblia sacra cum glossa ordinaria* 6 vols (Antverpiae 1634).

Notes to chapter 3

1. Pierre Belon, *L'histoire de la nature des oyseaux* (Paris 1555) 46.
2. Paul-Joseph Barthez, *Nouvelle méchanique des mouvements de l'homme et des animaux* (Carcassonne 1798) 190–230.
3. Iamblichus, *De mysteriis Aegyptiorum, Chaldaeorum, Assyriorum* III.xvi, trans. Thomas Taylor (London 1821).
4. Aristotle, *De incessu animalium* 706 a 25–29, 711 b 15–16.
5. Aristotle, *De incessu animalium* 709 b 10–11, 713 a 10. For other comments on the relationship of the bird's shape to the resisting medium, see 710 a 24–710 b 4.
6. Aristotle, *Physica* 254 b 31–33.
7. Aristotle, *Physica* 258 a 1–3.
8. Aristotle, *Physica* 259 b 6–31. In the *De motu animalium* he identifies the heart as the unmoved centre of the animal. See 698 b 5–8, 700 a 6–11, 701 b 25–33.
9. Lucretius, *De rerum natura* VI.830–39. For a denial of the absence of birds from Avernus, see pseudo-Aristotle, *De mirabilibus auscultationibus* 839 a 24–26: 'But to say that no bird flies over it is a lie; for those who have been there maintain that there is a large number of swans in it.'
10. Plutarch, *Titus Flamininus* X.6.
11. Plutarch, *Pompey* XXV.7. Birds falling dead from the air are common in poetic imagery. Cf. Dante, *Vita nuova* XXIII: 'Cader li augelli volando per l'are'; Agrippa d'Aubigné, *Stances* I.137–38: 'Ma presence fera dessecher les fontaines / Et les oiseaux tomber mortz à mes pieds. . . .'
12. Pliny the Elder, *Naturalis historia* X.liv.
13. Claudius Galen, *De tremore,* in *Omnia . . . opera* II (Lugduni 1550) 326–27.
14. Ulysses Aldrovandus, *Ornithologiae* I (Bononiae 1599) 30.
15. Marin Mersenne, *Correspondance* IV, ed. Cornelis de Waard (Paris 1955) 117–18.
16. Jean Antoine, l'Abbé Nollet, *Leçons de physique expérimentale* I (Paris 1743) 221–22.
17. Frederick II of Hohenstaufen, *De arte venandi cum avibus,* trans. Casey A. Wood and Marjorie F. Fyfe (Stanford 1943) 83a.
18. Pierre Belon, *L'histoire de la nature des oyseaux* (Paris 1555) 46.
19. Aristotle, *Physica* 254 b 12ff.
20. Aristotle, *De incessu animalium* 708 b 20ff; *De motu animalium* 698 b 8ff.
21. Belon, *L'histoire* 46.
22. For Aristotle on this point, see *De caelo* 313 a 14–313 b 23.
23. Extracting energy from the ship's underwater bow wave.

24. Belon, *L'histoire* 47.

25. Hieronymous Fabricius ab Aquapendente, *De alarum actione, hoc est volatu* 4, appended to his *De motu locali animalium secundum totum* (Patavii 1618) separately paginated 1–16.

26. Pierre Gassendi, 'De volatu animalium' in *Opera omnia* II (Lugduni 1658) 538a.

27. Fabricius, *De alarum actione* 8.

28. Hieronymus Cardanus, *Opus novum de proportionibus* (Basileae 1570) 25.

29. Fabricius, *De alarum actione* 9.

30. Fabricius, *De alarum actione* 14.

31. Giovanni Alfonso Borelli, *De motu animalium,* 2 vols (Romae 1680, 1681) I.285–326 and plates 11–13.

32. As, for example, in Diderot's *Encyclopédie* XVII.447–49, 'voler.'

33. Borelli, *De motu animalium* I.289.

34. MS Sloane 919 f. 17[r], English, January or February 16$\frac{73}{74}$.

35. MS Sloane 919 f. 17[r].

36. Borelli, *De motu animalium* I.310–11.

37. John Ray, *The Wisdom of God Manifested in the Works of the Creation* (London 1691) 106.

38. Ray, *Wisdom of God* 106.

39. Antoine Parent, *Essais et recherches* III (Paris 1713) 376–400.

40. Charles Louis de Secondat, Baron de Montesquieu, *Pensées* 758, in *Oeuvres complètes* I, ed. Roger Caillois (Paris 1949) 1207–08.

41. See Robert Shackleton, *Montesquieu: A Critical Biography* (Oxford 1961) 222.

42. François, Chevalier de Vivens, 'Du vol des oiseaux,' ed. Jules Duhem, *Mercure de France* 264 (15 November 1935) 30. Vivens later developed his idea of centripetal force in two books published anonymously: *Essai sur les principes de la physique* (n.p. 1746) and *Nouvelle théorie du mouvement, où l'on donne la raison des principes genéraux de la physique* (Londres [= Bordeaux] 1749).

43. Vivens, 'Du vol des oiseaux' 34.

44. 9 vols (Paris 1770–83).

45. Buffon, *Histoire naturelle* I.33.

46. Johann Esaias Silberschlag, 'Von dem Fluge der Vögel,' *Schriften der Berlinischen Gesellschaft Naturforschender Freunde* 2 (1781) 216–17.

47. Pierre-Jean-Etienne Mauduit de la Varenne, *Ornithologie* prefaced by 'Discours généraux sur la nature des oiseaux,' *Encyclopédie méthodique: histoire naturelle des animaux* (Paris and Liège 1782, 1784) I.ii, II.i.

48. Mauduit, *Ornithologie* I.356.

49. François Huber, *Observations sur le vol des oiseaux de proie* (Genève 1784).

50. Paul-Joseph Barthez, *Nouvelle méchanique des mouvements de l'homme et des animaux* (Carcassonne 1798) 190–230.

51. Aristotle, *De incessu animalium* 710 a 1ff.

52. Frederick II of Hohenstaufen, *De arte venandi cum avibus* 90b.

53. Fabricius, *De alarum actione* 11.

54. Francis Willughby, *The Ornithology of Francis Willughby . . . in Three Books* (London 1678) 4, trans. from the Latin ed. (Londini 1676) 3.

55. Borelli, *De motu animalium* I.311–12.

56. Silberschlag, 'Von dem Fluge der Vögel' 236–41.

57. Barthez, *Nouvelle méchanique* 194.

58. Mauduit, *Ornithologie* I.357.

59. Mauduit, *Ornithologie* I.355.

60. Barthez, *Nouvelle méchanique* 204.

Notes to chapter 4

1. Raffaele Giacomelli, *Gli scritti di Leonardo da Vinci sul volo* (Roma 1936); Arturo Uccelli, *I libri del volo di Leonardo da Vinci* (Milano 1952).
2. Charles Gibbs-Smith's treatment of them in his *Leonardo da Vinci's Aeronautics* (London 1967) is too brief to cover all the relevant material and is in any case inaccurate in one important particular. See later.
3. See Chapter 2.
4. MS K 3^r (1504-09). The translations are based on those in Edward MacCurdy, *The Notebooks of Leonardo da Vinci Arranged, Rendered into English and Introduced* 2 vols (London 1938).
5. MS F 41^v (c. 1508).
6. MS F 53^v (c. 1508).
7. *Codice Atlantico* $77^{r\text{-}b}$ (c. 1505).
8. *Codice Atlantico* $161^{r\text{-}a}$ (1505).
9. *Codice Atlantico* $381^{v\text{-}a}$ (c. 1485).
10. *Codice Atlantico* $214^{v\text{-}a}$ (c. 1505).
11. MS E 38^r (1513-15).
12. *Codice Atlantico* $395^{r\text{-}b}$ (c. 1515).
13. MS F 87^v (1508).
14. MS E 45^v (1513-15).
15. *Codice Atlantico* $45^{r\text{-}a}$ (c. 1515).
16. *Codice sul volo degli uccelli* $6(5)^v$ (1505).
17. *Codice Atlantico* $97^{v\text{-}a}$ (c. 1515).
18. MS E 36^r (1513-15).
19. MS E 46^r (1513-15).
20. *Codice sul volo degli uccelli* 18^r (1505).
21. See Arturo Uccelli, *I libri del volo di Leonardo da Vinci* (Milano 1952) 129-45.
22. MS E 42^v (1513-15).
23. MS E 37^r (1513-15).
24. MS E 37^r (1513-15).
25. MS E 43^r (1513-15).
26. E.g., *Codice Atlantico* $308^{r\text{-}b}$ (c. 1505).
27. See Colin J. Pennycuick, *Animal Flight* (London 1972) 58-59.
28. *Codice sul volo degli uccelli* $7(6)^r$ (1505).
29. *Codice sul volo degli uccelli* $7(6)^r$ (1505).
30. MS L 55^v (1499-1503).
31. MS E 39^v (1513-15).
32. *Codice sul volo degli uccelli* $17(16)^r$ (1505).
33. *Codice Atlantico* $97^{v\text{-}a}$ (c. 1515).
34. *Codice sul volo degli uccelli* $7(6)^r$ (1505).
35. MS K 12^r (1504-09).
36. *Codice sul volo degli uccelli (fogli mancanti)* $11(10)^r$ (1505).
37. *Codice sul volo degli uccelli (fogli mancanti)* $11(10)^v$ (1505).
38. MS K 5^v (1504-09).
39. MS K 7^r (1504-09).
40. *Codice Atlantico* $97^{v\text{-}a}$ (c. 1515).
41. *Codice Atlantico* $160^{r\text{-}b}$ (c. 1515).
42. *Codice sul volo degli uccelli* $13(12)^r$ (1505).
43. MS K 14^r (1504-09).
44. *Codice sul volo degli uccelli* $14(13)^v$ (1505).

45. MS K 59(10)^{r} (1504–09).
46. *Codice sul volo degli uccelli* 13(12)^{r} (1505). My translation.
47. MS L 56^{r} (1499–1503).
48. *Codice sul volo degli uccelli* 8(7)^{v} (1505).
49. *Codice sul volo degli uccelli* 8(7)^{v} (1505).
50. *Codice sul volo degli uccelli* 15(14)^{r} (1505).
51. *Codice Atlantico* 66$^{r\text{-}a}$ (c. 1505).
52. *Codice Atlantico* 66$^{r\text{-}a}$ (c. 1505).
53. *Codice Atlantico* 66$^{r\text{-}b}$ (c. 1505).
54. *Codice Atlantico* 97$^{v\text{-}a}$ (c. 1515).
55. MS L 55^{r} (1499–1503).
56. *Codice Atlantico* 66$^{r\text{-}b}$ (c. 1505).
57. MS K 7^{v} (1504–09).
58. MS K 8^{r} (1504–09).
59. MS E 44^{v}, 49^{r} (1513–15).
60. MS E 52^{v}, (1513–15).
61. See notes 1, 2.
62. Gibbs-Smith, *Leonardo da Vinci's Aeronautics* 19. See earlier.
63. Gibbs-Smith, *Leonardo da Vinci's Aeronautics* 22–23.
64. Gibbs-Smith, *Leonardo da Vinci's Aeronautics* 19.
65. Clive Hart, *The Dream of Flight: Aeronautics from Classical Times to the Renaissance* (London 1972) 134.
66. A few years later Leonardo sketched a variant design for the framework, allowing increased body movement. See Fig. 26 and Gibbs-Smith, *Leonardo da Vinci's Aeronautics* 26. I believe Gibbs-Smith is wrong in asserting that Leonardo has here adopted 'the unwise technique of exerting this shift from above, and hence making for a de-stabilising situation.' It seems that the whole of the pilot's body is placed beneath the wing surface.

Notes to chapter 5

1. Saint John Chrysostom, *Homiliae XXI de statuis* XV.5 [A.D. 387 or 388], trans. W. R. W. Stephens and others (New York 1889) 443.
2. John Wilkins, *The Discovery of a World in the Moone* (London 1638) 208.
3. Johan Daniel Major, *See=Farth nach der neüen Welt, ohne Schiff und Segel* (Kiel 1670) A4^{v}–B1^{r}.
4. Major, *See=Farth* B4^{r}.
5. Johann Caramuel Lobkovitz, *Mathesis biceps* I (Campaniae 1670) 740$^{a\text{–}b}$.
6. Noël-Antoine Pluche, *Spectacle de la Nature: or, Nature Display'd* I (London 1733) 30–31. Trans. from the French edition of 1732.
7. Nehemiah Grew, *Cosmologia sacra: or a Discourse of the universe as it is the Creature and Kingdom of God* (London 1701) 29.
8. William Derham, *Physico-theology* (London 1713) 309.
9. Joannes Ludovicus Hannemann [*Praeses, resp.* Georgius Matthias Hirsch], *Icarus in mare icarium praecipitatus ceu dissertatio qua hominem ad volandum esse ineptum ostenditur* (Kiloni 1709) 13–14.
10. Louis-Bertrand Castel, *Traité de physique sur la pesanteur universelle des corps* I (Paris 1724) 276.
11. Samuel Johnson, *The History of Rasselas, Prince of Abissinia* (1759), ed. Geoffrey Tillotson and Brian Jenkins (London 1971) 17.
12. Laurent Gaspard Gérard, *Essai sur l'art du vol aérien* (Paris 1784) 145–47.
13. Frederick Marriott, 'The Problem Solved,' *San Francisco News Letter and California Advertiser* (6 November 1880) 2.
14. Johann Esaias Silberschlag, 'Von dem Fluge der Vögel,' *Schriften der Berlinischen Gesellschaft Naturforschender Freunde* 2 (1781) 250–51.

15. Jean Molinet, 'Recollection des merveilleuses advenues' stanza 143 [c. 1500], in *Les faictz et dictz de Jean Molinet* I, ed. Noël Dupire (Paris 1936) 332.

16. Philippe d'Alcripe [= le Picard], *La nouvelle fabrique des excellens traits de vérité* [sixteenth century], ed. P. A. Gratet-Duplessis (Paris 1853) 22.

17. d'Alcripe, *La nouvelle fabrique* 178–79.

18. Pierre Jean Grosley, *Oeuvres inédites* I (Paris 1812) 84–88.

19. John Hacket, *Scrinia reserata: a Memorial Offer'd to the Great Deservings of John Williams, D.D.* ([London] 1693) 8.

20. Jean Juvénal des Ursins, *Histoire de Charles VI, roy de France* (1614) 2nd ed. (augmented) (Paris 1653) 71–72.

21. Mathieu d'Escouchy, *Chronique* II, ed. G. du Fresne de Beaucourt (Paris 1863) 149.

22. James Shirley, *Cupid and Death,* in *The Dramatic Works and Poems of James Shirley* VI, ed. Alexander Dyce (London 1833) 358.

23. See Clive Hart, *The Dream of Flight: Aeronautics from Classical Times to the Renaissance* (London 1972) 69–72.

24. Scipion Dupleix, *Histoire de Henry III* (1630) 5th ed. (Paris 1641) 8.

25. *Dictionnaire des origines, découvertes, inventions et établissemens* II (Paris 1777) 572–73.

26. Pierre Boaistuau, *Bref discours de l'excellence et dignité de l'homme* (Paris 1558) f. 20[r].

27. John Wilkins, *Mathematicall Magick* (London 1648) 207–08.

28. Hugues Le Roux and Jules Garnier, *Acrobats and Mountebanks,* trans. A. P. Morton (London 1890) 210.

29. Joseph Glanvill, *The Vanity of Dogmatizing: or Confidence in Opinions* (London 1661) 182.

30. Sagredo appears as one of the interlocutors in Galileo's *Dialogue Concerning the Two Chief World Systems* (1632), and in *Two New Sciences* (1638).

31. Tito Livio Burattini, *Ars volandi,* Paris, Bibliothèque nationale, MS Latin 11195, 1647, ff. 57[r–v].

32. Johann Caramuel Lobkovitz, *Mathesis biceps* I (Campaniae 1670) 741[a]. The Belgian *Ars volandi* of 1640 appears to be lost.

33. Robert Hooke, *Philosophical Collections* I (London 1679) 15.

34. Richard Waller, 'The Life of Dr. Robert Hooke,' prefaced to *The Posthumous Works of Robert Hooke* (London 1705) iv.

35. Waller, 'The Life of Dr. Robert Hooke,' iv.

36. Hooke, *The Posthumous Works* 56–57.

37. Hooke, *The Posthumous Works* 56.

38. Robert Hooke, *Micrographia* (London 1665) 198.

39. See *The Life and Work of Robert Hooke* II, ed. Robert William Theodore Gunther (Oxford 1930) 428. In the rest of this chapter, dates are given in New Style.

40. Marin Mersenne, *Questions inouyes, ou récréation des scavans* (Paris 1634) 1–5.

41. Marin Mersenne, *Correspondance* IV, ed. Cornelis de Waard (Paris 1955) 55.

42. Mersenne, *Correspondance* IX (Paris 1965) 524.

43. Mersenne, *Correspondance* X (Paris 1967) 87.

44. Mersenne, *Correspondance* XI (Paris 1970) 89.

45. Mersenne, *Correspondance* XI (Paris 1970) 435–36.

46. Henri Bouchot, *Jacques Callot: sa vie, son oeuvre et ses continuateurs* (Paris 1889) 231.

47. *Arliquiniana ou les bons mots, les histoires plaisantes & agréables* (Paris, Florentin & Pierre Delaulne et Michel Brunet, 1694) 85. Another edition, with different pagination, was published in Paris in the same year.

48. Royal Society, *Philosophical Transactions* 1.6 (6 November 1665) 99.

49. See Gottfried Wilhelm von Leibniz, *Otium Hanoveranum, sive miscellanea* (Lipsiae 1718) 185.

50. Johann Joachim Becher, *Närrische Weiszheit und weise Narrheit* (Franckfurt 1682) 27–28, 30–31, 148–49.

51. *Arliquiniana* 84–87.

52. Marin Mersenne, *Correspondance* XI (Paris 1970) 435–36.

53. Mersenne, *Correspondance* XII (Paris 1972) 392.

54. See Chapter 6.

55. Mersenne, *Correspondance* XII (Paris 1972) 394.

56. Marin Mersenne, *Novarum observationum physico-mathematicarum . . . tomus III* (Parisiis 1647) 73–74.

Notes to chapter 6

1. For Burattini's life in general, see Antonio Favaro, *Intorno alla vita ed ai lavori di Tito Livio Burattini, fisico agordino del secolo XVII* (Venezia 1896), and Antonio Favaro, 'Supplemento agli studi intorno alla vita ed alle opere di Tito Livio Burattini, fisico agordino del secolo XVII,' *Atti del Reale istituto veneto di scienze, lettere ed arti* 51.2 (1900) 855–60. For the date of birth, see Cornelis de Waard, 'Notes sur Stevin et Beeckman,' *Isis* 24.1 (December 1935) 125. For studies of the flying dragon, see Carl von Klinckowstroem, 'Tito Livio Burattini, ein Flugtechniker des 17. Jahrhunderts,' *Prometheus* 1100 (26 November 1910) 117–20, and Karolina Targosz, 'Jak wyglądał "latający smok" Tita Livia Burattiniego,' *Technika Lotnicza i Astronautyczna* 3 (1976) 37–40. A number of variant spellings of Burattini's name appear in the contemporary literature. In this chapter, dates are given in New Style.

2. John Greaves, *Pyramidographia: or a Description of the Pyramids in Ægypt* (London 1646) 86, 104.

3. Vienna, Nationalbibliothek, MS Hohendorf 7049, f. 447[r–v]. Letter from Des Noyers to Roberval, 4 December 1647.

4. Paris, Bibliothèque nationale, MS Latin 11195, ff. 50[r]–61[r], Italian-French-Polish, 1647.

5. Friedrich Hermann Flayder, *De arte volandi* ([Tübingen] 1627).

6. MS Latin 11195, f. 57[r].

7. The correct statements of principle had been clearly enunciated in Galileo's *Dialogue Concerning the Two Chief World Systems* (1632), trans. Stillman Drake (Berkeley and Los Angeles 1953) 223.

8. MS Latin 11195, f. 58[r]

9. MS Latin 11195, f. 58[r–v].

10. See Galileo Galilei, *Two New Sciences, Including Centers of Gravity & Force of Percussion,* trans. Stillman Drake (Madison 1974) 292–93.

11. These ideas are also heavily dependent on Galileo. See *Two New Sciences* 225–28. Cf. the demonstration of the thickness of the air proposed in the *Book of Sydrach* (see Chapter 2).

12. MS Latin 11195, f. 59[v].

13. Galileo Galilei, *Les méchaniques* [1593–c. 1600], trans. Marin Mersenne (Paris 1634) 17–26. The Italian original was posthumously published in 1649. Burattini's indebtedness to the book is specifically attested in a letter from Des Noyers to Roberval, 18 March 1648, MS Hohendorf 7049, f. 453[v].

14. MS Latin 11195 f. 60[r–v].

15. MS Hohendorf 7049, f. 450[r].

16. Christiaan Huygens, *Oeuvres complètes* III (La Haye 1890) 270. Letter from Des Noyers to Thévenot, 29 January 1648.

17. Paris, Bibliothèque nationale, MS n.a.fr. 6204, ff. 128[r]–129[r].

18. MS Hohendorf 7049, ff. 452[v]–453[r].

19. Paris, Bibliothèque nationale, MS n.a.fr. 6204, f. 130[r].

20. Huygens, *Oeuvres complètes* III.303.

21. MS Hohendorf 7049, f. 456[v].

22. Paris, Bibliothèque nationale, MS n.a.fr. 6206, f. 64[v].

23. MS Hohendorf 7049, f. 456[v].
24. Mère Marie Angélique de Sainte Madeleine, *Lettres* I (Utrecht 1742) 363. Letter to Queen Maria Louisa, 20 March 1648.
25. Cyrano de Bergerac, *Histoire comique des état et empire de la lune et du soleil* (1648–49), ed. Claude Mettra and Jean Suyeux (Paris 1962).
26. Cyrano, *Histoire comique* 181.
27. Johann Joachim Becher, *Närrische Weiszheit und weise Narrheit* (Franckfurt 1682) 165–66.

Notes to chapter 7

1. Emanuel Swedenborg, *Opera quaedam aut inedita aut obsoleta de rebus naturalibus* I, ed. Alfred H. Stroh (Holmiae 1907) 226.
2. *Daedalus hyperboreus* 4 (Oct.–Dec. 1716) 80–83.
3. Stifts- och Landesbibliotek, Linköping, Codex 14a.
4. See the following translations and commentaries: Carl Th. Odhner, 'Swedenborg's Flying Machine,' *New Church Life* 29.10 (October 1909) 582–91; Hugo Lj. Odhner and Carl Th. Odhner, *Suggestions for a Flying Machine* (Philadelphia 1910), reprinted in part in *The Aeronautical Journal* 14 (July 1910) 118–22; *Transactions of the International Swedenborg Congress* (London 1910) 45–46; Alfred Acton, *The Mechanical Inventions of Emanuel Swedenborg* (Philadelphia 1939) 20–26; *Machine att flyga i Wädret enligt utkast av Emanuel Swedenborg* (Stockholm 1960), translation reprinted from Acton.
5. Swedenborg's unit of length is the ell (= approx. 2 ft); his unit of weight is the *lispund* (= 18.75 lb).
6. Flap valves for wings are sketched in *Codice Atlantico,* f. 309[v-b] (1487–90). See Fig. 24.
7. In the printed version he says that the event occurred at Strängnäs.
8. See, for example, Clive Hart, *The Dream of Flight: Aeronautics from Classical Times to the Renaissance* (London 1972) 45–46.
9. *Angelic Wisdom Concerning the Divine Providence* (1764) I.20 (London 1949) 16–17.
10. *The Apocalypse Revealed* (1766) 2 vols (London 1970) section 245 (vol. I, pp. 216, 217).
11. *The Apocalypse Revealed,* section 561 (vol. II, p. 533).
12. *The Apocalypse Revealed,* section 757 (vol. II, p. 743).
13. *The Apocalypse Revealed,* section 757 (vol. II, pp. 744–45).

Notes to chapter 8

1. *The Whitehall Evening-Post; Or, London Intelligencer* 882 (3–5 October 1751) 1. A note at the head of the column states that 'The following Article is taken from one of our Daily Papers.' I have not been able to identify the source.
2. *Lettera scritta da uno di Londra ad un suo amico di Venezia sopra la Machina Volante, che con universale applauso vedesi colà guidata per aria dal famoso, e singolare Mecanico.* Although no printed copies appear to have survived, a manuscript copy, which may be a transcript from print, is preserved in the Biblioteca Civica, Bergamo, MS Gabinetto A, IV, 5, ff. 118[r]–119[r]. After the title, place and date are mentioned: 'In Venezia 1751.' A few running corrections are consistent with its being a transcript. The letter is reprinted, with inaccuracies, in Boffito, Venturini, and elsewhere. (See note 4.)
3. Clemente Baroni Cavalcabò, *L'impotenza del demonio di trasportare a talento per l'aria da un luogo all'altro i corpi umani* (Rovereto 1753) 108–09.
4. Francesco Milizia, *Le vite de' più celebri architetti d'ogni nazione et d'ogni tempo precedute da un saggio sopra l'architettura* (Roma 1768) 327. See also Giuseppe Boffito, *Il volo in Italia* (Firenze 1921) 163–71, and Galileo Venturini, *Da Icaro a Montgolfier* I (Roma 1928) 325–36.
5. Venturini, *Da Icaro a Montgolfier* 334–36.
6. See Clive Hart, *The Dream of Flight: Aeronautics from Classical Times to the Renaissance* (London 1972) 78–86.
7. Richard Owen Cambridge, *The Scribleriad: an Heroic Poem. In Six Books* (London 1751) IV. 125–56.

8. Besnier's famous attempt was widely discussed in European journals of the day.
9. Robert Paltock, *The Life and Adventures of Peter Wilkins. A Cornish Man* (London 1751).
10. Ed. N. M. Penzer (London 1926).
11. 'Dr. Musgrave's Machine: A Vision,' *The Oxford Magazine* 3.3 (September 1769) 109.
12. 'Dr. Musgrave's Machine: A Vision,' pp. 109–10.
13. The church, a collegiate institution, was destroyed during the revolution. For background information, see [Clément] Maxime de Montrond, *Essais historiques sur la ville d'Étampes* 2 vols (Paris 1836, 1837).
14. François de Ravaisson, ed., *Archives de la Bastille: documents inédits* XVII (Paris 1891) 219–22.
15. Most of the information given here is drawn from *Annonces, affiches, nouvelles et avis divers de l'Orléanois* 36, 39, 40 (4 and 25 September and 2 October 1772) 147–48, 161–62, 165–66. Additional details are taken from Laurent Gaspar Gérard, *Essai sur l'art du vol aérien* (Paris 1784) 40–45. Some of Gérard's information is apparently drawn from eyewitness reports.
16. News reports in the *Annonces* often appeared two or three weeks after the events. Failing to notice the sequence of dates, some earlier commentators have written garbled accounts. The first of these appeared in issue 43 of the Parisian journal *Affiches, annonces, et avis divers* (21 October 1772) 172, which summarised and briefly discussed Desforges's first published announcement. According to the *Affiches*, Desforges undertook the demonstration flight after the subscription money had been put up by a gentleman of Lyons. In a variant account, Gérard (pp. 41–42) says that the money was offered by a number of citizens of Lyons. As far as I can determine, this story, which is inconsistent with the known facts, is an apocryphal embellishment. (The *Affiches* went on to summarise Desforges's second announcement in issue 44, 28 October 1772, 175–76.)
17. *Annonces, affiches, nouvelles et avis divers de l'Orléanois* 40 (2 October 1772) 165–66.
18. Baron Friedrich Melchior von Grimm, *Correspondance* X, ed. M. Tourneaux (Paris 1879) 60–61. Letter dated 15 September 1772.
19. Ferdinand Galiani, *Correspondance inédite* II (Paris 1818) 86.
20. Galiani, *Correspondance* 86–87.
21. *Dictionnaire des origines* III (Paris 1777) 638.

Notes to chapter 9

1. The few facts that are known about Bauer are incorporated in a long historical novel by Peter Supf, *Der Himmelswagen: das Schicksal des Melchior Bauer* (Stuttgart 1953).
2. The manuscript, in which Bauer describes his flying machine and his attempts to find a patron, is known as 'Die Flugzeughandschrift des Melchior Bauer.' Dated 1764, it is in the Staatsarchiv, Weimar, catalogued a. Rep. A Greiz Rep 41 Nr. 12a. It consists of eight folios, written on both sides except for the last, f. 8^v, which is blank. The text, in a large, clear hand, is in eighteenth century German script, with occasional use of Roman for salutations and emphasis. There are diagrams and line drawings on ff. 4^v, 5^r, 5^v, 6^r, 6^v, 8^r. Folio numbers are included in the translations.
3. See Hubert c. Johnson, *Frederick the Great and His Officials* (New Haven and London 1975) 232–37.
4. Bauer, 'Die Flugzeughandschrift.'
5. The apostolic creatures: man, lion, ox, and eagle (Ezekiel 1:10).
6. The choice of 1364 as a date for comparison suggests very strongly that Bauer was writing in 1764, exactly four hundred years later.
7. Hieronymus Fabricius ab Aquapendente, *De alarum actione, hoc est volatu* (Patavii 1618) 14–16.

Notes to chapter 10

1. In *Oberrheinische Mannigfaltigkeiten* 8–10 (13, 20, 27 January 1783) 541–44, 545–55, 574–76.
2. *Die Kunst zu fliegen nach Art der Vögel* (Frankfurt und Basel 1784). As the text is so brief (46 pp.), I have not thought it necessary to cite page numbers.

3. *L'art de voler à la manière des oiseaux* (Basel 1784), new impression 1785; *A arte de voar a' maneira dos passaros* (Lisboa 1812). The French version is shortened and simplified. It contains errors of translation and interpretation showing that it cannot have been seen through the press in detail by Meerwein himself. One or two small modifications and improvements (e.g., the triangular structure for supporting the pilot's legs; see later) nevertheless suggest that some authorial second thoughts may have been included.

4. Meerwein's measurements are based on south German units, with a pound of 32 ounces, a foot of 10 inches, and an inch of 100 lines.

5. Johann August Schlettwein, *Bemühungen in der Naturkunde und anderen nützlichen Wissenschaften* (Jena 1756).

6. Sir Isaac Newton, *Correspondence* II, ed. H. W. Turnbull, F.R.S. (Cambridge 1960) 295.

Notes to chapter 11

1. Henry Goldwel, *A briefe declaration of the shews, devices, speeches, and inventions, done & performed before the Queenes Maiestie* (London 1581) [Bvi^r].

2. Dio Cassius, *Roman History* LXVI.22 (epit.).

3. Thomas of Cantimpré, *Miraculorum, & exemplorum memorabilium sui temporis, libri duo* II [mid-thirteenth century] (Duaci 1597) 448–49. See also 466–68.

4. See Etienne Baluze, *Capitularia regum francorum* 2 vols (Paris 1677) I.235, II.1131. Law of A.D. 789: *De auguriis vel aliis maleficiis.*

5. L'Abbé Montfaucon de Villars, *Comte de Gabalis, ou entretiens sur les sciences secretes* (Cologne ?1675) 133.

6. Stefano Breventano, *Trattato delle impressioni dell'aere, raccolto da varij autori di filosofia* (Pavia 1571) 8^v.

7. Saint Agobard, *Contra insulsam vulgi opinionem de grandine et tonitruis* II [c. 800], *PL* 104.148.

8. A late reference to the tradition of the *tempestarii* is included in Marc-Antoine le Grand's *Les avantures du voyageur aérien* (Paris 1724) 9: 'Je suis originaire du Païs où le vent trouve des Vendeurs & des Acheteurs, & où l'on peut faire deux cent lieuës en douze heures sans s'incommoder.' Cf. also Jean François Regnard's *Voyage de Laponie,* in *Oeuvres* I (Paris 1790) 149–50. During a visit to Lapland, c. 1680, Regnard learned that the Lapps believed they could use magic to control the winds.

9. Gervasius of Tilbury, *Otia imperialia* [1211], selected and ed. F. Liebrecht (Hannover 1856) 2–3, 62. For a study of this and other flying ships, see L. Gougaud, 'L'aéronef dans les légendes du moyen âge,' *Revue celtique* 41 (1924) 354–58.

10. Geoffroi de Vigeois, *Chronica Gaufredi coenobitae* XL, in *Novae bibliothecae manuscript. librorum* II, comp. P. Labbe (Paris 1657) 299–300.

11. See Kuno Meyer, 'The Irish Mirabilia in the Norse "Speculum Regale,"' *Ériu* 4 (1910) 12–13; Oscar Brenner, ed., *Speculum regale: ein altnorwegischer Dialog nach Cod. Arnamagn. 243 Fol. B und den ältesten Fragmenten* (München 1881) 44–45; Thomas Wright and James Orchard Halliwell, eds, *Reliquiae antiquae* II (London 1843) 106. For philological comments, see Jacob Grimm, *Deutsche Mythologie,* 4th ed., 3 vols (Berlin 1875–78) I.524–33, III.178–83.

12. Albert of Saxony, *Questiones . . . in octo libros Physicorum Aristotelis* IV.vi.2.3 [c. 1360](Parisiis 1516) 47^r.

13. Nicole Oresme, *Le livre du ciel et du monde,* ed. Albert D. Menut and Alexander J. Denomy, trans. Albert D. Menut (Madison Milwaukee and London 1968) 401, 403. Oresme's book, finished in 1377, is almost certainly later than Albert's.

14. Oresme, *Le livre du ciel et du monde* 405.

15. Further echoes of the idea are found in Francisco de Mendoça, *Viridarium sacrae, ac profanae* (Lugduni 1631) 117–18, and in Johann Caramuel Lobkovitz, *Mathesis biceps* I (Campaniae 1670) 734–45.

16. [Joseph Galien], *Mémoire touchant la nature et la formation de la grêle et des autres météores qui y ont rapport, avec une conséquence ultérieure de la possibilité de naviger dans l'air à la hauteur de la région de la grêle. Amusement physique et géométrique* (Avignon 1755) 51–87.

17. *L'art de naviger dans les airs* (Avignon 1757).

18. Joseph Galien, *Mémoire touchant la nature et la formation de la grêle* 76–77.

19. Jean François Cailhava d'Estendoux, *Arlequin Mahomet, ou le cabriolet volant,* in *Théâtre de M. Cailhava* II (Paris 1781) 1–88.

20. Act I, scene 7, pp. 26–27.

21. *Première suite du cabriolet volant, ou Arlequin cru fou, sultane et mahomet,* in *Théâtre de M. Cailhava* II (Paris 1781) 89–157.

22. Percy Bysshe Shelley, *Prometheus Unbound* IV, lines 206–17, ed. Lawrence John Zillman (New Haven and London 1968) 205.

Notes to appendix I

1. Erasmus Francisci, *Der Wunder=reiche Uberzug unserer Nider=Welt, oder Erd=umgebende Lufft=Kreys* (Nürnberg 1680) 370.

2. Joannes Ludovicus Hannemann, *Icarus in mare icarium praecipitatus ceu dissertatio qua hominem ad volandum esse ineptum ostenditur* (Kiloni 1709) 15.

3. (London 1966).

Notes to appendix II

1. Lynn White, Jr, 'The Invention of the Parachute,' *Technology and Culture* 9.3 (July 1968) 462–67; 'Medieval Uses of Air,' *Scientific American* 223.2 (August 1970) 100.

2. White, 'Medieval Uses of Air' 100.

3. Belief in the efficacity of the technique continued during the Middle Ages. See, for example, Angelomus Luxoviensis, *Commentarius in Genesin* [ninth century] *PL* 115.121.

Bibliography

I have divided the bibliography into three sections: A, manuscripts; B, primary and secondary printed sources; C, sources and further reading for modern ideas about flight and the air. Section C, a selective list of references, may assist the reader who wishes to compare the work of older thinkers with the propositions of recent flight theory.

Citations of Greek and Latin works without publication details should be understood to refer to the Loeb Classical Library, ed. T. E. Page, W. H. D. Rouse, and others (London and New York 1912–).

PG = J. P. Migne, *Patrologiae cursus completus, series graeca* (Paris 1857 etc.)
PL = J. P. Migne, *Patrologiae cursus completus, series latina* (Paris 1844 etc.)

I have adopted a spare bibliographic style, abbreviating long titles and minimizing punctuation. Arabic numerals placed immediately before parenthetical place and date refer to volumes or parts of journals and series. Roman numerals so placed refer to volumes of sets or to sections of subdivided works.

Brief notes on contents are sometimes included, especially for works that are not discussed in the text.

A. Manuscripts

i. Manuscripts relevant to the creation of the birds

I have listed a small selection of manuscripts containing illustrations of the Creation, Day Five (Genesis 1:20–23). Among them I have included a group of northern French *Bibles historiales* with related miniatures. The illustrations commonly take one of the following forms: (a) in the *Bibles historiales,* a separate, bordered miniature, often about 80 × 80 mm; (b) a mandorla (almond shape), lozenge, quadrilobe, or medallion in a set of seven or eight making up the initial I of Genesis (*In principio . . .*); (c) part of a composite scene showing the whole of the Creation sequence placed at the head of the text. Versions of most of the scenes described here are found in other manuscripts from the same ateliers. Those I have chosen are generally representative. Of particular interest are variations in the degree of freedom with which the birds are drawn, whether or not they are shown flying, and their pictorial relationship to God, the fishes, and the water. When the miniature has been reproduced in this book, the plate number is given in parentheses.

Amiens: Bibliothèque municipale.

MS 21, f. 7^{r}, fifth roundel. French. Thirteenth century. White birds flying on gold, God holding another in his clenched fist (Plate III).

MS 108, f. 2$^{r\text{-}a}$. French. 1197 and later. First days of Creation. The illustration is a crude replacement of a lost original, copies of which are extant in New York Public Library, Spencer Collection, MS 22, f. 5^{r}, and Harburg, Fürstlich Oettingen-Wallensteinische Bibliothek und Kunstsammlung, MS 1, 2, lat. 4°, 15, f. 4^{v}. See François Bucher, *The Pamplona Bibles,* 2 vols (New Haven and London 1970).

Arras: Bibliothèque municipale. (Old catalogue numbers are given in parentheses.)

MS 1 (3), vol. I, f. 7^{v}, fifth roundel. French. Twelfth–thirteenth century. God with birds, left, and fish, right, and with another bird about to be released. Badly damaged.

MS 299 (944^{1}), f. 6^{v}, fifth roundel. French. Thirteenth century. God holding a bird in his left clenched fist, with another bird flying freely.

MS 790 (919), f. 3^{v}, fifth roundel. French. Thirteenth century. Birds on the left of God balancing fish on the right.

Cambridge: Fitzwilliam Museum.

MS 11, f. 2^{v}. German. Fifteenth century. Biblia emblemata. Composite Creation scene, with two hastily sketched birds. This interesting ms contains many curious emblems, including birds and (f. 56^{r}) two depictions of the wheel of Ezekiel's vision.

Heidelberg: Universitätsbibliothek.

MS pal. germ. 164, f. 10^{r}. German. 1320. Jus feudale saxonicum (= the Sachsenspiegel of Eike von Repgow, thirteenth century). Left column, third of four vignettes: the Creation, with Adam, animals, two fish, and two birds flying straight up. The scene illustrates the statement 'When God created man, He gave him power over fishes, birds, and all wild creatures.'

MS pal. germ. 471, f. 56^{v}. German. 1425. Hugo von Trimberg, Der Renner. The whole of Creation depicted in one circular land- and cityscape. Nine birds are flying, six are in trees; one large phoenix is in flames. The scene is set simultaneously in both day and night. The birds are the only unenclosed creatures (Plate IX).

Klosterneuburg: Stiftsbibliothek.

MS 2, f. 4^{r}, fifth roundel from bottom. Southwest German. 1310–15. Creation of the birds and animals (no fish). Seven carefully drawn birds sit behind God, who faces six beasts. The roundel sequence is from bottom to top.

London: British Library.

MS Cotton Claud. B. IV, ff. 3^{v}, 6^{r}. English. Second half of the eleventh century. Aelfric's Old English metrical paraphrase of the Pentateuch and the Book of Joshua. Creation of the birds and fishes; naming of the animals (Plates I, II).

MS Royal 1.E.IX, f. 3^{v}. Latin Bible. End of the fourteenth century. Double Creation picture with God creating, on the right, the birds, fishes, and animals. Unusually, God is depicted as vastly bigger than his creatures.

MS Royal 17.E.VII, f. 5^{r}. French. 1357. Animals look on as God prepares to place a fish in a stream and to release a bird sitting freely in his hand. Two other birds sit in a tree. The birds, the fish, and the water are all given similar colouring, stressing their relationship.

MS Royal 19.D.III, f. 5^{v}. French. 1411. Creation of the birds and fishes. Two cockerels stand on the ground. God is surrounded by eight flying birds in a variety of beautiful attitudes (Plate VIII). At f. 3^{r} God the Geometer is shown against a blue heaven entirely filled with faintly lined-in four-winged seraphim. At f. 6^{r} the animals watching the creation of Adam include a winged serpent.

New York: Pierpont Morgan Library.

Illuminated book of Old Testament miniatures, f. 1^{v}. French. C. 1250. God with a large globe surrounded by the firmament. Five birds fly above four mushroom trees. The sun and the moon are above, fish and a whale swim in the water below. God gestures with both hands. Facsimile in *Old Testament Miniatures: A Medieval Picture Book with 283 Paintings from the Creation to the Story of David* with introduction and legends by Sydney C. Cockerell, preface by John Plummer (London 1969).

Oxford: Bodleian Library.

MS Bodley 270^{b}, ff. 3^{v}, 4^{r}. Northern French. Last quarter of the thirteenth century. Bible moralisée. Creation of the birds and fishes; creation of animals and birds. The moralising medallion on 4^{r} shows a contemplative man (symbolised by the birds) holding his hands to look like a pair of wings.

Paris: Bibliothèque de l'Arsenal.

MS 5056, f. 1^r. French. Thirteenth century. Birds, none of them flying, in medallion 4, with the animals; fish only in medallion 5.

MS 5059, f. 5^r. French. 1370. One bird sitting; two flying, one as if just released from God's hand (Plate IV).

MS 5212, f. 6^r. French. Fourteenth century. Small historiated initial D: God with one fish in his right hand, one bird with spread wings in his left; one other bird flying almost straight up.

Paris: Bibliothèque nationale.

MS franç. 8, f. 5^v. French. Fourteenth century. Two birds flying away at 45° on a checkered ground (Plate V).

MS franç. 152, f. 11^r, panel 5. French. Fourteenth century. Three birds soaring upwards, two fish apparently leaping upwards, suggesting expressions of joy.

MS franç. 6260, f. 26^r. French. Fifteenth century. Panel five of a composite Creation scene: birds fly as God creates fish and reptiles. Very primitive style.

MS franç. 15392, f. 3^v. French. Fourteenth century. Benign God with five birds arranged decoratively on trees.

MS franç. 20087, f. 3^v. French. Fifteenth century. God holds a cockerel, wings spread wanting to be released, by its legs and rump. Six other farmyard animals stand on the ground in a tamed, cultivated, rural parkland (Plate VII).

MS lat. 203, f. 4^v, fifth medallion. French. Thirteenth century. God has a bird, which is breaking through the frame, perched on his hand rather in the style of a falconer.

MS lat. 226, f. 4^v, fifth roundel. French. Thirteenth century. God releases a bird from his right hand while his left points to a sphere of water, suspended, with two fish in the middle, enclosed as in a bowl.

MS lat. 11549, f. 1^r, fifth roundel. French. Twelfth century. God holds the sphere of the world, divided into two halves: in water, at the bottom, are two fish; one bird flies in the air at the top.

Troyes: Bibliothèque municipale.

MS 59, f. 5^v. French. Fourteenth century. God holds one large green fish while three birds fly on a checkered ground.

Vienna: Österreichische Nationalbibliothek.

MS 2554, f. 1^{r}, third roundel, left. French (western Champagne). Thirteenth century. Bible moralisée. Creation of the birds and trees. Four birds seem to have been placed on mushroom trees by God's hand. Moralisation: birds are the people who hold to Holy Church.

ii. Other manuscripts

When illustrations have been reproduced in this book, the plate and figure numbers are given in parentheses.

Arras: Bibliothèque municipale. (Old catalogue numbers are given in parentheses.)

MS 42 (33). French. Seventeenth century. Traité de la nature. Vol. I, ff. 239^{r}–242^{v}: 'De l'air'; vol. II, ff. 171^{r}–217^{v}: 'Des vents.'

MS 897 (587). French. Fourteenth century. Le roman de la rose. F. 12^{r}: god of love with fiery wings; wings of other colours on ff. 11^{r}, 12^{v}.

Austin, Texas: Humanities Research Center.

MS Bede, ff. 7^{v}–8^{r}, 9^{r}. Bavarian. Second quarter of the eleventh century. De natura rerum. Air, wind, cloud, mist.

Bergamo: Biblioteca civica.

MS Gabinetto A, IV, 5, ff. 118^{r}–119^{r}. Italian. 1751. Lettera scritta da uno di Londra. Grimaldi.

Bologna: Biblioteca universitaria.

MS 2705, ff. 95^{v}–104^{v}. Italian. C. 1420. Giovanni da Fontana, Metrologum de pisce cane et volucre.

Cambridge: Fitzwilliam Museum.

MS 167, ff. 88^{r}, 89^{r}. French. Fifteenth century (?1486). Calendrier des bergers. Meteorological flying dragon; flying star.

Cambridge: University Library.

MS Dd. 12. 50, ff. 127^{r}–146^{r}. Sixteenth century. Latin treatise on birds.

MS Ii. 2. 27, ff. 4^{r}–111^{r}. Fourteenth century. Tractatus moralis super quatuor elementa.

Douai: Bibliothèque municipale.

MS 711, ff. 24^{v}–43^{r}. French. End of the thirteenth century. De natura animalium. Birds, various winged dragons, and (32^{r}) a winged Siren.

Downe, Kent: *Down House, Erasmus Darwin Room.*

Commonplace Book of Dr Erasmus Darwin, pp. 33, 38. English. 1777. Design for a mechanical bird.

Florence: Biblioteca Mediceo-Laurenziana.

MS Plut. 1.56, f. 13^{v}. Syrian/Mesopotamian. A.D. 586. Rabbula Gospels. Christ ascending, carried by the cherub wagon of Ezekiel's vision (Plate XII).

Graz: *Universitätsbibliothek.*

MS 1170, ff. 59^{r}–167^{r}. Latin. Seventeenth century (finished 5 March $16\frac{21}{22}$). De natura angelorum. An example of seventeenth century Thomist treatment of angels, including (122^{r}), a statement that angels assume bodies made from a mixture of air and the lower elements. Essentially the same material in MS 1163, ff. 1^{v}–69^{v}.

Heidelberg: *Universitätsbibliothek.*

MS pal. germ. 832. German. 1488. Astrolabium planum. Astrological moralisations about birds, passim.

Innsbruck: *Museum Ferdinandeum.*

MS 32009 (= 16.0.7), ff. [153^{r}]–[160^{r}]. German. C. 1440. Treatise, with diagrams, on the corporeal nature of the air.

Klosterneuburg: *Stiftsbibliothek.*

MS 125. Latin. Fifteenth century. 'Incipit Phisocosmus de proprietatibus rerum.' Ff. 8^{r}, 12^{r-v}, 15^{v}: various winged creatures; ff. 16^{v}–19^{v}: 'De aere,' etc.; ff. 429^{r}–452^{r}, 452^{v}–455^{v}: 'De avibus,' 'De minutis volatilibus,' from Albert.

Linköping: *Stifts- och Landesbibliotek.*

Codex 14a. Swedish. 1714. Swedenborg's 'Machina volatilis et Daedalea' (Figs. 34, 35).

London: *British Library.*

MS Harl. 679, ff. 28^{r}–29^{r}. Spanish. Seventeenth century. ?By Fr Sebastiano Bartadas. Commentary on the fifth day of Creation.

MS Lansdowne 396, ff. 90^{v}–92^{r}. English. ?Late sixteenth century. Notes for an exposition of Genesis to the death of Joseph.

MS Royal 2.C.VIII, f. 4. Thirteenth century. ?By Petrus Cantor. Glosses on Genesis 1:20. For other typical examples of thirteenth century commentaries, mainly based on the *Glossa ordinaria,* see Royal 2.F.XIII, ff. 10^{r}–11^{r}; 4.A.II, f. 8^{r}; 4.A.X, ff. 5^{v}–6^{r}; 4.C.X, ff. 4^{v}–5^{r}.

MS Royal 6.E.VI–VII. English. Mid-fourteenth century. A large encyclopaedia, preceded by pictures, four to a page, of Old Testament history and the life of Christ. See especially 6.E.VI, f. 1^{v}, creation of the birds; ff. 68^{r}–69^{r}, 'Aer'; ff. 89^{v}–94^{v}, angels; ff. 167^{v}–169^{r}, 'Aves'; f. 491^{r}, demons; f. 535^{v}, demons.

MS Royal 10.A.VII. English. Thirteenth century. Ff. 8^{r}, 111: 'avis,' 'volucres,' in a copy of some of the *Distinctiones* of William de Monte; ff. 150^{r}–161^{r}: 'De avibus.'

MS Royal 20.B.XVI, ff. 77^{r}–78^{v}. French. C. 1540. 'Les dictz daulcuns oyseaux.' Couplets.

MS Sloane 919, f. 17^{r}. English. 1672-74. Meteorological and scientific notes by John Conyers, apothecary of London. Passage on kites, ballistics, movements of the air, January or February $16\frac{73}{74}$ (Fig. 13).

MS Add. 15245, f. 3^{v}. French. Late fourteenth century. Saint Augustine, *De civitate Dei* XI. Creation of the birds and fishes (Plate VI).

MS Add. 34113, ff. 189^{v}, 200^{v}. ?Sienese. C. 1480. Protoparachute; conical parachute.

MS Add. 62708, pp. 5, 21, and last folio, verso. English. 1710-31. Picture Bible, with paraphrases, by Abraham Sheares. Creation scene; God flying with golden wings; wings of contemplation ascending to Heaven.

Munich: Bayerische Staatsbibliothek.

MS CLM 14399. German. 1160-70. Ambrose, Hexaemeron.

Orléans: Bibliothèque d'Orléans.

MS 976. French. Eighteenth–nineteenth centuries. Documents relevant to the history of the region. Section 82, ff. 9^{r}–12^{r} (item 1427): manuscript copies of the published Desforges correspondence.

Paris: Bibliothèque de l'Arsenal.

MS 85, f. 80^{v}. French. Thirteenth century. Ambiguous moralisations of the dove.

MS 5213. French. 1387. Traité de la nature des anges. Ff. 1^{v}–6^{r}: angelic nature and corporeality.

Paris: Bibliothèque nationale.

MS franç. 156, f. 4^{r}. French. Fourteenth century. Bible historiale. The elemental regions arranged in startling coloured concentric circles, like a target.

MS franç. 186. French. Fifteenth century. Le livre des angeles. At f. 15^{v} God is shown in glory, surrounded by coloured rings of angels.

MS franç. 1082, f. 103^{r-b}. French. 1377. Nicole Oresme. Le livre du ciel et du monde. Aerial ship floating on the surface of the atmosphere (Fig. 53).

MS franç. 1559, ff. 15^{r}, 16^{v}. French. Thirteenth century. Le roman de la rose. God of love with fiery wings.

MS franç. 6260, f. 156^{r}. French. Fifteenth century. Story of the Bible. Bat-winged demons falling through the air and into the earth (Plate X).

MS n.a.fr. 6204, ff. 128^{r}–130^{r}. French-Polish. 1648. Letters from Pierre des Noyers. Burattini.

MS n.a.fr. 6206, f. 21^{r-v}. French-Dutch. 6 April 1648. Letter to Marin Mersenne from Christiaan Huygens. Burattini. F. 64^{r-v}: French. 3/13 July 1648. Letter to Marin Mersenne from Theodore Haak. Burattini.

MS lat. 11108, f. 43^{v}. ?Irish. ?Twelfth century. Chronicles. Aerial ship.

MS lat. 11195, ff. 50^{r}–61^{r}. Italian-French-Polish. 1647. Ars volandi. Burattini (Figs. 32, 33).

Sibiu: Biblioteca raională.

MS II Varia 374, ff. 194^{r}, 258^{v}. German. Fourteenth, fifteenth, sixteenth centuries. Rüst= und Feuerwerksbuch. Rocket-carrying bird signed C[onrad]. H[aas]. Dated 1565. 'Fewer Taubenn,' carrying spherical bomb.

Vienna: Österreichische Nationalbibliothek.

MS 1179, ff. 43^{r}, second medallion, left; 224^{r}, fourth medallion, right. Thirteenth century. Latin moralised Bible. Doves as moralisations of simplicity and chastity.

MS 3068, f. 89^{r}. German. Fifteenth century. Liber de arte bellica. Air and clouds (Fig. 6).

MS Hohendorf 7049, ff. 447^{r}–471^{v}. French-Polish. 1647-49. Letters from Pierre des Noyers. Burattini, passim.

Weimar: Staatsarchiv.

MS a. Rep. A Greiz Rep 41 Nr. 12a. German. 1764. Melchior Bauer. Man-powered aircraft (Figs. 42–49).

B: Primary and secondary printed sources

Addison, Joseph. [Satirical letter and commentary.] *The Guardian* 112 (20 July 1713). Reprinted in *The Works* IV (London 1721) 180-82.

Aelian. *De natura animalium* II.2, II.31, IX.52, XII.4

Affiches, annonces, et avis divers (Paris) 43 (21 October 1772) 172; 44 (28 October 1772) 175-76. Desforges.

Agobard, Saint. *Contra insulsam vulgi opinionem de grandine et tonitruis* II [c. 800]. *PL* 104.148.

[Pseudo-] Aḥmad ibn Sīrīn. [Dream book.] *Artemidori Daldiani & Achmetis Sereimi F. oneirocritica . . .* (Lutetiae 1603) part II, 137-39. Flying dreams (chapters 161, 162), widely attributed in the Middle Ages and Renaissance to Aḥmad ibn Sīrīn (647-728) but probably by Isḥak ibn Ḥunain (sixth century A.D.). See A. Abdel Daïm, *L'oniromancie arabe d'après Ibn Sîrîn* (Damas 1958) 26.

d'Ailly, Pierre. *De impressionibus aeris* (Leipsic ?1495).

Alanus de Insulis. *Anticlaudianus* IV.5, 6 [c. 1183]. *PL* 210.525-27. Trans. James J. Sheridan and published as *Anticlaudianus or the Good and Perfect Man* (Toronto 1973) 127-29.

______. *De sex alis cherubim* [late twelfth century]. *PL* 210.265-80.

Albert of Saxony. *Questiones . . . in octo libros Physicorum Aristotelis* [c. 1360] IV.vi.2.3 (Parisiis 1516) 47^{r}.

Albertus Magnus. *De animalibus libri xxvi* XXIII [thirteenth century]. Ed. Hermann Stadler, II (Münster i. W. 1920) 1430-1514.

______. *De celo & mundo* [thirteenth century] (Venetijs 1495) 70^{v}, 73^{r-v}.

d'Alcripe [= Le Picard], Philippe. *La nouvelle fabrique des excellens traits de vérité* [sixteenth century]. Ed. P. A. Gratet-Duplessis (Paris 1853) 22, 178-79.

Aldrovandus, Ulysses. *Ornithologiae hoc est de avibus historiae libri xii* 3 vols (Bononiae 1599-1603).

d'Alembert, Jean le Rond. *Réflexions sur la cause générale des vents* (Berlin 1747).

Ambrose, Saint. *Hexaemeron libri sex* V.xii–xxv, VI.ix [fourth century]. *PL* 14.222-42, 272. Trans. John J. Savage and published in *Hexameron, Paradise, and Cain and Abel* (New York 1961) 191-226, 281. (*The Fathers of the Church: A New Translation* 42.)

Anania, Giovanni Lorenzo. *De natura daemonum libri iiii* (Venetiis 1581) 148-51 and passim.

Anastasius Sinaita, Saint. *Anagogiarum contemplationum in Hexaemeron ad Theophilum libri duodecim* V [seventh century]. *PG* 89.913–22. Creation of the birds and fishes.

Anaximenes. Fragments [sixth century B.C.]. Diels I.90–96. See also Fairbanks, *The First Philosophers of Greece* 17–22 and Kirk and Raven, *The Presocratic Philosophers* 143–62.

Anders Sunesøn. *Hexaëmeron libri duodecim* [c. 1200]. Ed. M. Cl. Gertz (Havniae 1892) I.391–424, II.708–23.

Angelomus Luxoviensis. *Commentarius in Genesin* [ninth century]. *PL* 115.121. Birds cannot fly unless the air is thick. Men climbing mountains keep damp sponges before their mouths.

Annonces, affiches, nouvelles et avis divers de l'Orléanois 36 (4 September 1772) 147–48; 39 (25 September 1772) 161–62; 40 (2 October 1772) 165–66. Desforges.

Anonymous. *De elementis.* Ed. Richard C. Dales, *Isis* 56.2 (Summer 1965) 174–89.

Apuleius, Lucius. *De deo Socratis* VIII.138–X.143 [second century A.D.]. In *Apulei platonici madaurensis opera quae supersunt* III, ed. Paulus Thomas (Stutgardiae 1970) 15–18. Trans. and published as *The God of Socrates* in *The Works . . . a new translation* (London 1899) 358–59.

Aquinas, Saint Thomas. *In Aristotelis libros De caelo et mundo, De generatione et corruptione, Meteorologicorum expositio.* Ed. R. M. Spiazzi (Torino 1952).

______. *In octo libros Physicorum Aristotelis expositio.* Ed. P. M. Maggiòlo (Taurini 1965). Trans. Richard J. Blackwell, Richard J. Spath, and W. Edmund Thirlkel and published as *Commentary on Aristotle's* Physics *by St. Thomas Aquinas* (New Haven 1963). See esp. commentary on 254 b 15ff (p. 505) for aerial bodies that naturally move upwards.

______. *Summa theologiae* [1265–73]. Ed. P. Caramello 3 vols (Torino 1952–56). Trans. and published as *The 'Summa Theologica'. . . literally translated by fathers of the English Dominican Province* 27 vols (London 1911–22).

Arias Montanus, Benedictus. *Naturae historia* (Antverpiae 1601) 293–312. Flying creatures, with a theory of flight (298).

Ariotti, Piero E. 'Christiaan Huygens: Aviation Pioneer Extraordinary.' *Annals of Science* 36.6 (November 1979) 611–24.

Aristotle [and pseudo-Aristotle]. *The Works of Aristotle. Translated into English under the Editorship of J. A. Smith . . . W. D. Ross.* 12 vols (Oxford 1908–52).

Arliquiniana ou les bons mots, les histoires plaisantes & agréables. Recueillies des conversations d'Arlequin (Paris, Florentin & Pierre Delaulne et Michel Brunet, 1694) 84–87. Desson. Another edition, with different pagination, was published in Paris in the same year.

Artemidorus Daldianus. *Oneirocritica* II.68 [late second century A.D.]. Trans. with a commentary by Robert J. White and published as *The Interpretation of Dreams* (Park Ridge, N.J. 1975) 132–33. For an earlier translation, see *The Judgement, or Exposition of Dreames* (London 1606) 103–05.

d'Aubigné, Agrippa. *Stances* I.137–38. In *Oeuvres,* ed. Henri Weber and others (Paris 1969). (Bibliothèque de la Pléiade.) Birds falling dead.

Auger, Léon. *Un savant méconnu: Gilles Personne de Roberval (1602–1675)* (Paris 1962) 97–98. Burattini.

Augustine, Saint, Bishop of Hippo. *De civitate Dei contra paganos libri viginti duo* VIII.14–16, X.21–22, XXI.10. *PL* 41.239–41, 298–300, 724–25. Trans. Rev. Marcus Dods, M.A., and published as *The City of God* 2 vols (Edinburgh 1872) I.325–30, 411–13, II.434–36.

______. *De divinatione dæmonum liber unus* III.7. *PL* 40.581–92.

______. *De genesi ad litteram libri duodecim* III.vi, vii, ix, x. *PL* 34.282–83, 284–85.

______. *De genesi ad litteram, imperfectus liber* XIV, XV. *PL* 34.237–41.

______. *De genesi contra Manichæos* I.15. *PL* 34.184.

______. *Enarrationes in Psalmos* 103 [104].I(a).12–14. *PL* 37.1346–48. Trans. H. M. Wilkins and published as *Expositions on the Book of Psalms* V (Oxford 1853) 80–82. Wings of the wind.

Avitus, Alcimus Ecdicius. *De mundi initio* [c. 500]. Ed. Abraham Schippers (Kampen 1945). Lines 32–35 on the creation of birds.

Bachelard, Gaston. *L'air et les songes: essai sur l'imagination du mouvement* (Paris 1943).

Bacon, Francis. *Sylva sylvarum: or A Naturall Historie. In Ten Centuries* VII.681, IX.823–24, IX.886 (London 1627) 170, 216, 235.

Baluze, Etienne. *Capitularia regum francorum* 2 vols (Paris 1677) I.235, II.1131. Law of A.D. 789: *De auguriis vel aliis maleficiis.*

Baroni Cavalcabò, Clemente. *L'impotenza del demonio di trasportare a talento per l'aria da un luogo all'altro i corpi umani* (Rovereto 1753) esp. 92–117: 'Si dimostra, che il Demonio non può instruire gli Uomini al volo.'

Barthez, Paul-Joseph. *Nouvelle méchanique des mouvements de l'homme et des animaux* (Carcassonne 1798) 190–230.

Bartholomaeus Anglicus. *De proprietatibus rerum* XII.1 [c. 1250]. Trans. John Trevisa (13$\frac{98}{99}$) and published as *On the Properties of Things* I, ed. M. C. Seymour (Oxford 1975) 596–97.

Bartoli, Danielo, S.J. *De' simboli trasportati al morale* II.xiv (Venetia 1677) 470–71. Comment on Diodorus Siculus II.xiii: children carried by large birds in Ethiopia.

Basil, Saint. *Homilia VIII in hexaemeron* [fourth century]. *PG* 29.163–88. Trans. Sister Agnes Clare Way and published as 'Homily 8: On the Hexaemeron' in *Exegetical Homilies* (Washington D.C. 1963) 120–34. (The Fathers of the Church: A New Translation 46.)

______. *The Letters* CLXXXVIII.xv.

Baumé, Antoine. *Chymie expérimentale et raisonnée* I (Paris 1773) 62–69, 'Sur l'air'; III (Paris 1773) 693–98, 'Appendix sur l'air fixe.'

Becher, Johann Joachim. *De nova temporis dimetiendi ratione* (Londini 1680) 4, 22.

______. *Närrische Weiszheit und weise Narrheit* (Franckfurt 1682) 27–28, 30–31, 148–49, 164–68. Desson; flight attempts.

Beda, Venerabilis. *De natura rerum* XXV, XXVI. *PL* 90.244–47.

______. *In Pentateuchum commentarii* I.i. *PL* 91.199. Paraphrase of Origen on separation of the waters.

Bellarminus, Robertus, Cardinal. *De ascensione mentis in deum per scalas rerum creatarum* (Antverpiae 1615) 74–87.

Belon, Pierre. *L'histoire de la nature des oyseaux, avec leurs descriptions, & naïfs portraicts retirez du naturel* (Paris 1555).

Benedetti, Giovanni Battista. *Diversarum speculationum mathematicarum, & physicarum liber* (Taurini 1585) 361. Clouds.

Berefelt, Gunnar. *A Study on the Winged Angel: The Origin of a Motif.* Trans. from the Swedish by Patrick Hort, M.A. (Stockholm 1968).

Berga, Antonius. *Paraphrasis eorum quae in quarto libro operis Meteorologici habentur* (In Monte Regale 1565).

Bernardus Silvestris. *De mundi universitate libri duo* II.vii. Ed. C. S. Barach and J. Wrobel (Innsbruck 1876). Trans. Winthrop Wetherbee and published as *The 'Cosmographia' of Bernardus Silvestris* (New York and London 1973) 105–08.

Biblia sacra cum glossa ordinaria 6 vols (Antverpiae 1634).

Biliński, Bronisław. *Galileo Galilei e il mondo polcco* (Wrockław-Warszawa-Kraków 1969) 113–117. Burattini.

Boaistuau, Pierre. *Bref discours de l'excellence et dignité de l'homme* (Paris 1558) 20[r]. Appended to *Le theatre du monde* (Paris 1558). Latin translation by Laurentius Cuperus published as *Tractatus de excellentia et dignitate hominis,* appended to *Theatrum mundi minoris, sive humanae calamitatis oceanus* 2nd ed. (Antverpiae 1589) 239.

Bochart, Samuel. *Hierozoicon . . . De animalibus scripturae* I.1.iii, II.1, 2. In *Opera omnia* 3 vols (Lugduni Batavorum 1712) II.13-22, III.1-356.

Bodin, Jean. *Universae naturae theatrum* (Lugduni 1596) 363-81. Birds. Air, winds, demons, passim.

Boffito, Giuseppe. 'La posizione di Aristotele nella storia dell'aeronautica.' *Rivista di filologia e di istruzione classica* 48.2 (1920) 258-66.

______. *Il volo in Italia* (Firenze 1921).

______. 'L'aeronautica nelle città italiane. Venezia: Tito Livio Burattini (1615-82) costruttore d'una macchina volante.' *Rivista aeronautica* 4 (1928) 198-201.

______. *Biblioteca aeronautica italiana illustrata. Precede uno studio sull'aeronautica nella letteratura nell'arte e nel folklore* 2 vols (Firenze 1929, 1937).

Bollandus, Joannes. *Af-beeldinghe van d'eerste eeuwe der Societeyt Iesu* (t'Antwerpen 1640) 86, 290, 296, 396, 410, 504. Emblems with aeronautical content.

The Book of Leinster, formerly Lebar na Núachongbála [late twelfth–early thirteenth century] V, ed. Richard I. Best and others (Dublin 1967) 274[a].36–38, p. 1204. Aerial ships.

Borelli, Giovanni Alfonso. *De motionibus naturalibus a gravitate pendentibus* (Regio Julio 1670) 254–63. Compressibility of the air.

______. *De motu animalium* 2 vols (Romae 1680, 1681) I.285–326 and plates 11–13.

Bouchot, Henri. *Jacques Callot: sa vie, son oeuvre et ses continuateurs* (Paris 1889) 230–31. Desson.

Bowen, Ivor. 'John Williams of Gloddaeth.' *Transactions of the Honourable Society of Cymmrodorion* (London 1927–28). Life, but lacks mention of his flight.

Boyle, Robert. *New Experiments Physico-Mechanicall, Touching the Spring of the Air, and its Effects* (Oxford 1660).

______. *Tracts: containing . . . Suspicions about some Hidden Qualities of the Air* (London 1674). Mixed nature of air and variety of phenomena produced by it.

______. *The General History of the Air* [part pre-1677] (London 1692).

Bradstreet, Anne. 'The Four Elements' (1650 and 1678). *The Works of Anne Bradstreet,* ed. Jeannine Hensley (Cambridge, Mass. 1967) 18–32. Air: 29–32.

Brenner, Oscar, ed. *Speculum regale: ein altnorwegischer Dialog* (München 1881) 44–45. Flying ship at Clonmacnois.

Breventano, Stefano. *Trattato de l'origine delli venti* (Venetia 1571).

______. *Trattato degli elementi* (Pavia 1571) 6^{v}–10^{r}.

______. *Trattato delle impressioni dell'aere, raccolto da varij autori di filosofia* (Pavia 1571) 5^{r}, 8^{v}, 9^{r}. Aerial 'dragons,' aerial ships.

Browne, Sir Thomas. *Religio Medici* I.30 [1643]. In *The Works of Sir Thomas Browne* I, ed. Geoffrey Keynes, new ed. (London 1964) 40–41.

Bruno Astensis, Saint. *Expositio in Genesim* I [eleventh or twelfth century]. *PL* 164.155.

Bruyn, Johannes de. *Disputatio physica de aëre* (Lugduni Batavorum 1654).

______ [*Praeses, resp.* G. de Ridder]. *Disputationis physicae de corporum levitate et gravitate. Pars quarta* (Ultrajecti 1668) A4^{r}.

Buccaferrea, Ludovicus. *In quartum meteororum Aristotelis librum* (Venetiis 1563) 146.

Bucher, François. *The Pamplona Bibles: A facsimile compiled from two picture Bibles with martyrologies by King Sancho el Fuerte of Navarra (1194–1234) Amiens manuscript Latin 108 and Harburg MS 1, 2, lat. 4°, 15* 2 vols (New Haven and London 1970) esp. I, illus. 15b and II, plate 7.

Büchner, Georg Heinrich. *Merkwürdige Beyträge zu dem Weltlauf der Gelehrten* III (Langensalza 1766) 453–524, 'Ars volandi frustra tentata. Oder die Thorheit der Menschen, wie Vögel durch die Luft zu fliegen'; 525–66 'Seltsamer Lebenslauf eines Rotterdamer Bürgers, Adriaen Baartjens genannt, welcher durch mancherley sich selbst zugezogene Unglücksfälle in die gröste Armuth verfallen und endlich im gemeinen Gefängniß elender Weise gestorben ist'; 569–648 'Nauta aereus male compositus. Oder gezeigte Unmöglichkeit mit einem Schiffe durch die Luft zu seegeln.'

Buffon, Georges Louis Leclerc, comte de. *Histoire naturelle des oiseaux* 9 vols (Paris 1770–83) I.1–60, III.151–54.

Buonaventura, Federigo. *De causa ventorum motus* (Urbini 1592).

______. *Anemologiae pars prior, id est de affectionibus, signis, causisque ventorum ex Aristotele, Theophrasto, ac Ptolemęo tractatus* (Urbini 1593).

Burattini, Tito Livio. Letter about Egyptian mummies, trans. from Italian and appended to Melchisedech Thévenot's French translation of John Greaves, *Pyramidographia* (London 1646): *Relations de divers voyages curieux* (Paris 1663) [I].xxv.

[______.] Letters by Jerzy B. Cynk and Charles H. Gibbs-Smith in *Flight International* 83.2826 (9 May 1963) 695; 83.2831 (13 June 1963) 935–36; 84.2834 (4 July 1963) 26.

Burgundius, Horatius. 'De volatu.' In Michel Giuseppe Morei, *Arcadum carmina* II (Romae 1756) 1–8.

Burnet, John. *Early Greek Philosophy* 3rd ed. (London 1920).

Cailhava d'Estendoux, Jean François. *Arlequin Mahomet, ou le cabriolet volant* (1770); *Première suite du cabriolet volant, ou Arlequin cru fou, sultane et mahomet* (1770). In *Théâtre de M. Cailhava* II (Paris 1781) 1–88, 89–157.

Calcidius. *See* Den Boeft, J.

Cambridge, Richard Owen. *The Scribleriad: an Heroic Poem. In Six Books* (London 1751) esp. IV.125–56.

Campanella, Thomas. *De sensu rerum et magia, libri quattuor* III.vi–x, IV.vi (Francofurti 1620) 216–35, 280–83. The air, flight.

Cantor, G. N. and Hodge, M. J. S., eds. *Conceptions of Ether: Studies in the History of Ether Theories 1740–1900* (Cambridge 1981). With an introduction on earlier ideas.

Caramuel Lobkovitz, Johann. *Mathesis biceps* I (Campaniae 1670) 740–62 and plate XVIII.

Cardanus, Hieronymus. *De subtilitate libri xxi* (Lugduni 1554) 400, 401–02. Flying fish; manucaudiata.

______. *De rerum varietate libri xvii* (Basileae 1557) 269, 287, 440. Flying fish; manned flight; Archytas.

______. *Commentarii, in Hippocratis de aere, aquis et locis opus* (Basileae 1570) 30b.

______. *Opus novum de proportionibus* (Basileae 1570) 23–24, 25. Voluntary motion.

Castel, Louis-Bertrand. *Traité de physique sur la pesanteur universelle des corps* I (Paris 1724) 276.

Caus, Salomon de. *Les raisons des forces mouvantes* (Francfort 1615) 1v. Air is cold, dry, and light.

Cavalcabò. *See* Baroni Cavalcabò, Clemente.

Cavendish, Margaret, Duchess of Newcastle on Tyne. *Poems, and Fancies* (London 1653) 7, 31–32, 33, 34–35, 105–06, 142–43. Air, wind, birds; esp. 142–43: 'Similizing Birds to a Ship.'

[Charbonnet, Pierre-Mathias.] *Eloge prononcé par La Folie devant les habitans des petites-maisons* (Avignon 1761). Bacqueville.

Chauvin, Stephanus. *Lexicon rationale sive thesaurus philosophicus ordine alphabetico digestus* (Rotterodami 1692). 'Aer'; 'avis'; 'draco volans'; 'volatus.'

Chrysostom, Saint John. *Homiliae XXI de statuis ad populum Antiochenum habitae* XI.4, XV.3, 5 [A.D. 387 or 388]. *PG* 49.124–25, 157, 160. Trans. W. R. W. Stephens and others and published as *The Homilies on the Statues to the People of Antioch* (New York 1889) 416, 441, 443. (A Select Library of the Nicene and Post-Nicene Fathers of the Christian Church 9.)

Cicero, Marcus Tullius. *De natura deorum* I.x, II.x, II.xxvi, II.xxxix. The air.

Clausenius, Petrus. *Disputationum physicarum de aere prima* (Havniae 1712).

Columbus, Ferdinand. *The Life of the Admiral Christopher Columbus by His Son Ferdinand.* Trans. and ann. Benjamin Keen (London 1960) 147, 243.

Cordier, Henri. 'Un précurseur dans l'aviation.' *La revue hebdomadaire* 3.3 (March 1918) 329–32. Bacqueville.

Cornford, Francis MacDonald. *Plato's Cosmology: The Timaeus of Plato translated with a running commentary* (London 1937).

Le courrier d'Avignon [= *Le courrier historique, politique, littéraire,* etc.] 58 (Tuesday 20 July 1784) 236. Flying attempts by M. Ariès, of Embrun.

Cumont, Franz. 'A propos de Properce, III, 18, 31 et de Pythagore.' *Revue de philologie de littérature et d'histoire anciennes* ns 44.1 (January 1920) 75–78. Infernal regions in the air.

______. *After Life in Roman Paganism* (New Haven 1922).

Cuvier, Georges Léopold Chrétien Frédéric Dagobert de, Baron. *Lectures on Comparative Anatomy* I, trans. William Ross (London 1802) 534–42. 'Of Flying.'

Cyrano de Bergerac. *Histoire comique des état et empire de la lune et du soleil* [1648–49]. Ed. Claude Mettra and Jean Suyeux (Paris 1962) 181.

Dales, Richard C. 'A Twelfth-Century Concept of the Natural Order.' *Viator* 9 (1978) 179–92.

Danaeus, Lambertus. *Physices christianæ pars altera* II.ix, V.xxvi–xxxvi. In *Opuscula omnia theologica* (Genevae 1583) 281–83, 365–68.

Dante Alighieri. *Opere.* Ed. Fredi Chiappelli (Milano 1965). Esp. 34 (*La vita nuova* XXIII), 548 (*Inferno* XXIX).

Darwin, Erasmus. *The Botanic Garden* I (London 1791) canto I.289–96 and additional note I; canto IV.

Daudin, François Marie. *Traité élémentaire et complet d'ornithologie, ou Histoire naturelle des oiseaux* 2 vols (Paris 1800).

Defoe, Daniel. *The Consolidator: or, Memoirs of Sundry Transactions from the World in the Moon* (London 1705) 36–54. Satirical account of a flying machine.

Den Boeft, J. *Calcidius on Demons (Commentarius CH. 127–136)* (Leiden 1977).

Derham, William. *Physico-theology: or, a demonstration of the being and attributes of God, from his works of creation* (London 1713) 7n, 308–09, 372–96, 405–06. Human flight; birds; flight of insects.

Descartes, René. [Letters to Mersenne 30 July 1640, 30 August 1640.] See Marin Mersenne, *Correspondance* IX, X.

Desforges, Pierre, Abbé d'Etampes. *Avantages du mariage, et combien il est nécessaire & salutaire aux prêtres & aux évêques de ce tems-ci d'épouser une fille chrétienne* 2 vols (Bruxelles 1758). Reprinted 1760, 1768. Italian translation 1770. Condemned to be burned 3 October 1758.

Desmarets de Saint Sorlin. *Ariane* 2 pts (Paris 1632) I.v, 289-90. Trans. and published as *Ariana* (London 1636) 98. Parachute jump.

De Waard, Cornelis. 'Notes sur Stevin et Beeckman.' *Isis* 24.1 (December 1935) 123-25. Burattini 125.

Dictionnaire des origines, découvertes, inventions et établissemens (Paris 1777) II. 571-73, III.638. 'Machines merveilleuses'; 'voler.'

Diderot, Denis. *Oeuvres complètes* XVIII (Paris 1876) 484-86. Bacqueville.

______ and d'Alembert, Jean le Rond. *Encyclopédie, ou dictionnaire raisonné des sciences, des arts et des métiers* 17 vols (Paris 1751-65) XI.433-39, XVII.447-49. 'Oiseau' (by Louis de Jaucourt), and 'voler' (unattributed). Also 'oiseau' in *Supplément* IV (Paris 1777).

Diel, Paul. *Le symbolisme dans la mythologie grecque: étude psychanalytique* (Paris 1952) 61-115.

Diels, Hermann. *Die Fragmente der Vorsokratiker* 7th ed. Ed. Walther Kranz 3 vols (Berlin-Charlottenburg 1954).

Dio Cassius. *Roman History* LXVI.22-23 (epit.). Flying giants.

Diodorus Siculus. *Bibliothecae historicae libri xv* I.12.7, II.58.5, IV.77.8-9.

Diogenes Laertius. *Vitae* VII.147 (Zeno), VIII.32 (Pythagoras). Hera and the air; spirits of the air.

Diogenes of Apollonia. Fragments [fifth century B.C.]. Diels II.51-69. Trans. and commentary in Freeman, *The Pre-Socratic Philosophers*, 279-84, Kirk and Raven, *The Presocratic Philosophers,* 427-45. Primacy of the air.

Pseudo-Dionysius. *De coelesti hierarchia* XV [?fifth century]. *PG* 3.331-32. Angel wings.

'Dissertazione contro l'Operetta del Signor Clemente Baroni, intitolata *L'Impotenza del Demonio di trasportare a talento per l'aria da un luogo all' altro i Corpi umani.' Nuova raccolta d'opusculi scientifici e filologici* I (Venezia 1755) 129-91.

Doppelmayr, Johann Gabriel. *Historische Nachricht von den Nürnbergischen Mathematicis und Künstlern* (Nürnberg 1730) 300–01. Johann Hautsch.

Dorisi, Joannes. *Curiosae quaestiones de ventorum origine et de accessu maris ad littora & portus nostros, & ab ijsdem recessu* (Parisiis 1646) 1–119.

Dracontius. *Hexaemeron* I [late fifth century; said to have been emended by Saint Eugenius]. *PL* 87.373. Birds.

du Bartas, Guillaume de Salluste, Sieur. *Works* II [late sixteenth century]. Ed. with introduction, commentary, and variants by Urban Tigner Holmes, Jr and others (Chapel Hill 1935–40) 357–75. Birds.

Duhem, Jules. 'Une théorie de la locomotion aérienne.' *Mercure de France* 263 (1 November 1935) 515–45. Bacqueville, Rousseau, Vivens.

______. 'Un essai de vol à voile d'après le plus ancien des recueils connus sous le nom d'*Arlequiniana*.' *Bulletin du bibliophile* (July 1939) 297–302. Desson.

______. 'Le saut de Lavini.' *Bulletin du bibliophile* (March 1940) 42–45. Parachutes.

______. *Histoire des idées aéronautiques avant Montgolfier* (Paris 1943).

______. *Musée aéronautique avant Montgolfier* (Paris 1943).

______. 'Biblioteca aeronautica vetustissima: inventaire des écrits laissés par l'antiquité classique sur le pouvoir du vol.' *Bulletin du bibliophile* (1953) 259–76, (1954) 39–47, 69–81, (1955) 11–19, 179–86, (1956) 19–28, (1958) 8–15, (1959) 147–54, (1962) 73–85; *Thalès: recueil des travaux de l'Institut d'Histoire des Sciences et des Techniques de l'Université de Paris* 8 (1955) 1–31.

______. *Histoire de l'arme aérienne avant le moteur* (Paris 1964).

Dupleix, Scipion. *Histoire de Henry III. Roy de France et de Pologne* (1630) 5th ed. (Paris 1641) 8.

Ellinger, Adamus. *Ex physicis disputationem publicam de aere* [Wittenberg 1659] B[1]^{r}.

Erckenbrecht, Fredericus Casimirus. *Stricturae physicae de vi aëris elasticae* (Hanoviae 1697).

Eriugena, Johannes Scotus. *De divina praedestinatione* XIX [mid-ninth century]. *PL* 122.436–38.

______. *De divisione naturae* III.36–40 [mid-ninth century]. *PL* 122.727–42.

Erythraeus, Janus Nicius [= Giovanni Vittorio Rossi]. *Pinacotheca imaginum illustrium, doctrinae vel ingenii laude, virorum, qui, auctore superstite, diem suum obierunt* (Coloniae Agrippinae 1643) 123.

d'Escouchy, Mathieu. *Chronique* II. Ed. G. du Fresne de Beaucourt (Paris 1863) 149–50.

Esmeijer, Anna C. *Divina quaternitas: A Preliminary Study in the Method and Application of Visual Exegesis* (Amsterdam 1978). Symbolism and numerology of air and birds, passim.

Evliyâ Çelebi [1611–83]. *Seyahatnâme* I (Istanbul A.H. 1314 [= 1896]) 670. Flight of Hezârfen Ahmed Çelebi.

Fabricius ab Aquapendente, Hieronymus. *De alarum actione, hoc est volatu.* In *De motu locali animalium secundum totum* (Patavii 1618) separately paginated 1–16.

Fairbanks, Arthur. *The First Philosophers of Greece* (London 1898).

Favaro, Antonio. *Intorno alla vita ed ai lavori di Tito Livio Burattini, fisico agordino del secolo XVII* (Venezia 1896) esp. 55–57, 72–75. (*Memorie del Reale istituto veneto di scienze, lettere ed arti* 25.8.)

______. 'Nuove contribuzioni alla storia delle scienze nel decimosettimo secolo: Tito Livio Burattini.' *Atti del Reale istituto veneto di scienze, lettere ed arti* 54 (1896) 110–16.

______. 'Supplemento agli studi intorno alla vita ed alle opere di Tito Livio Burattini, fisico agordino del secolo XVII.' *Atti del Reale istituto veneto di scienze, lettere ed arti* 59.2 (1900) 855–60.

Flayder, Friedrich Hermann [*resp.* to Johann Oswald and Johann Ulrich Brehizer]. *De arte volandi* ([Tübingen] 1627).

Fludd, Robert. *Utriusque cosmi maioris scilicet et minoris metaphysica, physica atque technica historia* 4 vols (Francofurti 1617–21). Air passim.

Fonteny, Jacques de. 'L'oeuf de Pasques ou pascal' [1616]. In *Variétés historiques et littéraires* V, ed. Edouard Fournier (Paris 1856) 59–74.

Francisci, Erasmus. *Der Wunder=reiche Uberzug unserer Nider=Welt, oder Erd=umgebende Lufft=Kreys* (Nürnberg 1680).

Frederick II of Hohenstaufen. *De arte venandi cum avibus* [c. 1250]. Facsimile of Biblioteca Apostolica Vaticana, MS Pal. Lat. 1071, with commentary by Carl Arnold Willemsen 2 vols (Graz 1969). Trans. Casey A. Wood and Marjorie F. Fyfe and published as *The Art of Falconry* (Stanford 1943).

Freeman, Kathleen. *Ancilla to the Pre-Socratic Philosophers* (Oxford 1948).

______. *The Pre-Socratic Philosophers: A Companion to Diels, 'Fragmente der Vorsokratiker'* (Oxford 1946).

Frescarode, Jeremias. *Dissertatio philologico-physica de ventis* (Trajecti ad Rhenum 1702).

Fulgentius, Saint, Bishop of Ruspe [467-533]. *De trinitate liber unus* IX. *PL* 65.505.

Fulke, William. *A Goodly Gallerye with a Most Pleasaunt Prospect, into the garden of naturall contemplation, to behold the naturall causes of all kynde of Meteors* (Londini 1563).

Furetière, Antoine. *Dictionnaire universel* III (A la Haye et a Rotterdam 1690) 'Voler.' Birds of paradise, Burattini.

Gaietanus de Thienis. *Liber Aristotelis de celo et mundo* II.1 (Padua ?1475). Air-filled bladders.

Galen, Claudius. *De tremore, & palpitatione, convulsionéque, ac rigore, liber* [second century A.D.]. In Cl. Galeni *Omnia . . . opera in latinam linguam conversa* II (Lugduni 1550) 321-42.

Galiani, Ferdinand. *Correspondance inédite* II (Paris 1818) 86-87.

[Galien, le Père Joseph.] *Mémoire touchant la nature et la formation de la grêle et des autres météores qui y ont rapport, avec une conséquence ultérieure de la possibilité de naviger dans l'air à la hauteur de la région de la grêle. Amusement physique et géométrique* (Avignon 1755) esp. 51-87. Second ed. published as *L'art de naviger dans les airs. Amusement physique et géometrique, précedé d'un mémoire sur la nature & la formation de la grêle, dont il est une conséquence ultérieure* (Avignon 1757).

Galilei, Galileo. *Dialogue Concerning the Two Chief World Systems* (1632). Trans. Stillman Drake (Berkeley and Los Angeles 1953) 223.

______. *Les méchaniques* [1593–c. 1600]. Trans. Marin Mersenne (Paris 1634) V, VI, 17-26. (The Italian original was posthumously published in 1649.)

______. *Two New Sciences, Including Centers of Gravity & Force of Percussion.* Trans. Stillman Drake (Madison 1974) 225–28, 292–93. Resistance of air to motion (1638); force of percussion (c. 1635).

Galon, V. *Machines et inventions approuvées par l'Académie royale des sciences* 7 vols (Paris 1735–77) I.71–72 and fig. 18, III.33–41 and figs. 152, 153. Huygens's machine for measuring the moving force of the air; sail-driven chariots (1714).

Garnerus. *Gregorianum* II [twelfth century]. *PL* 193.65–84. Symbolic meanings of birds.

Gassendi, Pierre. *Syntagmatis philosophici pars secunda, quae est physica* III.ii.11.vi. In *Opera omnia* II (Lugduni 1658) 537–40. 'De volatu animalium.'

Geoffroi de Vigeois. *Chronica Gaufredi coenobitae* XL. In Philippe Labbe, comp., *Novae bibliothecae manuscript. librorum* II (Paris 1657) 299–300.

Gérard, Laurent Gaspard. *Essai sur l'art du vol aérien* (Paris 1784).

Gervasius of Tilbury. *Otia imperialia* [1211]. Selected and ed. F. Liebrecht (Hannover 1856) 2–3, 62, 261. Aerial ship.

Gesner, Conrad. *Historiae animalium liber III, qui est de avium natura* (Tiguri 1555).

______. *Historiae animalium liber IIII, qui est de piscium & aquatilium animantium natura* (Tiguri 1558) 514, 516, 1279, 1291. Flying fish.

Giacomelli, Raffaele. 'La scienza dei venti di Leonardo da Vinci.' *Atti del convegno di studi Vinciani* (Florence 1953) 374–400.

______. *See also* Leonardo da Vinci.

Gibbs-Smith, Charles H. *A Directory and Nomenclature of the First Aeroplanes 1809 to 1909* (London 1966).

______. *Leonardo da Vinci's Aeronautics* (London 1967).

Gille, Bertrand. 'La conquête de l'air au Moyen Âge et à la Renaissance.' *La recherche* 12.119 (February 1981) 182–91. Derivative, outdated, inaccurate.

Glanvill, Joseph. *The Vanity of Dogmatizing: or Confidence in Opinions* (London 1661) 182.

Glorieux, P., comp. *Autour de la spiritualité des anges* (Tournai 1959). (*Monumenta Christiana selecta* 3.) Compendium, from Genesis to Aquinas, of passages about the constitution of angels and demons.

Goldwel, Henry. *A briefe declaration of the shews, devices, speeches, and inventions, done & performed before the Queenes Maiestie* (London 1581) [Bvi[r]].

Gombrich, Sir Ernst Hans. 'The Form of Movement in Water and Air.' In *Leonardo's Legacy: An International Symposium,* ed. C. D. O'Malley (Berkeley and Los Angeles 1969) 171–204.

Gougaud, L. 'L'aéronef dans les légendes du moyen âge.' *Revue celtique* 41 (1924) 354–58.

Goulard, Jean François Thomas. *Cassandre mécanicien, ou le bateau volant, comédie parade, en un acte et en vaudevilles* (Paris 1783). Parody of a flying machine with nacelle, rudder, and two wings worked by springs.

Goulart, Simon. *Thrésor d'histoires admirables et mémorables de nostre temps* I (Genève 1620) 33, 34, 47–64, 549–54. Compendium of various apparitions.

Gower, John. *Confessio amantis* VII.254–374 [c. 1390]. In *The English Works of John Gower* II, ed. G. C. Macaulay (London 1901) 240–43. (Early English Text Society Extra Series 82.) Lines making conventional Aristotelian points about the air.

Grant, Edward. *Much Ado About Nothing: Theories of Space and Vacuum from the Middle Ages to the Scientific Revolution* (Cambridge 1981).

Graziani, Girolamo. *Il conquisto di Granata* [1650]. In *Parnasso italiano* 39 (Venezia 1789) 1, 175. Cuts of flying ships.

Greaves, John. *Pyramidographia: or a Description of the Pyramids in Ægypt* (London 1646) 86, 104. Burattini.

Gregory I, Pope, Saint, the Great. *Homiliarum in Ezechielem prophetam libri duo* I.iii.1–2 [sixth century]. *PL* 76.806.

Gregory of Nyssa, Saint. *In hexaemeron. PG* 44.61–124, esp. 81–88. The waters above the firmament; creation of the air.

Grew, Nehemiah. *Cosmologia sacra: or a Discourse of the Universe as it is the Creature and Kingdom of God* (London 1701) 28, 29.

Grimm, Friedrich Melchior von, Baron. *Correspondance littéraire, philosophique et critique par Grimm, Diderot, Raynal, Meister, Etc.* Ed. M. Tourneaux, 16 vols (Paris 1877–82) V.102–03, X.60–61.

Grimm, Jacob. *Deutsche Mythologie* 4th ed. 3 vols (Berlin 1875–78) I.524–33, III.178–83.

Grosley, Pierre Jean. *Oeuvres inédites* I. Ed. L. M. Patris-Debreuil (Paris 1812) 84–88. Bolori.

Grosses vollständiges Universal Lexicon aller Wissenschafften und Künste IX, XVIII (Halle und Leipzig 1735, 1738). 'Flüge=Kunst'; 'Lufft=Schiff=Kunst.'

Grosseteste, Robert. *Hexaëmeron* VI [1230–35]. Ed. Richard C. Dales and Servus Gieben (London 1982) 185–98. (*Auctores britannici medii aevi* VI.) Day five.

Guibertus Abbas. *Moralium geneseos* I [eleventh or early twelfth century]. *PL* 156.48–51.

Hacket, John, Bishop of Lichfield and Chichester. *Scrinia reserata: a Memorial Offer'd to the Great Deservings of John Williams, D.D.* ([London] 1693) 8.

[Hale, Sir Matthew.] *An Essay Touching the Gravitation, or Non-gravitation of Fluid Bodies, and the Reasons thereof* (London 1673) 83–88.

Hales, Stephen. *Vegetable Staticks* (London 1727) chapter VI: 'A specimen of an attempt to analyze the Air. . . .'

Hall, Marie Boas [= Boas, Marie]. *Robert Boyle and Seventeenth-Century Chemistry* (Cambridge 1958) 181–204. 'A Digression on Air.'

Hammerschmidt, Karl. *Die Ornithologie des Aristoteles* (Speier 1897).

Hamond, George. *Πνευματολογια: or, a Discourse of Angels* (London 1701) 59–60. Movement of angels.

Hannemann, Joannes Ludovicus [*Praes., resp.* Georgius Matthias Hirsch]. *Icarus in mare icarium praecipitatus ceu dissertatio qua hominem ad volandum esse ineptum ostenditur* (Kiloni 1709).

Harding, Francis. 'In artem volandi.' In *Musarum anglicanarum analecta* (Oxon. 1692) 77–81. Poem.

Harriott, Rosemary. *Poetry and Criticism before Plato* (London 1969) 83–91. Poetry and flight.

Harrison, Thomas P. *They Tell of Birds: Chaucer, Spenser, Milton, Drayton* (Austin 1956).

Harsdörffer, Georg Philipp. *Delitiae mathematicae et physicae . . . zweyter Theil* XII.xi (Nürnberg 1651) 475–76. First part by Daniel Schwenter.

Hart, Clive. *The Dream of Flight: Aeronautics from Classical Times to the Renaissance* (London 1972).

Hauksbee, Francis. *Physico-Mechanical Experiments on Various Subjects* (London 1709).

Hautsch, Ernst. 'Der Nürnberger Zirkelschmied Johann Hautsch (1595–1670) und seine Erfindungen.' *Mitteilungen des Vereins für die Geschichte der Stadt Nürnberg* 46 (1955) 533–56, esp. 545–46.

Havenreuter, Johann Ludwig. *Commentarii . . . in libros octo Physicorum Aristotelis philosophorum principis* (Francofurti 1604).

______. *Commentarii . . . in Aristotelis philosophorum principis Meteorologicorum libros quatuor* (Francofurti 1605).

______. *Commentarii . . . in Aristotelis philosophorum principis De cœlo, Generatione, & corruptione libros* (Francofurti 1605).

The Havenreuter works are good examples of standard sixteenth and seventeenth century Aristotelianism, with Greek texts, translations, and commentaries.

Heggen, Alfred. 'Die „ars volandi" in der Literatur des 17. und 18. Jahrhunderts.' *Technikgeschichte* 42.4 (1975) 327–37.

Heninger, Simeon Kahn, Jr. *A Handbook of Renaissance Meteorology* (Durham, N.C. 1960).

Hennig, Richard. 'Beiträge zur Frühgeschichte der Aeronautik.' *Beiträge zur Geschichte der Technik und Industrie* 8 (1918) 100–16.

______. 'Zur Vorgeschichte der Luftfahrt.' *Beiträge zur Geschichte der Technik und Industrie* 18 (1928) 87–94.

Hero of Alexandria. *Pneumatica et automata.* Ed. Wilhelm Schmidt (Lipsiae 1899). (*Heronis Alexandrini opera quae supersunt omnia* 1.) Trans. Joseph G. Greenwood, ed. Bennett Woodcroft, and published as *The Pneumatics of Hero of Alexandria* (London 1851). Facsimile with introduction by Marie Boas Hall (London and New York 1971).

Hesiod. *Opera et dies* 547–56. *Aer.*

______. *Theogonia* 123–25. Night.

Hewitt, Barnard, ed. *The Renaissance Stage: Documents of Serlio, Sabbattini and Furttenbach translated by Allardyce Nicoll, John H. McDowell, George R. Kernodle* (Coral Gables 1958). Cloud machines passim.

Heywood, Thomas. *The Hierarchie of the Blessed Angells* (London 1635) 505–07.

Higden, Ranulph. *Polychronicon* II.v [1327]. Ed. Churchill Babington and Joseph Rawson Lumby (London 1865–86) II.236–37. (*Rerum britannicarum medii aevi scriptores.*)

Hildegard of Bingen, Saint [1098–1179]. *Liber divinorum operum simplicis hominis* I.iv.lviii, I.iv.lxxx. *PL* 197.847, 861–62. The body needs the soul to move it, just as the bird needs air to fly. The air as honey surrounding the earth imagined as a piece of honeycomb, symbolising the soul and the body, respectively. Birds elsewhere, passim, esp. II.v.xxxviii, *PL* 197.937–39.

______. *Subtilitatum diversarum naturarum creaturarum* VI. *PL* 197.1287–1312. De avibus.

Hill, Thomas. *The Moste pleasaunte Arte of the Interpretacion of Dreames* (London 1576) [39[r]].

Hippolytus. *Philosophoumena* [c. A.D. 200]. Trans. Rev. J. H. MacMahon and published as *The Refutation of All Heresies* (Edinburgh 1868) 30–40.

Hoffmann, Immanuel. *Die Anschauungen der Kirchenväter über Meteorologie. Ein Beitrag zur Geschichte der Meteorologie* (München 1907). (Münchener geographische Studien 22.)

Hooke, Robert. *Micrographia* (London 1665) esp. 172–74, 195–98, 217–40.

______. 'An Account of the Sieur Bernier's [*sic*] Way of Flying.' *Philosophical Collections* I.1 (London 1679) 14–29. Followed by further notes, with translations from Lana.

______. *The Posthumous Works of Robert Hooke* (London 1705). Prefaced by Richard Waller, 'The Life of Dr. Robert Hooke.'

______. *The Life and Work of Robert Hooke.* Ed. Robert William Theodore Gunther, 2 vols (Oxford 1930) 9, 114, 247, 428, 517, 518, 523. (Early Science in Oxford 6, 7.)

Horst, Johann Daniel. *Physica hippocratea* (Francofurti 1682) 51–58. 'De volatilibus. Resp. Joh. Daniele Molthero.'

Huber, François. *Observations sur le vol des oiseaux de proie* (Genève 1784).

Hugo of Saint Victor. *De bestiis et aliis rebus* [early twelfth century]. *PL* 177.13–164. Birds passim.

Hunter, John, F.R.S. 'An Account of certain Receptacles of Air, in Birds, which communicate with the Lungs, and are lodged both among the fleshy Parts and in the hollow Bones of those Animals.' *Philosophical Transactions* 64 (1774) 205–13.

Huygens, Christiaan. *Oeuvres complètes* 22 vols in 23 (La Haye 1888–1950) III.270, 302–03, VII.357, 359.

Iamblichus. *De mysteriis Aegyptiorum, Chaldaeorum, Assyriorum* III.xvi. Trans. Thomas Taylor and published as *On the Mysteries of the Egyptians, Chaldeans, and Assyrians* (London 1821) 156–57.

Isidor, Saint, of Seville. *Etymologiarum, sive originum, libri xx* XII.iv, vi, vii, viii. Ed. W. M. Lindsay, 2 vols (Oxonii 1911) II, unpaginated.

Jaeger, Werner. *Aristotle: Fundamentals of the History of His Development* 2nd ed. Trans. Richard Robinson (Oxford 1948) 144–48. Fire animals.

Johnson, Hubert C. *Frederick the Great and His Officials* (New Haven and London 1975) 232–37.

Johnson, Samuel. *The Rambler* 67 (6 November 1750); *The Rambler* 199 (11 February 1752). In *The Works* III.357, V.272, ed. W. J. Bate and Albrecht B. Strauss (New Haven and London 1969). Vanity of attempts at flying.

______. *The History of Rasselas, Prince of Abissinia* (1759), ed. Geoffrey Tillotson and Brian Jenkins (London 1971) 14–18: 'A dissertation on the art of flying.'

Jonstonus, Joannes. *Historiae naturalis* II (Francofurti ad Moenum 1650) 6, 169–71. Flight, *manucaudiata.*

______. *An History of the Wonderful Things of Nature* (London 1657) 39–42, 80–81, 167–205. Air, winds, birds.

Le journal des sçavans (12 December 1678). In ed. of 1679 = VI.459–60, 460–64. Danti, Besnier.

Journal politique de Bruxelles [= pt 2 of *Mercure de France*] (18 October 1783) 124–27. Montgolfiers; a glide of about 500 yards by an anonymous experimenter.

______. (12 June 1784) 76–79. Leibniz, Lana, Montgolfiers.

Juvénal des Ursins, Jean. *Histoire de Charles VI, roy de France* (1614). 2nd ed. (augmented) (Paris 1653) 71–72.

Kahn, Charles H. *Anaximander and the Origins of Greek Cosmology* (New York and London 1960). Primaeval air, passim.

Keller, Alex. 'Kepler, the Art of Flight and the Vision of Interplanetary Travel as the Next Great Invention.' *Actes du XIIIe congrès international d'histoire des sciences, 1971* (1974) 70–79.

Kepler, Johannes. *Somnium, seu opus posthumum de astronomia lunari* (Francofurti 1634). Trans. with a commentary by Edward Rosen and published as *Kepler's Somnium: The Dream, or Posthumous Work on Lunar Astronomy* (Madison Milwaukee and London 1967).

Keysler, Johann Georg. *Neüeste Reise durch Teütschland, Böhmen, Ungarn, die Schweitz, Italien und Lothringen* 2 vols (Hannover 1740) I.252. 2nd ed. (1751) trans. and published as *Travels through Germany, Bohemia, Hungary, Switzerland, Italy, and Lorrain* 4 vols (London 1756, 1757) I.218–19n. Besnier, Burattini, etc.

Kircher, Athanasius. *Arca Noë* (Amstelodami 1675) 74–94. Birds of the Ark.

Kirk, Geoffrey Stephen. *Heraclitus: The Cosmic Fragments Edited with an Introduction and Commentary* (Cambridge 1954).

Kirk, Geoffrey Stephen and Raven, John Earle. *The Presocratic Philosophers: A Critical History with a Selection of Texts* (Cambridge 1957).

Klinckowstroem, Carl von. 'Tito Livio Burattini, ein Flugtechniker des 17. Jahrhunderts.' *Prometheus* 1100 (26 November 1910) 117–20.

______. 'Luftfahrten in der Literatur.' *Zeitschrift für Bücherfreunde* ns 3.2. (1912) 250–64.

Knight, Richard Payne. *A Discourse on the Worship of Priapus, and its Connection with the Mystic Theology of the Ancients.* New ed. with 'An Essay on the Worship of the Generative Powers During the Middle Ages of Western Europe' [by Thomas Wright and others] (London 1865) 119–21; plate XXV, between 146 and 147; plate XXVI, between 152 and 153. Winged phalluses.

Kolb, Gwin J. 'Johnson's "Dissertation on Flying" and John Wilkins' *Mathematical Magick.*' *Modern Philology* 47.1 (August 1949) 24–31.

Konrad von Megenberg. *Das Buch der Natur* [1349/50]. Ed. Franz Pfeiffer (Stuttgart 1861) 253. Flying fish.

Kramp, Christian. *Geschichte der Aerostatik, historisch, physisch und mathematisch ausgeführt* 2 pts with appendix volume (Strasburg 1784–86). Esp. II.235–347: a history of aviation.

Lactantius Firmianus [c. A.D. 260–c. 320]. *Divinarum institutionum libri septem. PL* 6.111–822. Angels and demons, passim.

[Lana.] *Franz Lana und Philipp Lohmeier von der Luftschiffkunst.* Trans. anon. (Tübingen 1784) sig. $\chi 5^{v}$–[$\chi 6^{r}$]. Schweikart.

Larson, Orville K. 'Bishop Abraham of Souzdal's Description of *sacre rappresentazioni.' Educational Theatre Journal* 9 (1957) 208–13.

Latini, Brunetto. *Li livres dou trésor* I.c, cvi, cvii [c. 1265]. Ed. Francis J. Carmody (Berkeley and Los Angeles 1948) 83–84, 90–93. (University of California Publications in Modern Philology 22.)

Laurentius, Joannes Christophorus. *Ex physicis de aere* (Wittebergae 1662).

Lavoisier, Antoine. *Essays Physical and Chemical* (1774). Trans. Thomas Henry (London 1776) 210. Includes a quotation from Antoine Baumé (1773): air is one substance.

Lawrence, Henry. *Of our Communion and Warre with Angels* ([Amsterdam] 1646) 15. Angels assume bodies from thickened air.

Le Clert, Louis. [Report of an address on Bolori.] *Mémoires de la Société académique d'agriculture, des sciences, arts et belles-lettres du département de l'Aube* 74 (Troyes 1910) 431.

[Le Grand, Marc-Antoine.] *Les avantures du voyageur aérien. Histoire espagnole* (Paris 1724) 9. Late reference to the sellers and buyers of winds.

Leibniz, Gottfried Wilhelm von. *De elevatione vaporum.* In *Opera omnia* II.2. (Genevae 1768) 82–86.

______. *Hypothesis physica nova* (Londini 1671) 24–26. On the air, Lana, etc.

______. *Opuscules et fragments inédits.* Ed. L. Couturat (Paris 1903) 491 [passage c. 1702–04]. Definition of flight.

______. *Otium Hanoveranum, sive miscellanea* (Lipsiae 1718) 185. Desson.

Lenglet Dufresnoy, Nicolas. *Recueil de dissertations, anciennes et nouvelles, sur les apparitions, les visions & les songes* II.ii.135 (Avignon 1751). 'Voler facilement avec des aîles, *liberté, richesses & dignité.'*

Lenoble, Robert. *Mersenne ou la naissance du mécanisme* (Paris 1943) 492–93.

Leonardo da Vinci. *Il Codice Atlantico di Leonardo da Vinci . . . Trascrizione diplomatica e critica di Giovanni Piumati* 8 vols (Milano 1894–1904).

______. *Codice sul volo degli uccelli.* Transcription and translation ed. G. Piumati and C. Ravaisson-Mollien (Parigi 1893).

______. *Codice sul volo degli uccelli.* Trans. Ivor B. Hart in his *The Mechanical Investigations of Leonardo da Vinci* 2nd ed. (Berkeley and Los Angeles 1963) 194–235.

______. *Gli scritti di Leonardo da Vinci sul volo,* ed. Raffaele Giacomelli (Roma 1936).

______. *I fogli mancanti al codice di Leonardo da Vinci su'l volo degli uccelli nella Biblioteca di Torino.* Facsimile and transcription ed. E. Carusi (Torino 1926).

______. *I libri del volo . . . nella ricostruzione critica di Arturo Uccelli* (Milano 1952).

______. *Manuscrit B (2173 et 2184) de l'Institut de France. Traduction française de Francis Authier . . . Transcriptions du Dr Ing. Nando de Toni . . . Introduction d'André Corbeau* 2 vols (Grenoble 1960). With facsimile.

______. *Les manuscrits de Léonard de Vinci . . . facsimilés . . .* 6 vols (Paris 1881–91). The manuscripts of the Institut and BN Ash. 2038, 2037. With transcriptions and French translations, ed. C. Ravaisson-Mollien.

______. *The Notebooks of Leonardo da Vinci Arranged, rendered into English and Introduced* 2 vols, ed. and trans. Edward MacCurdy (London 1938).

Le Roux, Hugues and Garnier, Jules. *Acrobats and Mountebanks.* Trans. A. P. Morton (London 1890) 210, 240.

Levis, Howard C. *The British King Who Tried to Fly* (London 1919). Bladud.

Lewis, C. S. *The Discarded Image: An Introduction to Medieval and Renaissance Literature* (Cambridge 1964). Various aerial creatures passim.

Libellus de natura animalium: A Fifteenth Century Bestiary. Facsimile with an introduction by J. I. Davis (London 1958) $C2^{v}$–$C3^{r}$. Salamanders.

Liebeschütz, Hans. *Das allegorische Weltbild der heiligen Hildegard von Bingen* (Leipzig/Berlin 1930).

Lipstorp, Daniel. *Specimina philosophiae Cartesianae* (Lugduni Batavorum 1653) 94–207. 'De aere.'

[Lister, Martin.] 'Some Observations Concerning the odd Turn of some Shell-snailes, and the darting of Spiders.' *Philosophical Transactions* 50 (16 August 1669) 1011–16. Airborne spiders, floating on web.

Listonai, Mr. de [pseud. Daniel Jost de Villeneuve]. *Le voyageur philosophe dans un pais inconnu aux habitans de la terre* I (Amsterdam 1761) 69. Description of a flying ship.

Lobkovitz. *See* Caramuel Lobkovitz, Johann.

Locatelli, Giuseppe. 'La manica attraversata a volo da un frate italiano nel 1751.' *Corriere della sera* (Milan) 34.214 (4 August 1909) 3. Grimaldi.

Loomis, Roger Sherman. 'Alexander the Great's Celestial Journey.' *The Burlington Magazine* 32 (April 1918) 136–40, (May 1918) 177–85.

Lucretius. *De rerum natura* I.335–97, V.273–80, VI.738–839.

Luedecke, Heinz. *Vom Zaubervogel zum Zeppelin* (Berlin 1936).

Luther, Martin. *The Creation: A Commentary on the First Five Chapters of the Book of Genesis* (1544). Trans. Henry Cole (Edinburgh 1858) 72–75. Creation of the birds.

Lydgate, John. 'A Pageant of Knowledge' [first half of the fifteenth century]. *The Minor Poems of John Lydgate* II, ed. Henry Noble MacCracken (London 1934) esp. pp. 731, 736. (Early English Text Society Original Series 192.) Stanzas on the affects of fickle air on men.

McMullin, Ernan. *The Concept of Matter in Greek and Medieval Philosophy* (Notre Dame 1963).

Macrobius. *In somnium Scipionis* I.vi.26–27, 36–39, I.xi.8, I.xvii.15 [early fifth century]. Trans. with an introduction and notes by William Harris Stahl and published as *Commentary on the Dream of Scipio* (New York and London 1952) 105, 107, 132, 158.

______. *Saturnalia* I.xv.20, III.iv.8 [early fifth century]. Ed. Nino Marinone (Torino 1967) 226, 380. Trans. Percival Vaughan Davies (New York and London 1969) 103, 201–02.

Maierus, Michael. *Tractatus de volucri arborea, absque patre et matre, in insulis Orcadum, forma anserculorum proveniente* (Francofurti 1619).

Maignan, Emanuel. *Cursus philosophicus* [1653] (Lugduni 1673) 433–48. 'De aëre elemento.'

Maiolus, Simon. *Dies caniculares: hoc est colloquia tria et viginti physica, nova et penitus admiranda ac summa iucunditate concinnata* I (Moguntiae 1607) 9, 183–218, 301–02, 334–35, 762.

Major, Johann Daniel. *See=Farth nach der neüen Welt, ohne Schiff und Segel* (Kiel 1670) A4v–B1r, K1v.

Mansuy, Abel. *Le monde slave et les classiques français aux XVIe–XVIIe siècles* (Paris 1912) 203–29. Burattini.

Marie Angélique, de Sainte Madeleine, Mère [=Jacqueline Marie Angélique Arnaud]. *Lettres de la révérende Mère Marie Angélique Arnaud* I (Utrecht 1742) 363. Burattini.

Marius. *De elementis* [twelfth century]. Ed. and trans. Richard C. Dales and published as *Marius: On the Elements* (Berkeley Los Angeles and London 1976).

Marius Victorinus. *Commentariorum in Genesin libri tres* I [fourth century]. *PL* 61.941.

Marriott, Frederick. 'The Problem Solved.' *San Francisco News Letter and California Advertiser* (6 November 1880) 2.

Martelli, Pietro Jacopo. *Degli occhi di Gesù, libri sei* (Roma 1710). Flying ship. The ship is illustrated in the edition of the poem published in his *Opere* VI (Bologna 1729) following p. [70].

______. 'Del volo: dialogo.' In *Opere* V (Bologna 1723) 371–449, and frontispiece.

Martianus Capella. *De nuptiis philologiae et mercurii libri VIIII* II.149–68, VII.732 [c. A.D. 420]. Ed. Adolfus Dick, rev. Jean Préaux (Stutgardiae 1969) 64–69, 368. Trans. William Harris Stahl and Richard Johnson with E. L. Burge and published as *Martianus Capella and the Seven Liberal Arts* II: *The Marriage of Philology and Mercury* (New York 1977) 50–55, 277–78.

Martius, Johannes Nicolaus. *Dissertatio . . . de magia naturali* (Erfordiae 1700) 6–7. Doubts whether flight will ever be possible.

Marvell, Andrew. *The Poems and Letters of Andrew Marvell* I. Ed. H. M. Margoliouth. 3rd ed. (Oxford 1971).

Massip, M. 'Une victime de l'aviation au onzième siècle.' *Mémoires de l'Académie des sciences, inscript. et belles-lettres de Toulouse* 10ème série 10 (1910) 199–217. Eilmer of Malmesbury.

Massuet, Pierre. *Elémens de la philosophie moderne* 2 vols (Amsterdam 1752) I.221: birds landing; I.421–32: the air and the atmosphere; II.433–81: measurements of air, experiments with air.

Mauduit de la Varenne, Pierre-Jean-Etienne. *Ornithologie,* prefaced by 'Discours généraux sur la nature des oiseaux.' *Encyclopédie méthodique: histoire naturelle des animaux* (Paris and Liège 1782, 1784) I.ii, II.i.

Meerwein, Carl Friedrich. *Der Mensch: sollte der nicht auch zum Flügen gebohren seyn?* 3 pts. *Oberrheinische Mannigfaltigkeiten* 8–10 (13, 20, 27 January 1783) 541–44, 545–55, 574–76. Expanded version published in one volume as *Die Kunst zu fliegen nach Art der Vögel* (Frankfurt und Basel 1784). French version published as *L'art de voler à la manière des oiseaux* (Basle 1784). Portuguese translation published as *A arte de voar a' maneira dos passaros* (Lisboa 1812).

Mendoça, Francisco de. *Viridarium sacrae, ac profanae* (Lugduni 1631) 91, 117–18, 310–12.

Le mercure hollandois, contenant les choses les plus remarquables de toute la terre, arrivées en l'an 1673. jusqu'à l'an 1674 (Amsterdam 1678) 98–99. Flight of Bernouin on 15 January $16\frac{72}{73}$.

Mersenne, Marin. *Questions inouyes, ou récréation des scavans* (Paris 1634) 1–5.

______. *Les questions theologiques, physiques, morales, et mathematiques* (Paris 1634) 171–74. Clouds.

______. *Correspondance.* Ed. Cornelis de Waard (Paris 1932–) IV.54–56, 117–18, IX.53, 484, XI.89, 435–36, XII.392.

______. *Novarum observationum physico-mathematicarum . . . tomus III* (Parisiis 1647) 73–74.

Méry, Jean. 'Observations sur la peau du pélican.' *Mémoires de l'académie royale des sciences depuis 1666 jusqu'à 1699* X (Paris 1730) 433–38. Memoir dated 31 December 1693 of observations made in 1686. Air sacs, interpreted as greatly reducing specific gravity and aiding the ability to fly very high.

Meyer, Kuno. 'The Irish Mirabilia in the Norse "Speculum regale."' *Ériu* 4 (1910) 1–16. Flying ships.

Milizia, Francesco. *Le vite de' più celebri architetti d'ogni nazione et d'ogni tempo precedute da un saggio sopra l'architettura* (Roma 1768) 326–27. Trans. Mrs Edward Cresy and published as *The Lives of Celebrated Architects, Ancient and Modern* II (London 1826) 151–52. Guidotti, Danti, Grimaldi.

Millet, Gabriel. 'L'ascension d'Alexandre.' *Syria: revue d'art oriental et d'archéologie* 4 (1923) 85–133.

Minor, Jacob. 'Die Luftfahrten in der deutschen Literatur.' *Zeitschrift für Bücherfreunde* ns 1.1 (1909) 64–73.

Mirowski, Andreas. *Theoria ventorum* (Wirceburgi 1596).

Mittendorffius, Bernhardus. *Disputatio physica-mathematica de ventis insolentibus et imprimis eo, qui circa proxime praeteritum 9 Dec. totam ferme Europam perflasse creditur. Cum appendice de recenti cometa* (Wittebergae 1661).

Mizauld, Antoine. *Ephemerides aeris perpetuae: seu popularis & rustica tempestatum astrologia, ubique terrarum & vera, & certa* (Lutetiae 1554).

______. *Le mirouer de l'air* (Paris 1548).

Molinet, Jean. 'Recollection des merveilleuses advenues,' stanza 143 [c. 1500]. In *Les faictz et dictz de Jean Molinet* I, ed. Noël Dupire (Paris 1936) 332.

Molsdorf, Wilhelm. *Christliche Symbolik der mittelalterlichen Kunst* (Leipzig 1926) 123-36. 'Engel und Teufel.'

Mongés, J. A. 'Mémoire sur l'imitation du vol des oiseaux.' *Observations sur la physique, sur l'histoire naturelle et sur les arts* II ed. l'Abbé Rozier (July 1773) 140-44. Read at the Académie de Lyon, 11 May 1773.

Montesquieu, Charles Louis de Secondat, Baron de. *Pensées* 758. In *Oeuvres complètes* I, ed. Roger Caillois (Paris 1949) 1207-08. (Bibliothèque de la Pléiade.) Reflections on bird flight and the possibility of human flight.

Montrond, [Clément] Maxime de. *Essais historiques sur la ville d'Étampes* 2 vols (Paris 1836, 1837). Topographical and historical background for Desforges.

Morhof, Daniel. *Polyhistor literarius, philosophicus et practicus* II.2.iv.4, II.2.xviii.1-9 (Lubecae 1714) 289-90, 354-60. Lana, the air.

Morris, Ralph [pseud.]. *A Narrative of the Life and Astonishing Adventures of John Daniel, A Smith at Royston in Hertfortshire, for a Course of Seventy Years* (London 1751). Flying machine, with cut opposite p. 178. Reprinted as vol. I of the *Library of Impostors,* ed. N. M. Penzer (London 1926).

Munsterberg, Peggy. Introduction to *The Penguin Book of Bird Poetry*. Ed. Munsterberg (London 1980) 25-99. Best essay on the subject.

[Musgrave, Dr Samuel.] 'Dr. Musgrave's Machine: A Vision. With a Copper-plate descriptive of the same.' *The Oxford Magazine* 3.3. (September 1769) 108-10.

Musschenbroek, Pieter van. *Beginsels der Natuurkunde.* Trans. Pierre Massuet and published as *Essai de physique* (Leyden 1739) 629-706, 878-914. Air, wind.

Nagler, Georg Kaspar. *Neues allgemeines Künstler=Lexicon* III (München 1836) 353, 363-64. 'Derson, Nicolaus,' 'Deson, N.'

Nashe, John. *Lenten Stuffe* 192. In *The Works* III, ed. Ronald B. McKerrow, revised F. P. Wilson (London 1958).

Neckam, Alexander. *De naturis rerum* I.iii [c. 1200]. Ed. Thomas Wright (London 1863) 21. (*Rerum britannicarum medii aevi scriptores.*)

Newton, Sir Isaac. *Correspondence* II. Ed. H. W. Turnbull, F.R.S. (Cambridge 1960) 295. Letter to Boyle of 28 February 16$\frac{78}{79}$. Gravity caused by gradations in the *aether* content of bodies.

Nicolson, Marjorie Hope. *Voyages to the Moon* (New York 1948).

Nieuwentijt, Bernard. *Het regt Gebruik der Werelt Beschouwingen* (Amsterdam 1717). Trans J. Chamberlayne and published as *The Religious Philosopher: or, the Right Use of Contemplating the Works of the Creator* 2 vols (London 1724) I.181–234, II.335–44. Air, wind, birds, bird flight.

Nollet, Jean Antoine, l'Abbé. *Leçons de physique expérimentale* I (Paris 1743) 220–30.

O'Brien, D. *Empedocles' Cosmic Cycle: A Reconstruction from the Fragments and Secondary Sources* (Cambridge 1969) esp. 189–95, 301–13.

Observator [pseud.]. 'On the Flight of Birds.' *The Monthly Magazine* 56 (1 March 1800) 126–28. (Letter of 11 January 1800.) Birds fly through the direct action of the will, counteracting gravity.

Oemich, Joannes Georgius. *Vim aeris elasticam* (Gryphiswaldiae [1689]).

Oresme, Nicole, *Le livre du ciel et du monde* [1377]. Ed. Albert D. Menut and Alexander J. Denomy. Trans. with an introduction by Albert D. Menut (Madison Milwaukee and London 1968) 293–95, 401–05, 681–91, 705, 719.

Origen [A.D. 185 or 186–253] *De principiis* I.viii.1. Trans. G. W. Butterworth and published as *Origen on First Principles* (London 1936) 67. Original text taken from the quotation in Leontius Byzantinus *De sectis* X.v [c. 535]. *PG* 86(1).1263–66.

______. *Homiliae in Genesim* I.8. *PG* 12.152–53.

Owen, Hugh and Blakeway, John Brickdale. *A History of Shrewsbury* II (London 1825) 409–11n. Robert Cadman, a 'flying' funambulist.

Paduanius, Fabricius. *Tractatus duo, alter de ventis, alter perbrevis de terrae motu* (Bononiae 1601).

Palladius, Bishop of Helenopolis. *Historia lausiaca* XXVII [A.D. 419 or 420]. *PL* 73.1126. Variant version trans. Robert T. Meyer and published as *Palladius: The Lausiac History* (Westminster, Md. and London 1965) 76. (*Ancient Christian Writers* 34.)

Paltock, Robert. *The Life and Adventures of Peter Wilkins* (London 1751).

Paracelsus, Philippus Aureolus [Theophrastus Bombastus von Hohenheim]. *Philosophia de generationibus et fructibus quatuor elementorum* I.i-xii [c. 1525]. In *Achter Theil der Bücher und Schrifften des . . . Paracelsi*, ed. Johannes Huserus (Basel 1590) 55-64.

______. 'De elemento aeris' [part I of a fragmentary *Buch meteororum* of uncertain date]. In *Achter Theil der Bücher und Schrifften des . . . Paracelsi*, ed. Johannes Huserus (Basel 1590) 283-96.

______. *Liber de nymphis, sylphis pygmæis et salamandris, et de cæteris spiritibus* [1529-32]. In *Neundter Theil der Bücher und Schrifften des . . . Paracelsi*, ed. Johannes Huserus (Basel 1590) 45-78.

Paré, Ambroise. *Des monstres et prodiges* (1573). Ed. Jean Céard (Genève 1971) 114-17, 126-30. (*Travaux d'humanisme et Renaissance* 115.) Flying fish; 'Des monstres volatiles.'

Parent, Antoine, *Essais et recherches de mathematique et de physique* III (Paris 1713) 376-400. Flight.

Parmelee, Alice. *All the Birds of the Bible: Their Stories, Identification and Meaning* (New York 1959).

Parthenius, Josephus Marianus [=Giuseppe Maria Mazzolari]. *Electricorum libri vi* (Romae 1767) II.68-70.

Pasch, Georg. *Schediasma de curiosis hujus seculi inventis* (Kiloni 1695) [2]12-49 (wrongly paginated).

Patch, Howard Rollin. *The Other World According to Descriptions in Medieval Literature* (Cambridge, Mass. 1950).

Paucton, Alexis Jean Pierre. *Théorie de la vis d'Archimède* (Paris 1768) 210-13. Convertiplane.

Pechlin, Johannes Nicolaus. *De aeris et alimenti defectu, et vita sub aquis, meditatio* (Kiloni 1676) 34-50.

Peletier, Jacques. *La Savoye* (Anecy 1572) 16-17. Air, clouds.

Pena, Jean. *Euclidis optica & catoptrica* (Parisiis 1557) bbij^{r-v}. Comets and thickened air.

Pestel, Johannes [*Resp., Praeses* P. Lohmeier]. *Exercitationum physicarum de paradoxis gravitatis & levitatis prima* (Rinthelii 1678) 25–29. Flying eggs; manned flight.

Peter Abelard. *Expositio in Hexaemeron* [early twelfth century]. *PL* 178.756.

Petrus Lombardus. *Sententiarum libri quatuor* II.vi–viii [c. 1160]. *PL* 192.662–69. Angels and demons.

Pezzi, Giuseppe. 'La meccanica del volo nell'opera di Leonardo e nel "De motu animalium" di Gian Alfonso Borelli.' *Minerva medica* 63.38 (1972) 2184–88.

Philo Judaeus. *De gigantibus* II.7–11.

______. *De opificio mundi* VII, IX.

______. *De plantatione* III.xii.

______. *De somniis* I.xxii, xxiii.

Philoponus, Johannes. *De mundi creatione* II.viff [sixth century]. Ed. Walther Reichardt (Lipsiae 1897) 69ff. In Latin in *Bibliotheca veterum patrum* 12 (Venetiis 1778) 502ff.

Physiologus latinus versio Y XLV [A.D. fourth–fifth century]. Ed. Francis J. Carmody (Berkeley and London 1941) 132. (University of California Publications in Classical Philology 12.7.) Salamander.

Pico della Mirandola, Giovanni. *Heptaplus* 18. In *Opera omnia* I (Basileae 1557).

______. *De elementis* 170. In *Opera omnia* II (Basileae 1573).

Plato. *Cratylus* 404C. Hera and the air.

______. *Phaedrus* 246D–E. Wings.

______. *Sophist* 220B. Flying and swimming.

______. *Symposium* 202E–203A. Creatures of the air.

______. *Theaetetus* 197C ff. Aviary of the soul.

______. *Timaeus* 31B–32C, 39E–40A, 56B, 56D–57C, 58C–D, 91D–92C. Trans. Cornford, *Plato's Cosmology.*

Pliny the Elder. *Naturalis historia* X.xii, liv–lv, lxxxvi, XI.xlii, XXIX.xxiii.

Pluche, Noël-Antoine. *Spectacle de la Nature: or, Nature Display'd. Being discourses on such particulars of natural history as were thought most proper to excite the curiosity, and form the minds of youth* I (2) (London 1733-36) 1-76. Trans. from the French ed. in 8 vols (Paris 1732-34). Birds.

Plutarch. *De communibus notitiis adversos Stoicos* 1085C–E. Activity of air and fire.

______. *De facie quae in orbe lunae apparet* V, 921F–922F. Shadowy air on the moon.

______. *De primo frigido* 948A–949F. Stoic argument that air is cold.

______. *Platonicae quaestiones* VI, 1004C–D. Wings of the soul.

______. *Titus Flamininus* X.6. (*Vitae.*) Rending of the air.

______. *Pompey* XXV.7. (*Vitae.*) Rending of the air.

Porta, Giovanni Baptista della. *De aeris transmutationibus libri iiii* (Romae 1614 [colophon dated 1610]).

Powell, Thomas. *Humane Industry: or, a History of most Manual Arts* (London 1661) 27-35. Wind-operated automata, including Archytas.

Prescheur, Franciscus David [*resp.*; *Praes.* Philippus Lohmeier]. *Exercitatio physica de artificio navigandi per aerem* (Rinthelii 1676). Plagiarised from Lana.

Procopius of Gaza. *Commentarii in Genesim* [fifth–sixth century]. *PG* 87.99-106.

Psellus, Michael [1018–c. 1078]. *Dialogus de energia, seu operatione daemonum e graeco translatus* (Parisiis 1577).

[Pseudo-Sydrach] [c. 1250]. Versions used: *Het Boek van Sidrac* [early fourteenth century] I, ed. J. F. J. van Tol (Amsterdam 1936); *Il libro di Sidrach* [fourteenth century], ed. Adolfo Bartoli (Bologna 1868); *Das Buch Sidrach. Nach der Kopenhagener mittelniederdeutschen Handschrift v. J. 1479*, ed. H. Jellinghaus (Tübingen 1904); *Sydrak efter Haandskriftet ny kgl. saml. 236 4*[to] [second half of fifteenth century], ed. Gunnar Knudsen (København 1921-32); *The History of Kyng Boccus, & Sydracke how he confoundyd his lerned men . . .* (London ?1510).

Ptolemaeus, Claudius. *Hypotheses planetarum* II.7. Trans. from the Arabic into German in *Claudii Ptolemaei opera quae exstant omnia* II, ed. J. L. Heiberg (Lipsiae 1897) 119-20. Planetary motion likened to that of birds.

Ptolemaeus Lucensis. *Exaemeron* VII.vi [thirteenth century]. Ed. P. F. Pius-Thomas Masetti (Senis 1880) 94-95. (*Biblioteca Tomistica* I.1.)

Rabanus Maurus [c. 776-856]. 'De avibus,' 'De apibus.' In *De universo* VIII.6, 7. *PL* 111.240-58. Symbolic and analogical significance of birds.

______. *Commentariorum in Genesim libri quatuor* I.6. *PL* 107.456-57.

______. *De magicis artibus. PL* 110.1095-1110. Aerial demons.

Ralegh, Sir Walter. *The History of the World* [1614]. Ed. C. A. Patrides (London and Basingstoke 1971) 95-96. Creation of air from water.

Ravaisson, François, ed. *Archives de la Bastille: documents inédits* XVII (Paris 1891) 219-22. Desforges.

Ray, John. *The Wisdom of God Manifested in the Works of the Creation* (London 1691) 106-07, 111-12. Birds.

______. *Synopsis methodica avium & piscium* (Londini 1713).

Réau, Louis. *Iconographie de l'art chrétienne* 3 pts (Paris 1955-59) II.36-37, III.3, 1499. Angel wings.

Régis, Pierre Silvain. *Cours entier de philosophie, ou système général selon les principes de M. Descartes* (1690). New ed. in 3 vols (Amsterdam 1691) II.617-19. Bird flight, summarising Borelli.

Regnard, Jean François. *Voyage de Laponie.* In *Oeuvres* I (Paris 1790) 149-50. Visit to Lapland, c. 1680. Lapps using magical powers to sell the winds.

Remigius Antissiodorensis. *Commentarius in Genesim* I [ninth century]. *PL* 131.56, 58.

Rémy, Nicholas. *Daemonolatreiae libri tres* I.24, 25, 29 (Lugduni 1595).

Rey, Jean. *Essays . . . sur la recherche de la cause pour laquelle l'estain & le plomb augmentent de poids quand on les calcine* (Bazas 1630) 23-47.

Reyher, Samuel. *Dissertatio de aere* (Kiliae 1670).

______ and Laudenbach, Joannes. *Disputatio de ventis* (Lipsiae [1657]).

Richard, l'Abbé. *Histoire générale de l'air et des météores* 10 vols (Paris 1770–71).

Rohault, Jacques. *Traité de physique,* 4 pts in 1 (Paris 1671) III.148–55, 256–64. Air; wind.

Romanus, Adrianus. *Ventorum secundum recentiores distinctorum usus* (Wirceburgi 1596).

Ronsard, Pierre de. *Les amours* I.xxxi: 'Legers Daimons, qui tenez de la terre' (1553). In *Oeuvres complètes* I, ed. Hugues Vaganay (Paris 1923) 36–37.

______. 'Les daimons.' In *Oeuvres complètes* VI, ed. Hugues Vaganay (Paris 1923) 59–68.

Rossi, Medoro Ambrozio. *Novelle della repubblica letteraria per l'anno MDCCLII* 8 (Venezia 1752) 62–63. Grimaldi; sceptical report dated 19 February 1752.

?Rousseau, Jean-Jacques. *Le nouveau Dédale* (1742). Ed. Pierre-Paul Plan, *Mercure de France* 87.4 (1 October 1910) 577–97.

______. *Le nouveau Dédale* (1742). Ed. Charles Wirz, *Annales de la Société Jean-Jacques Rousseau* 38 (1975) 155–239.

Royal Society. *Philosophical Transactions* 1.6 (6 November 1665) 99. Notes on Desson.

Rupert of Deutz. *Commentarii in Genesim* I.xlviii–l [c. 1124]. *PL* 167.239–41.

Saint Romuald, Pierre de. *Trésor cronologique et historique* 3 vols (Paris 1669) I.12–15; III.583. Creation of birds and fish. Italian flying from the Tour de Nesles c. 1550.

Salerne, François. *L'histoire naturelle, éclaircie dans une de ses parties principales, l'ornithologie.* Trans. from the Latin by John Ray (Paris 1767).

Sauval, Henri. *Histoire et recherches des antiquités de la ville de Paris* II (Paris 1724) 544–47. Funambulists, etc.

Savérien, Alexandre. *Histoire des progrès de l'esprit humain* (Paris 1766) 273–74. Archytas, Regiomontanus, Torriano.

Saxl, Fritz. 'Continuity and Variation in the Meaning of Images.' In his *Lectures* I (London 1957) 1–12. Winged and wingless creatures 7–12.

Scaliger, Julius Caesar. *Exotericarum exercitationum liber quintus decimus, De subtilitate, ad Hieronymum Cardanum* (Lutetiae 1557) 294ʳ, 299ʳ–310ᵛ. Flying fish, birds.

Schlettwein, Johann August. *Bemühungen in der Naturkunde und anderen nützlichen Wissenschaften* (Jena 1756).

Schneweis, Emil. *Angels and Demons According to Lactantius* (Washington, D.C. 1944). (Catholic University of America Studies in Christian Antiquity 3.)

Shelley, Percy Bysshe. *Prometheus Unbound*, ed. Lawrence John Zillman (New Haven and London 1968) esp. 205.

______. *Physica curiosa, sive mirabilia naturae et artis* II (Herbipoli 1662) 1101–1253. Birds.

Schulze, Hans-Georg and Stiasny, Willi. *Flug durch Muskelkraft* (Frankfurt 1936) 15–35. Prehistory of flight. Undocumented, not wholly accurate.

Segneri, Paolo, the Elder. *L'incredulo senza scusa* I.XII.vii (Bologna 1690) 48.

Seneca the Younger. *Naturales quaestiones* II.4–11; III.14, 15; IV.10; V.

Servius Grammaticus [fourth century A.D.]. *In Vergilii carmina commentariorum* [*Aeneid*] II.63. Ed. Arthur F. Stocker and Albert H. Travis (Oxonii 1965) 27–28. (The 'Harvard Edition' III.) Creatures of the air.

Severian. *In mundi creationem* IV [fourth century]. *PG* 56.457–72.

Shackleton, Robert. *Montesquieu: A Critical Biography* (Oxford 1961) 220, 221–23.

Shelley, Percy Bysshe. *Prometheus Unbound*, ed. Lawrence John Zillman (New Haven and London 1968) esp. 205.

Shirley, James. *Cupid and Death* [26 March 1653]. In *The Dramatic Works and Poems of James Shirley* VI, ed. Alexander Dyce (London 1833) 358.

Sidrach. *See* [Pseudo-Sydrach].

Silberschlag, Johann Esaias. 'Von dem Fluge der Vögel.' *Schriften der Berlinischen Gesellschaft Naturforschender Freunde* 2 (1781) 214–70 and plates VIII, IX.

Simplicius [sixth century A.D.]. *In Aristotelis de caelo libros commentaria.* Ed. J. L. Heiberg in *Commentaria in Aristotelem graeca* VII (Berlin 1894). Passage about acceleration through the air trans. Drabkin in Morris R. Cohen and Israel E. Drabkin *A Source Book in Greek Science* (New York 1948) 209–11.

Solmsen, Friedrich. *Aristotle's System of the Physical World: A Comparison with His Predecessors* (Ithaca 1960).

Steneck, Nicholas H. *Science and Creation in the Middle Ages: Henry of Langenstein (d. 1397) on Genesis* (Notre Dame and London 1976). Useful general commentaries, passim.

Sternberger, Gottfridus. *Ἀνεμολογια, sive dissertatio de ventis* (Lipsiae [1654]).

La storia dell' anno MDCCLI (Amsterdam n.d.) [=*La storia degli anni* 20] 225–29. Reprint of Italian letter about Grimaldi.

Stresemann, Erwin. *Ornithology from Aristotle to the Present.* Trans. Hans J. and Cathleen Epstein, ed. G. William Cottrell (Cambridge, Mass. and London 1975).

Sturm, Johannes Christophorus. *Collegium experimentale, sive curiosum* 2 parts (Norimbergae 1676, 1685) I.56–66, II.96–106. Lana.

Supf, Peter. *Der Himmelswagen: das Schicksal des Melchior Bauer* (Stuttgart 1953). Novel.

______. *Das Buch der deutschen Fluggeschichte: Vorzeit, Wendezeit, Werdezeit* (Stuttgart 1956).

Swan, Conrad. 'Heraldica Aeronautica: Some Notes an [*sic*] Aviation History.' *Aerospace* 2.7 (August/September 1975) 19–25.

Swan, John. *Speculum mundi, or a Glasse representing the Face of the World* (London 1644) 375, 382–419. Flying fish, flying creatures.

Swedenborg, Emanuel. *Angelic Wisdom Concerning the Divine Providence* (1764) I.20 (London 1949) 16–17, translated from the Latin for the Swedenborg Society. Wings.

______. *The Apocalypse Revealed* (1766) 2 vols (London 1970) sects. 245, 561, 757, translated from the Latin for the Swedenborg Society. Birds, wings.

______. *Machine att flyga i Wädret* (Stockholm 1960). Translations reprinted from next item.

______. *The Mechanical Inventions of Emanuel Swedenborg.* Trans. with a commentary by Alfred Acton (Philadelphia 1939) 20–26.

______. *Opera quaedam aut inedita aut obsoleta de rebus naturalibus* I. Ed. Alfred H. Stroh (Holmiae 1907) 224–29. Letters proposing *inter alia* a flying machine.

______. *Suggestions for a Flying Machine.* Trans. Hugo Lj. Odhner and Carl Th. Odhner (Philadelphia 1910). Variant of 'Swedenborg's Flying Machine' (below).

______. 'Suggestions for a Flying Machine.' *The Aeronautical Journal* 14.55 (July 1910) 118–22. Reprinted from preceding item.

______. 'Swedenborg's Flying Machine.' Trans. with a commentary by Carl Th. Odhner. *New Church Life* 29.10 (October 1909) 582–91.

______. 'Utkast til en *Machine* at flyga i wädret.' *Daedalus hyperboreus* 4 (Oct.–Dec. 1716) 80–83.

[______.] *Transactions of the International Swedenborg Congress held in Connection with the Celebration of the Swedenborg Society's Centenary, London, July 4 to 8, 1910* (London 1910) 45–46 and Fig. 1. Flying machine.

Sydrach. *See* [Pseudo-Sydrach].

Synesius, Bishop of Cyrene [c. 373–c. 414]. *De insomniis* 12 (154). *PG* 66.1318. Flying in dreams.

Targosz, Karolina. 'Jak wyglądał "latający smok" Tita Livia Burattiniego.' *Technika Lotnicza i Astronautyczna* 3 (1976) 37–40.

Theophilus of Antioch. *Ad Autolycum* II.13 [late second century A.D.]. Ed. and trans. Robert M. Grant (Oxford 1970) 46–49. 'Spirit borne over the water.'

Thierry of Chartres. *Tractatus de sex dierum operibus* [1130–40]. In *Commentaries on Boethius by Thierry of Chartres and His School*, ed. Nikolaus M. Häring (Toronto 1971) 553–75. (Pontifical Institute of Mediaeval Studies, Studies and Texts 20.)

Thomas of Cantimpré. *Liber de natura rerum* [mid-thirteenth century]. Ed. H. Boese I (Berlin and New York 1973) 173–231, 243, 246, 263–64, 411–12. Birds; flying fish; air.

______. *Miraculorum, & exemplorum memorabilium sui temporis, libri duo* II [mid-thirteenth century] (Duaci 1597) 448–49, 466–68. Demons and the air.

Thyraud, Jacques. *Histoire des hommes volants* (Paris 1977). Unreliable; undocumented. Many interesting illustrations.

Topsell, Edward. *The Fowles of Heaven or History of Birdes* [early seventeenth century]. Ed. Thomas P. Harrison and F. David Hoeniger (Austin, Tx. 1972). Based on Aldrovandus. Pp. xxiv–xxix include a useful essay by the editors on 'Renaissance Ornithology.'

Townshend, Aurelian and Jones, Inigo. *Albions Triumph* (London $16\frac{31}{32}$) B1r.

Tuerchner, Balthasar Tobias. *Cosmographia elementaris, propositionibus physico-mathematicis proposita* (Pragae 1673) H2r–K2v. 'Aero-graphia.'

Uccelli, Arturo. *See* Leonardo da Vinci.

Vairus, Leonardus. *De fascino libri tres* II.xiii (Paris 1583). Ed. used: (Venetiis 1589) 173-84. Demons.

Vallesius, Franciscus. *De iis quae scripta sunt physicè in libris sacris, sive de sacra philosophia. Liber singularis* (Lugdun 1595) 42-45, 265-67, 277-78, 464-6[5]. Birds, winds.

Van Cleve, Thomas Curtis. *The Emperor Frederick II of Hohenstaufen, Immutator Mundi* (Oxford 1972).

Venturini, Galileo. *Da Icaro a Montgolfier* 2 vols (Roma 1928).

Verduc, Jean Baptiste. *Nouvelle osteologie* (Paris 1689) 340-90. Dependent on Borelli.

Vermiglioli, Giovanni Battista. *Biografia degli scrittori Perugini e notizie delle opere loro* I (Perugia 1828) 371n. Danti.

Vicq d'Azyr, Félix. *Traité d'anatomie et de physiologie* tome premier (no more published) (Paris 1786, 1789) 33-42. Birds.

Villars, l'Abbé Montfaucon de. *Comte de Gabalis, ou entretiens sur les sciences secretes* (Paris 1670). Ed. used: (Cologne ?1675) 115-33. Sylphs, flying lovers, Charlemagne and the storm makers.

Villette, Jeanne. *L'ange dans l'art d'occident du XIIème au XVIème siècle, France, Italie, Flandre, Allemagne* (Paris 1940).

Vivens, François, Chevalier de. 'Du vol des oiseaux' (1742). Ed. Jules Duhem, *Mercure de France* 264 (15 November 1935) 25-41.

______. *Essai sur les principes de la physique* (n.p. 1746).

______. *Nouvelle théorie du mouvement, où l'on donne la raison des principes genéraux de la physique* (Londres [=Bordeaux] 1749).

Vossius, Isaac. *De motu marium et ventorum liber* (Hagae-Comitis 1663).

Vulson, Marc de. *Traité des songes et des visions nocturnes* (Paris 1690) 42-44. Dreams of the air.

Wagenseil, Christian Jakob. *Versuch einer Geschichte der Stadt Augsburg* IV.2 (Augsburg 1822) 485-87. Salamo Idler, the flying shoemaker.

Wandalbertus Prumiensis. *De creatione mundi per ordinem dierum sex* [ninth century]. *PL* 121.637.

Waterlow, Sarah. *Nature, Change, and Agency in Aristotle's 'Physics'* (Oxford 1982).

Weicker, Georg. *Der Seelenvogel in der alten Litteratur und Kunst: eine mythologisch-archaeologische Untersuchung* (Leipzig 1902).

White, Lynn, Jr. 'The Invention of the Parachute.' *Technology and Culture* 9.3 (July 1968) 462–67.

______. 'Medieval Uses of Air.' *Scientific American* 223.2 (August 1970) 92–100.

White, Terence Hanbury. *The Book of Beasts, Being a Translation from a Latin Bestiary of the Twelfth Century* (London 1954) 101–61, 199. Birds; flying fish.

The Whitehall Evening-Post; Or, London Intelligencer 882 (3–5 October 1751) [1]. Note on Andero Grimale Volante. Flying machine alleged to have crossed the Channel.

Wicbodius. *Quaestiones super librum Genesis* [?eighth century]. *PL* 96.1106–07.

Wiedeburg, Johann Bernhard. *Matheseos Biblicae specimen quintum* (Jenae 1729) 54–56, 57. In his *Mathesis Biblica septem speciminibus comprehensa* (Jenae 1730 etc.). Flight by repeated jumps against resistant air; turns made by differential flapping. The nature of the air.

Wild, Friedrich. *Drachen im Beowulf und andere Drachen. Mit einem Anhang: Drachenfeldzeichen, Drachenwappen und St. Georg* (Wien 1962). (*Sitzungsberichte der Österreichische Akademie der Wissenschaften,* Phil.-Hist. Kl., 238.5.)

Wilhelm, Balthasar, S.J. *Die Anfänge der Luftfahrt, Lana-Gusmão. Zur Erinnerung an den 200. Gedenktag des ersten Ballonaufstieges (8. Aug. 1709–8. Aug. 1909)* (Hamm i. W. 1909).

______. 'Schweikart und Mohr, zwei schwäbische Flieger aus alter Zeit.' *Illustrierte aeronautische Mitteilungen* 13 (1909) 441–45.

Wilkins, John. *Mathematicall Magick; or, the wonders that may be performed by mechanicall geometry* (London 1648) 191–223.

______. *The Discovery of a World in the Moone. Or, a discourse tending to prove, that 'tis probable there may be another habitable world in that planet* (London 1638) 208–[11] [misnumbered 208, 107, 208, 209].

William of Conches. *De philosophia mundi libri quatuor* [twelfth century]. *PL* 172.41–102, esp. 172.55, 172.75–84. Creation of the birds and fishes; five regions of air. Migne wrongly attributes the work to Honorius Augustodunensis.

Willughby, Francis. *Ornithologiae libri tres* (Londini 1676). English trans. published as *The Ornithology of Francis Willughby . . . in Three Books* (London 1678).

Worcester, Edward Somerset, second Marquis of. *A Century of the Names and Scantlings of such Inventions as at present I can call to mind to have tried and perfected* . . . (London 1663) 54–55. Invention 77.

Wotton, Edward. *De differentiis animalium libri decem* (Lutetiae Parisiorum 1552) 103^r–136^r.

Wray, John. 'A Confirmation of what was formerly Printed in Numb. 50 of these Tracts, about the manner of Spiders projecting their Threds.' *Philosophical Transactions* 65 (14 November 1670) 2103–05. *See* [Lister], 'Some Observations.'

Wright, Thomas and Halliwell, James Orchard, eds. *De rebus Hiberniae admirandis,* from MS Cotton. Titus, D.xxiv, f. 74^v [thirteenth century], in *Reliquiae antiquae* II (London 1843) 106.

Wright, Thomas. *See also* Knight, *A Discourse on the Worship of Priapus.*

Yapp, Brunsdon. *Birds in Medieval Manuscripts* (London 1981).

Zachariä, August Wilhelm. *Die Elemente der Luftschwimmkunst* (Wittenberg 1807).

______. *Geschichte der Luftschwimmkunst, von 1783 bis zu den Wendelsteiner Fallversuchen* (Leipzig 1823).

Zanardus, Michael. *Disputationes de universo elementari* (Venetiis 1619) III.II.xiii–xxvi.

Zuccolo, Vitale. *Dialogo delle cose meteorologiche* (Venetia 1590).

C: Sources and further reading for modern ideas about flight and the air

Bellairs, A. D'A. and Jenkin, C. R. 'The Skeleton of Birds.' *Biology and Comparative Physiology of Birds* I, ed. A. J. Marshall (New York and London 1960) 289–93. Air sacs.

Bennett, Leon. 'Clap and Fling Aerodynamics—An Experimental Evaluation.' *Journal of Experimental Biology* 69 (1977) 261–72.

Berger, M. and Hart, J. S. 'Physiology and Energetics of Flight.' *Avian Biology* IV, ed. Donald S. Farner and James R. King (New York and London 1974) 415–77.

Brown, R. H. J. 'Flight.' *Biology and Comparative Physiology of Birds* II, ed. A. J. Marshall (New York and London 1961) 289-305. A useful summary, though now a little out of date.

Clancy, Laurence Joseph. *Aerodynamics* (London 1975). A mathematical introduction.

Dalton, Stephen. *The Miracle of Flight* (Maidenhead 1977). Contains a readable and reliable account of the aerodynamics of bird and insect flight.

Davis, J. Michael. 'The Coordinated Aerobatics of Dunlin Flocks.' *Animal Behaviour* 28.3 (August 1980) 668-73.

Gray, Sir James. *Animal Locomotion* (London 1968) 79-81, 186-87, 194-240, 296-98.

Greenewalt, Crawford H. 'The Flight of Birds.' *Transactions of the American Philosophical Society* n.s. 65.4 (July 1975) 1-67. Includes a mathematically based summary of the aerodynamics of bird flight.

Handbook of Aviation Meteorology 2nd ed. (London 1971). An introduction produced by the Meteorological Office, London.

Kermode, Alfred Cotterill. *Flight Without Formulae* 4th ed. (London 1970). An elementary guide using no mathematics.

______. *Mechanics of Flight* 8th ed. (London 1972). A simple mathematical introduction.

Lasiewski, Robert C. 'Respiratory Function in Birds.' *Avian Biology* II, ed. Donald S. Farner and James R. King (New York and London 1975) 288-89, 296-98. Air sacs.

Lighthill, Sir Michael James. 'A Note on "Clap and Fling" Aerodynamics.' *Journal of Experimental Biology* 73 (1978) 279-80.

O'Connell, D. J. K., S.J. *The Green Flash and Other Low Sun Phenomena* (Amsterdam and New York 1958).

Pennycuick, Colin J. *Animal Flight* (London 1972). (The Institute of Biology's Studies in Biology 33.) A brief introduction.

______. 'Mechanics of Flight.' *Avian Biology* V, ed. Donald S. Farner and James R. King (New York and London 1975) 1-75.

Rüppell, Georg. *Vogelflug* (München 1975). Trans. Marguerite A. Biederman-Thorson and published as *Bird Flight* (New York 1977). Indifferently reproduced photographs; first class text.

Salt, G. W. and Zeuthen, Erik. 'The Respiratory System.' *Biology and Comparative Physiology of Birds* I, ed. A. J. Marshall (New York and London 1960) 372–74. Air sacs.

Selous, Edmund. *Thought-Transference (or What?) in Birds* (London 1931).

Tucker, Vance A. and Parrott, G. Christian. 'Aerodynamics of Gliding Flight in a Falcon and Other Birds.' *Journal of Experimental Biology* 52 (1970) 345–67.

Van Tyne, Josselyn and Berger, Andrew J. *Fundamentals of Ornithology* 2nd ed. (New York 1976) 381–98.

Vaughan, Terry A. 'Adaptations for Flight in Bats.' *About Bats: A Chiropteran Biology Symposium,* ed. Bob H. Slaughter and Dan W. Walton (Dallas, Tx. 1970) 127–43.

______. 'Flight Patterns and Aerodynamics.' *Biology of Bats* I, ed. William A. Wimsatt (New York and London 1970) 195–216.

Wallace, George J. and Mahan, Harold D. *An Introduction to Ornithology* 3rd ed. (New York and London 1975) 106–17.

Weis-Fogh, Torkel. 'Quick Estimates of Flight Fitness in Hovering Animals, Including Novel Mechanisms for Lift Production.' *Journal of Experimental Biology* 59 (1973) 169–230. Theories of lift in nonsteady conditions, description of the 'clap-fling' movement of wings.

Withers, Philip C. and Timko, Patricia L. 'The Significance of Ground Effect to the Aerodynamic Cost of Flight and Energetics of the Black Skimmer (*Rhyncops Nigra*).' *Journal of Experimental Biology* 70 (1977) 13–26.

Wu, Theodore Yao-tsu, Brokaw, Charles J., and Brennen, Christopher, eds. *Swimming and Flying in Nature* II (New York and London 1974) 729–1000.

Index

Designer:	Steve Renick
Compositor:	Interactive Composition Corporation
Text:	10/12 Fournier
Display:	Fournier
Printer:	Malloy Lithographing, Inc.
Binder:	John H. Dekker & Sons